Life On Mars and Venus

Algae, Lichens, Fungi, Fossils, Minerals, Microbial Mats, Stromatolites, Metazoans, Evolution, Skulls, Bones, Methane, Martians and the Interplanetary Transfer of Life

Life On Mars and Venus

Algae, Lichens, Fungi, Fossils, Minerals, Microbial Mats, Stromatolites, Metazoans, Evolution, Skulls, Bones, Methane, Martians and the Interplanetary Transfer of Life

Rhawn Gabriel Joseph, Rudolf Schild.
L. Graham, Burkhard Büdel, Patrick Jung, G. J. Kidron,
K. Latif, R. A. Armstrong, H. A. Mansour, J.G. Ray, G. J. P. Ramos,
L. Consorti, R. Dass, G. Bianciardi, N. Cantasano, V. Rizzo,

Copyright © 2019, 2020 Rhawn Gabriel Joseph, Ph.D.
Published by: Cosmology Science Publishers, Cambridge, MA

ISBN: 978-0-9716445-6-4

Life On Mars and Venus

Algae, Lichens, Fungi,
Fossils, Minerals, Microbial Mats, Stromatolites,
Metazoans, Evolution, Skulls, Bones,
Methane, Martians
and the Interplanetary
Transfer of Life

Contents

I. LIFE ON VENUS: THE INTERPLANETARY TRANSFER OF BIOTA FROM EARTH

Rhawn Gabriel Joseph
Astrobiology Associates of Northern California
San Francisco, Palo Alto, Santa Clara, Santa Cruz, California, USA

ABSTRACT

Evidence and observations favoring the hypothesis that Venus is habitable, and the celestial mechanisms promoting the interplanetary transfer of life, are reviewed. Venus may have been contaminated with Earthly life early in its history via interplanetary transfer of microbe-laden bolide ejecta; and this seeding with life may have continued into the present via spacecraft and due to radiation pressure and galactic winds blowing microbial-laden dust ejected from the stratosphere via powerful solar winds, into the orbit and atmosphere of Venus. Venus may have had oceans and rivers early in its history until 750 mya, and, hypothetically, some of those species which, theoretically, ßcolonized the planet during that time, may have adapted and evolved when those oceans evaporated and temperatures rose. Venus may be inhabited by a variety of extremophiles which could flourish within the lower cloud layers, whereas others may dwell 10 m below the surface where temperature may be as low as 200°C--which is within the tolerance level of some hyperthermophiles. Speculation as to the identity of mushroom-shaped specimens photographed on the surface of Venus by the Russian probe, Venera 13 support these hypotheses.

KEY WORDS: Venus, Extremophiles, Thermophiles, Atmosphere, Subsurface, Microorganisms, Solar Winds; Radiation Pressure, Galactic Winds, Stratosphere, Fungi, Panspermia, Lithopanspermia Interplanetary Transfer of Life.

1. Overview: Hypotheses, Observations, Evidence

Observations and evidence for the possibility of life on the subsurface and clouds of Venus, and celestial mechanics promoting the transfer of life from Earth, are reviewed, and images of specimens resembling fungal mushroom-shaped organisms photographed by the Russian Space Probe Venera 13, on the surface of Venus are presented (Figures 1-4) with the caveat that similarities in morphology are not proof of life.

A number of scientists have hypothesized that Earth may be seeding this solar system and Venus with microbes buried within ejecta following impact by meteors (Cockell, 1999; Gladman et al. 1996; Melosh 2003; Mileikowsky et al. 2000; Gao et al., 2014; Schulze-Makuch et al. 2004), and via microbes attached to space craft (Joseph 2018). Extremophiles attached to dust particles which are lofted into Earth's upper atmosphere and via powerful solar winds blown into space (Joseph & Schild 2010) may also be transported by radiation pressure and galactic winds into the atmosphere of Venus (Arrhenius, 1908, 1918). Atmospheric contamination by Earthly life would account for data indicating microbes may be dwelling within and attached to particles in the lower cloud layers of Venus (Bullock 2018; Limaye et al. 2018; Grinspoon & Bullock, 2007).

In support of the solar-galactic-wind microbe-laden-dust hypothesis is evidence of an interplanetary (Misconi & Weinberg, 1978) circumsolar ring of dust (Krasnopolsky & Krysko 1979) which appears to follow (Leinert & Moster, 2007) and to have possibly originated external to the orbit of Venus (Altobeli et al. 2003; Jones et al. 2013; Leinert & Moster, 2007; Russell & Vaisberg, 1983) and which penetrates the planet's atmosphere (Carrillo-Sanchez et al. 2020). As to dust originating from Earth, it is probable that microbes may be adhering to some of these particles (Arrhenius, 1908, 1918).

Cockell (1999) contends that atmospheric conditions at higher altitudes would freeze but not necessarily kill microorganisms. As these organisms fall toward the surface they would encounter environments between the middle and lower cloud layers that are conducive to biological activity (Limaye et al. 2018). Moreover, the turbulence experienced may enhance the viability of some species (Gibson 2000; Gibson & Thomas, 1995) including plankton which have been hypothesized to be dwelling in the clouds of Venus (Konesky, 2009; Schulze-Makuch et al. 2004).

The Venusian subsurface may also provide a livable habitat for photoautotrophic, lithobiontic, poikilohydric, and fungal hyperthermophiles which have evolved and adapted to these conditions. For example, temperatures at 10 m below the surface may range from 305°C to 200°C--which is within the tolerance level of some thermophiles (Baross & Deming, 1983; Kato & Qureshi, 1999). It is well established that extremophiles are "capable of living in extreme environments such as highly acidic or alkaline conditions, at high salt concentration, with no oxygen, extreme temperatures (as low as -20 degrees C and as high as 300 degrees C), at high concentrations of heavy metals and in high pressure environments" (Kato & Takai 2000).

Organisms deep beneath the surface may also have access to moisture drawn from subterranean sources before completely evaporating. In support of this hypothesis, when the Russian probe Venera 13 landed on Venus and in so doing blew away surrounding top soil, specimens resembling fungi were later photographed by Venera, as reported here (Figure 1, 2, 3, 4).

2. Terrestrial Analogs: Life In The Clouds

Over 1,800 different types of bacteria and other species thrive and flourish within the troposphere, the first layer of the Earth's atmosphere (Brodie et al. 2007). These denizens of the air include fungi and fungal spores, lichens, algae, protozoa, diatoms, plants, pollen, seeds, and invertebrates including insects, spiders, mites, and nematodes (Adhikari et al. 2006; Diehl, 2013; Fröhlich-Nowoisky e al. 2009; Griffin 2004; Polymenakou, 2012); many of which become attached to dust and other particles (Möhler et al. 2007; Sattler et al. 2001).

Due to tropical storms, monsoons, and seasonal upwellings of columns of air (Dehel, et al., 2008; Deleon-Rodriguez et al., 2013; Randel et al., 1998, 2010), microbes, spores, fungi, along with dust, dirt, water, methane, and other gases may be transported to the stratosphere (Griffin 2004; Imshenetsky et al. 1978). Many organisms eventually fall back to the surface where they may negatively impact (Griffin et al. 2007; Polymenakou 2012) or promote the growth and metabolic activity of other species (Jones et al. 2008).

Microorganisms and spores have been recovered at heights of 40 km (Soffen 1965), 61 km (Wainwright et al., 2010) and up to 77-km (Imshenetsky, 1978). These include Mycobacterium and Micrococcus, and fungi Penicillium notatum Circinella muscae, and Aspergillus niger, 77-km above Earth's surface (Imshenetsky, 1978). These atmospheric fungi and microorganisms play a major role in organic compound degradation and the alteration and chemical composition of the atmosphere (Amato, et al. 2007; Ariya et al. 2002; Côté, et al. 2008).

Clouds and wind are an effective transport mechanism for the dispersal of bacteria, virus particles, algae, protozoa, lichens, and fungi including those which dwell in soil and water and which may be transported by air thousands of miles across land and oceans (Prospero et al. 2005; Polymenakou 2012) or lofted into the stratosphere (Griffin 2004; Imshenetsky et al. 1978) where they may continue to thrive.

3. Atmospheric Transfer Of Biota From Earth

How often or how much of this atmospheric biota is lofted into space, is unknown. However, under extreme conditions, and if solar winds strike with sufficient force, then water molecules, surface dust (Schroder & Smith, 2008), along with air-borne bacteria, spores, fungi, lichens, algae, and other microbes may be ejected into space (Joseph & Schild 2010), where, as demonstrated experimentally, this biota may easily survive (Novikova et al. 2016 Setlow 2006; Horneck, et al. 1994, 2002, Nicholson et al. 2000; Onofri et al. 2012; McLean & McLean 2010; Kawaguchi et al 2013).

For example, between September 22-25, 1998, NASA's Ultraviolet Imager aboard the Polar spacecraft, detected and measured a series of coronal mass ejections and a powerful solar wind which created a shock wave that struck the magnetosphere with sufficient force to cause helium, hydrogen, oxygen, and other gases to gush from Earth's upper atmosphere into space (Moore and Horwitz, 1998). When the CME struck on Sept. 24, the pressure jumped to10 nanopascals, whereas normally the pressure is around 2 or 3 nanopascals. Thus it could be predicted that airborne microbes attached to dust particles were also swept from the stratosphere into space; and those survivors which were eventually deposited in the atmosphere and clouds of Venus may have begun to flourish and multiply.

As determined by Leinert and Moster, (2007) interplanetary dust accumulates outside the orbit of Venus, and the dust ring is similar to that found along Earth's orbit. Krasnopolsky and Krysko (1979) based on data from the Venera 9 and 10 orbiters, observed a dust layer at heights of 100-700 km above the surface of Venus, and which had a ring formation. According to the data provided by Carrillo-Sánchez et al. (2020) these interplanetary dust particles are mass contributors to the atmosphere of Venus.

The Ulysses space probe was equipped with dust detectors and reported 374 impacts, beginning within the orbit of Venus at 0.7 AU (Grun et al. 1992). At 2 AU Ulysses detected fluxes in the dust density "compatible with a population of interplanetary dust particles moving on low to moderately eccentric (e = 0.1 to 0.5) and low inclination (i = 0 deg. to 30 deg.) orbits" thereby indicating at least two dust particle populations in different orbital trajectories (Grun et al. 1992). It is not likely that those dust particles, with a radii smaller than .03 mm, detected between and within the orbits of Venus and Mars, originated from the outer solar system beyond Mars (Altobelli et al. 2005; Landgraf et al., 2000). Competition between gravitation and radiation pressure determines how deep dust can penetrate such that the smallest are believed to be filtered by radiation pressure before entering the inner solar system (Altobelli et al. 2005). Therefore, the smallest dust particles following or ringing Venus (Leinert & Moster, 2007; Krasnopolsky & Krysko 1979), most likely originated within the orbits of Mars, Earth and Venus.

As determined by Leinert and Moster, (2007) these inner planetary dust particles range in size from 0.1 mm to 0.001 mm. Most bacteria range from 0.002 to 0.0002 mm in diameter (Evans, 2016), whereas fungal spores (e.g. Penicillium, Cladosporium, Aspergillis) range from 0.016 to 0.0036 in diameter (Reponen et al. (2001). And microbes in Earth's upper atmosphere are often attached to particles of dust (Möhler et al. 2007; Sattler et al. 2001) which may range in size from 0.077 to 0.002 mm (Arimoto, et al. 1997; Betzer et al. 1988; Maring et al. 2000; Tegen & Lacis 2012).

Galactic winds and radiation pressure can propel dust particles at speeds ranging from 300 and 3,000 km/sec (Chandrasekhar, 1989; Heckman et al. 2017; Ishibashi & Fabian 2015, Murray et al. 2011; Wibking et al. 2018). According to the calculations of Nobel laureate Arrhenius (1908, 1918) if microbes ranging in size from 0.002-0.0002 mm in diameter are attached to particles ranging from 0.02-0.002 mm in diameter, or if fungal spores ranging in size from 0.0036 to 0.0016 mm are attached to particles ranging from 0.036 to 0.016 mm in diameter, and "if we assume that the spore starts with zero velocity" and originate from Earth, then due to the sun's gravity, galactic winds and radiation pressure, it would take "forty days" to travel from Earth to Venus and would then fall into the atmosphere of Venus with a velocity of approximately two kilometres per second.

As to organisms which may survive the descent to the surface of Venus--especially during the planet's early history--they may have also remained viable and may have evolved over time (Cockell, 1999; Schulze-Makuch et al. 2004). In this regard, and as reported here, there are mushroom-shaped specimens on the surface of Venus; the source of which may be the lower clouds (Limaye et al. 2018) which in turn may have been seeded from microbe-laden interplanetary dust (Arrhenius,1908, 1918; Joseph & Schild 2010), bolide particles originating on Earth (Cockell, 1999; Gladman et al. 1996; Grinspoon & Bullock, 2007; Melosh 1988, Schulze-Makuch et al. 2004) or contamination by space craft (Joseph 2018). However, due to the poor quality of the Venera 13 photos, it is unknown if these are in fact Venusian fungi or mushroom-shaped anomalies. One can only hypothesize and speculate.

In support of the spacecraft-Earth-to-Venus-transfer-of-life scenario is evidence that viable fungi and bacteria not only survived heat-treatment sterilization of spacecraft (La Duc et al. 2014; Venkateswaran et al. 2012; Puleo et al. 1977) but may have survived the journey from Earth to Mars, as successive photos taken months apart depict what appears to be masses of fungal-bacterial organisms growing on the NASA's Mars rovers Curiosity and Opportunity (Joseph 2018; Joseph et al., 2019). There is also evidence (but no proof) that fungi, lichens, algae and microbes may have, at different times in that planet's history, colonized the surface of Mars (Dass 2017; Joseph 2014, 2016; Joseph et al. 2019; Krupa 2017; Levin & Straat, 2016; Noffke 2015; Rabb 2018; Rizzo, & Cantasano 2009, 2016; Roffman 2019; Ruffi & Farmer, 2016; Small 2015). These putative Martian organisms include mushroom- and lichen-shaped specimens similar to those reported here (Figures 1-4). In fact, 15 Martian specimens resembling "puff balls" emerged from beneath the surface and dramatically increased in size over a period of three days (Joseph et al. 2019); observations which may indicate the rovers are stimulating the growth of these specimens, or they are the source. Hence, probes from Earth may have also transported innumerable microbes to Venus.

Similarities in morphology are not proof of life, and there is no definitive proof of life on Mars or Venus. Moreover, several scientists have taken issue with the findings of Joseph et al (2019) and questioned if fungi and lichens could flourish on the Red Planet (Armstrong 2019; Kidron 2019). On the other hand, simulation studies have shown that these and other species can survive in a Mars-like environment (de Vera et al. 2014, 2019; De la Torre Noetzel et al. 2017; Onofri et al. 2012, 2019; Pacelli et al. 2016; Sanchez et al. 2012; Schuerger et al., 2017; Selbman et al. 2015; Zakharova et al. 2014).

Life-simulation studies conducted with Venus-like environments also supports the possibility of life (Seckbach & Libby, 1970; Seckbach et al. 1970). Moreover, many scientists have argued that a variety of terrestrial life forms could survive on or below the surface and especially within the cloud layers (Cockell, 1999; Grinspoon & Bullock, 2007; Ksanfomality 2013; Limaye et al. 2018; Schulze-Makuch et al. 2004). Thus, there is reason to suspect that biota already adapted to life in the clouds of Earth, and which survived a journey to Venus, may flourish and multiply.

4. Transfer Of Life By Spacecraft

Several space craft and balloons have passed through the atmosphere and crashed or landed on Venus (Marov et al. 1998). Since all attempts to completely sterilize space craft have failed (La Duc et al. 2014; Venkateswaran et al. 2012; Puleo et al. 1977), these vehicles were likely transporting billions of microorganisms to Venus--beginning with the Venera 3 Soviet probe on March 1, 1966, followed by the Venera 5 and 6 in May of 1969 and Venera 7 (12/15/70), Venera 8 (7/22/1972), Venera 9 (10/22/1975), and two Pioneer Space craft in 1978, as well as probes from the European Space Agency and Japan (Marov et al. 1998).

As based on published studies on survival following heat treatment sterilization of spacecraft and equipment, at a minimum, billions of organisms, or their spores, per spacecraft, were likely transported to Venus (as well as Mars). For example, immediately after sterilization, 350 to 500 distinct colonies (on average) consisting of millions of organisms, including fungi, were found per square meter on the outer surfaces of the Viking Landers, rovers, and other spacecraft (La Duc et al. 2014; Venkateswaran et al. 2012; Puleo et al. 1977). As to species which were not or could not be cultured, and those masses of fungi and bacteria growing within the equipment's interior, the number of survivors is unknown. Many of the survivors of these failed attempts at sterilization, including fungi, can also survive long duration exposure to space (Novikova et al. 2016 Setlow 2006; Horneck, et al. 1994, 2002, Nicholson et al. 2000; Onofri et al. 2012; McLean & McLean 2010; Kawaguchi et al 2013) well in excess of the four months it takes a space craft to reach Venus (Marov et al. 1998). Moreover, after clean-room sterilization, and prior to launch, these crafts and much of this equipment was then exposed to the biota-laden external environment.

Following the explosion of the space shuttle Columbia, during its reentry from space, and subsequent crash in 2003, viable organisms, including the nematode, Caenorhabditis elegans (Szewczyk et al., 2005) and Microbispora sp (McLean et al., 2006) survived. Likewise, it can be predicted that spaceships crashing on Venus (or Mars) would still be laden with viable microorganisms.

5. Bolide Ejection & Spores: Transfer Of Life From Earth To Venus

Microbe-laden debris ejected into space following terrestrial impact by asteroids, comets, and meteors, may have repeatedly transferred life from Earth to Venus and Mars. Many species of microbe have evolved the ability to survive a violent hypervelocity impact, shock pressures of 100 GPa, and extreme acceleration and ejection into space including the vacuum and frigid temperatures of an interstellar environment; the cosmic rays, gamma rays, UV rays, and ionizing radiation they would encounter; and the descent through the atmosphere and the crash landing onto the surface of a planet (Burchell et al. 2004; Horneck et al. 2001a.b, Horneck et al. 1994; Mastrapaa et al. 2001; Nicholson et al. 2000, 2004; Mitchell & Ellis 1971).

It is reasonable to assume that Venus and Mars, beginning in the early asteroid-comet-meteor bombardment phase 3.8 bya (Cockell 1999; Gladman et al. 1996) may have been repeatedly and continually seeded with life from Earth, attached to dust, ejecta, and more recently, via space craft. In the early history of Mars, Earth and Venus, these planets may have repeatedly exchanged life--the ultimate source of which is as yet unknown.

When a meteor, asteroid, or comet, strikes this planet, the surface area of ejecta may be heated to temperatures in excess of 100 C upon impact (Artemieva and Ivanov 2004, Fritz et al. 2005), which is well within the tolerance range of thermophiles (Baross & Deming, 1983; Kato & Qureshi, 1999; Stetter, 2006). Moreover, because the ejecta-surface is acting as a heat shield, the interior may never be heated above 100°C (Burchell et al. 2004; Horneck et al. 2002). Spores can survive post shock temperatures of over 250°C (Horneck et al. 2002). Thus, innumerable microbes may remain viable despite violent impact-induced ejection into space.

Meteors at least ten kilometers across will also punch a hole in the atmosphere before striking the planet (Van Den Bergh, 1989) and will eject tons of dust, rocks, and boulders (Beech et al. 2018; Gladman et al. 1996; Melosh, 2003) through that hole into space (Van Den Bergh, 1989), along with any adhering microbes, fungi, algae, and lichens (Gladman et al. 1996; Melosh, 2003; Mileikowsky et al., 2000) before air can rush back in to completely fill the gap (Van Den Bergh, 1989). Thus, following the initial impact, ejected debris and adhering organisms would not be subject to extreme heating as they pass through the atmosphere.

There are currently 200 known terrestrial impact craters (Earth Impact Database, 2019), and this planet may have been struck thousands of times (Melosh, 1989), resulting in the ejection of millions of rocks, boulders and tons of debris into space (Beech et al. 2018; Gladman et al. 1996; Melosh, 1989, 2003; Van Den Bergh, 1989). Many scientists have argued that some of this debris and any adhering microbes likely landed on Mars and Venus (Beech et al. 2018; Gladman et al. 1996; Melosh, 2003; Davies, 2007; Fajardo-Cavazosa et al. 2007; Hara et al. 2010; Schulze-Makuch, et al. 2005).

Given that microbes can survive the shock of a violent impact and hyper velocity launch ejecting them into space (Mastrapaa et al. 2001; Burchell et al. 2001, 2004; Nicholson, et al. 2006; Hazael et al. 2017; Horneck et al. 2008), as well as the descent to the surface of a planet (Burchell et al. 2001; Horneck et al. 2002; McLean & McLean 2010), the interplanetary transfer of viable microorganisms, via bolides, within our Solar System, is overwhelmingly likely (Mileikowsky et al., 2000; Beech et al. 2018), beginning, possibly, soon after life appeared on Earth over 3.8 bya.

Over the course of the last 550 million years there have been a total of 97 major impacts, leaving craters at least 5 kilometers across (Earth Impact Database, 2019), and it's been estimated that in consequence, approximately "1013 kg of potentially life-bearing matter has been ejected from Earth's surface into the inner solar system" (Beech et al. 2018) during this time frame. Consider, for example, the Chicxulub crater, formed approximately 66 Mya, which has a 150 km diameter (Alvarez, et al., 2008). It's been estimated, given an 25 km/s impactor velocity, that up to 5.5 x 1012 kg of debris may have been ejected into space (Beech et al. 2018), along with unknown volumes of water, and perhaps millions of trillions of organisms buried within this ejecta. According to calculations by Beech et al (2018), given an impact velocity greater than 23 km/s, this microbial-laden ejecta could have entered the orbits of and intercepted Venus, Mars and other planets within a few weeks or months.

Studies have demonstrated that bolide ejecta provides nutrients that can sustain trillions of microorganisms, including algae and fungi, perhaps for thousands of years (Mautner 1976, 2002). However, ejecta may orbit in space for millions of years before impacting another planet (Gladman et al. 1996; Melosh, 2003). Microbes and fungi, however, are well adapted to survive life in space for even tens of millions of years; accomplished via the formation of spores. Cano and Borucki (1995) have reported that spores, embedded in amber, may remain viable for 25- to 40-million-years, whereas Vreeland and colleagues (2000) have re-animated 250 million-year-old halotolerant bacterium from a primary salt crystal.

Specifically, in the absence of water, nutrients, or under extreme life-neutralizing conditions, microbes and fungi may instantly react by forming a highly mineralized enclosure consisting of heat or cold shock proteins which wrap around their DNA and which alters the chemical and enzymatic reactivity of its genome making it nearly impermeable to harm (Setlow and Setlow 1995; Marquis and Shin 2006; Sunde et al., 2009). "In the dormant stage a spore has no metabolism and resists cycles of extreme heat and cold, extreme desiccation including vacuum, UV and ionizing radiation, oxidizing agents and corrosive chemicals" (Nicholson et al. 2000).

Space experiments and the Long Duration Exposure Facility Mission have shown that bacteria and fungal spores can easily survive the vacuum of space and constant exposure to solar, UV, and cosmic radiation with just minimal protection (Horneck 1993; Horneck et al.1995; Mitchell and Ellis, 1971). Moreover, survival rates increase significantly from 30% to 70% if coated with dust, or embedded in salt or sugar crystals (Horneck et al. 1994). Spores buried beneath the surface of ejecta could easily survive in space for millions of years.

Although the full spectrum of UV rays are deadly against spores, a direct hit is unlikely, even if unprotected while traveling through space. A few meters of surface material provides a significant degree of protection for those buried deep inside (Horneck et al. 2002). Although high-energy radiation or particles which strike a meteor from Earth may create secondary radiation, studies have shown that as the thickness of surrounding material increases beyond 30 cm, the dose rate and lethal effects of heavy ions, including secondary radiation, depreciates significantly (Horneck et al. 2002).

It's been estimated that because microbes and spores are so small that even when bombarded with photons and deadly gamma and UV radiation, they can drift in space for up to a million years before being struck (Horneck et al. 2002). If shielded by 2 meters of meteorite, even after 25 million years in space, a substantial number of spores would survive (Horneck et al. 2002). B. subtilis spores can even survive a direct hit (Horneck et al. 2002).

Further, many species form colonies with those in the outer layers creating a protective crust, which blocks out radiation and protects those in the inner layers from the hazards of space (Nicholson et al. 2000). Therefore, even if dwelling just beneath the surface of ejecta, colonies of living microbes provide their own protection.

Of the 60,556 meteorites so far documented, 227 originated on Mars, and 360 are from the Moon (Meteoritical Bulletin Database, 2019). Meteors from Venus have not yet been identified. However, given the estimated 1000 impact craters detected by the Magellan spacecraft (Schaber et al. 1992) clearly Venus has been repeatedly impacted by meteors which survived descent through the atmosphere without vaporization.

It can be predicted, depending on size, that the surface of ejecta from Earth, would be subject to extremely high temperatures for only a few seconds upon striking the atmosphere of Venus. And if the bolide fragments or explodes, it could be predicted that it may shed spores into the Venusian cloud layers. A variety of organisms remain viable even following atmospheric explosions (Szewczyk et al., 2005), and despite reentry speeds of up 9700 km h-1 (McLean et al., 2006), and high velocity impact onto a planet's surface (Burchell et al. 2001; Horneck et al. 2002; McLean & McLean 2010). Therefore, microbial-infested bolides from Earth could have repeatedly contaminated Venus throughout its history. And those survivors which could adapt, would likely go forth, multiply, and then evolve.

6. Can Life Survive On Venus? Terrestrial Analogs
It has been argued that a variety of terrestrial organisms could tolerate the high temperatures and dwell in the atmosphere of Venus (Clarke et al., 2013; Cockell, 1999; Grinspoon & Bullock, 2007; Konesky, 2009). These include thermophilic photothrophs (Arrhenius, 1918; Cockell. 1999), algae (Sekbach & Libby 1970) and acidophilic microbes (Schulze-Makuch et al. 2004). Sagan and Morowitz (1967) have hypothesized that even complex multi-cellular organisms swim between the thick layers of clouds.

Venus, at ground level has a temperature of 465°C (Avduevsky et al. 1971), with a surface atmospheric pressure of 9.5 MPa (Bougher et al. 1997; Donahue & Russell, 1997). Although there are trace amount of H_2O and water vapors, the atmosphere is 96.5% CO_2, 3.5% nitrogen, and subject to a high sulfur cycle due to active volcanism (Donahue and Hodges, 1992; Barstow et al., 2012).

Several terrestrial species, such as algae and Cyanidium caldarium can survive environments which are 100% CO_2 (Seckbach & Libby, 1970; Seckbach et al. 1970). Chlorella will continue to grow at 0.6 atm and can survive CO2 of 1atm, (Pirt & Pirt, 1980). The biological activity of some species of fungi and archaea is inhibited by the presence of oxygen and increases with increased levels of carbon dioxide (Lenhart, et al. 2012). Therefore, the high levels of CO2 in the atmosphere of Venus would not preclude colonization by a variety of microorganisms and fungi.

Numerous species of prokaryotes and eukaryotes easily survive pressures of 9.5 MPa at a depth of 950 m beneath the sea, including elephant seals and sperm whales; whereas microbes, such as obligate barophiles and shewanella and Moritella (Deming et al. 1988; Kato, 1999; Kato et al. 1998ab, 2010; Robb et al. 2007) flourish within the Mariana trench at a depth of 10,898 m (108 Mpa).

Terrestrial microbes and communities of hyperthermophiles have also been discovered up to 3.5 km below ground (Biddle et al., 2008; Chivian et al., 2008; Doerfert et al., 2009; Moser et al., 2005; Gohn et al., 2008). Baccilus infernus, have been discovering thriving at depths of 2,700 meters where the weight and pressure is 300 x that of the surface (Boone et al. 1995). Genomic analysis of the bacterium, Desulforudis audaxviator--discovered 2.8 km beneath the surface-- indicates this species metabolizes minerals in place of sunlight and "is capable of an independent life-style well suited to long-term isolation from the photosphere deep within Earth's crust" (Chivian et al., 2008). Species similar to these may also thrive at great depths or just beneath the surface of Venus, despite the high temperatures.

Some species of hyperthermophiles continue to grow at a temperature of 122°C (Robb et al. 2007), and have been discovered thriving adjacent to 400°C thermal vents (Stetter, 2006)--the most resilient of which include Pyrolobus fumarii (Baross & Deming, 1983) and Methanopyrus kandleri (Kurr et al. 1991). However, there are no known terrestrial species which can survive direct exposure to temperatures above 300°C (Kato & Qureshi, 1999; Kato & Takai, 2000).

Based on data from Earth, unless any hypothetical Venusian organisms have evolved heat-related adaptive features enabling them to survive surface temperatures of 465°C, then the most likely habitats are the clouds or beneath the ground where subsurface temperatures may be significantly cooler --as is the case on Earth (Davis & Schubert 1981; Smerdon et al. 2004; Al-Temeemi & Harris, 2001).

7. Venusian Water Vapors And Habitability

In addition to atmospheric water vapors (Donahue & Hodges, 1992; Barstow et al., 2012)--which could sustain life dwelling in the clouds of Venus--the atmosphere contains high levels of deuterium with a deuterium-to-hydrogen ratio which is approximately 150 times that of terrestrial water (Donahue et al. 1997). The D/H ratio indicates that Venus at one time had rivers and shallow oceans (Donahue et al. 1997), thereby making early Venus habitable (Way et al. 2016, 2019). What became of this water is unknown.

The D/H ratio leaves open the possibility there are tremendous reservoirs of water buried deep beneath the surface, and which, like terrestrial underground water under high temperature, arid surface conditions (Al-Sanad & Ismael, 1992), may seep upward only to evaporate. Hence, subsurface water evaporation may account for the trace amounts of cloud-water and contribute to the D/H ratio.

Therefore, Venus may have had lakes and oceans for billions of years--making it more conducive to life (Kasting 1988; Colin & Kasting 1992; Donahue et al. 1997; Way et al. 2016, 2019)--which then may have slowly evaporated over the following years, the remainder possibly remaining deep beneath the soil. Hence, early in its history Venus could have become home to Earthly archae, bacteria, and fungi which may have evolved, or migrated to the clouds or beneath the soil. As summed up by Cockell (1999), the possible "existence of hot surface oceans on Venus may have favored the successful evolution or transfer of early life."

On the surface of Venus there is no evidence of water vapor. Given that sea water will completely evaporate at 420°C it is not likely there is any surface water. The only source of moisture would be deep beneath the surface or within the clouds. Any species which colonized the surface of Venus early in its history, and in order to survive under present day conditions, would have had to evolve as temperatures rose and surface water was lost (Schulze-Makuch et al 2004); or they would have had to migrate to the clouds or beneath the surface.

However, even present day organisms transferred from Earth may remain viable. Cockell (1999) has argued: "Neither the pressure (9.5 MPa) nor the high carbon dioxide concentrations (97%) represent a critical constraint to the evolution of life on the surface or in the atmosphere. The most significant constraints to life on the surface are the lack of liquid water and the temperature (464°C). In the lower and middle cloud layers of Venus, temperatures drop and water availability increases, generating a more biologically favorable environment. However, acidity and the problem of osmoregulation in hygroscopic sulfuric acid clouds become extreme and probably life-limiting. If it is assumed that these constraints can be overcome, considerations on the survival of acidophilic sulfate-reducing chemoautotrophs suspended as aerosols in such an environment show that Venus does come close to possessing a habitable niche."

In support of Cockell's (1999) argument is evidence indicating that microbes may have colonized the cloud layers of Venus (Limaye et al. 2018), and the observations, presented in this report, of mushroom-shaped surface features, the latter of which are similar to those observed on Mars (Dass, 2017; Joseph 2016, Joseph et al. 2019). Moreover, Grinspoon (1997) and Grinspoon and Bullock (2007), have argued that bright radar surface signatures may be evidence that living organisms have colonized portions of the Venusian surface.

8. Life In The Clouds Of Venus

In 1967, Sagan and Morowitz, (1967) presented a scenario where multi-cellular organisms, up to 4 cm in diameter may dwell in the upper Venusian atmosphere and metabolize and generate hydrogen as propellants and a means of floatation.

At this writing, the only evidence remotely suggestive of multi-cellular organisms have been photographed at ground level (Ksanfomality (2013) and as reported here.

Venus has three cloud layers which circle the planet at an altitude of 48-68 kms and which are composed of H2SO4 aerosols, with an acid concentration ranging from 98% in the lower atmosphere to 81% in the upper atmosphere (Bougher et al. 1997; Donahue et al. 1997). The clouds of Venus also contain high levels of deuterium and trace amounts of water (Donahue and Hodges, 1992; Barstow et al., 2012) which could sustain life.

According to Limaye et al. (2018): "The lower cloud layer of Venus (47.5–50.5 km)" provides "favorable conditions for microbial life, including moderate temperatures and pressures (60°C and 1 atm), and the presence of micron-sized sulfuric acid aerosols," which could sustain sulfate-reducing microbes.

Konesky (2009) argues in favor of the life-cloud hypothesis, and states "there exists a habitable region in the atmosphere, centered at approximately 50 km, where the temperature ranges from 30 to 80°C and the pressure is one bar." He's suggested that organisms similar to plankton may dwell in the upper atmosphere. Likewise, Schulze-Makuch et al. (2004) hypothesized that Venusian clouds, 48 to 65 km above the surface, could harbor aeroplankton (which metabolize sulphur), and acidophilic microbes who employ sulfur allotropes as photo-protective pigments which enable them to engage in photosynthesis.

The upper cloud layers of Venus are subject to wind speeds of up to 355 km/hour (100 meters a second) and then increase to more than 700 km/hour in the lower layers (Blamont et al. 1986). Although increasing turbulence can have an inhibitory influence on the growth and metabolism of a variety of species, the opposite is true of phytopklanton and diatoms. Turbulence can increase photosynthetic activity (Gibson 2000; Gibson & Thomas, 1995; Thomas, Vernet, Gibson, 1995), thus supporting the plankton hypotheses of Konesky (2009) and Schulze-Makuch et al. (2004).

9. Radiation And Uv Absorption

Schulze-Makuch and colleagues (2004) have argued "that the cloud deck of the Venusian atmosphere may provide a plausible refuge for microbial life.... Here we make the argument that such an organism may utilize sulfur allotropes present in the Venusian atmosphere, particularly S8, as a UV sunscreen, as an energy-converting pigment, or as a means for converting UV light to lower frequencies that can be used for photosynthesis. Thus, life could exist today in the clouds of Venus."

Fungi, lichens, and prokaryotes can survive long-term exposure to gamma, and solar UV radiation and remain viable (Horneck et al. 2002; McLean & McLean 2010; Nicholson et al. 2000; Novikova et al 2016; Onofri et al. 2012; Sato et al. 2011; Tugay et al. 2006; Sancho et al. 2007; Raggio et al. 2011). A variety of terrestrial species thrive in and are attracted to highly radioactive environments (Becket et al. 2008; Dadachova et al. 2007; Tugay et al. 2006; Wember & Zhdanova 2001), including fungi, lichens and numerous species of microbe. UV rays would have no detrimental effects on these species.

Vesper et al. (2008) and Novikova et al. (2016; Novikova 2009) reported that fungi are invigorated and grow rapidly within the International Space Station as a consequence of the heightened radiation levels. Fungi also flourish on the outskirts and along the walls of the highly radioactive Chernobyl nuclear power plant (Dighton et al. 2008; Zhdanova et al. 2004). Fungi (Wember & Zhdanova 2001; Zhdanova et al. 2004) and radiation-tolerant bacteria (Moseley & Mattingly 1971; Ito et al. 1983) will seek out and grow towards sources of radiation which serve as an energy source for metabolism (Dighton et al. 2008; Tugay et al. 2006). These organisms can easily repair radiation-damaged DNA due to a redundancy of genes with repair functions (White et al. 1999).

Ground and low cloud level radiation may not be nearly as intense as first believed as an unknown UV absorber has been detected in the cloud layers (Krasnopolsky, 2017; Markiewicz et al., 2014; Rossi et al., 2015); thus making Venus more hospitable to life (Limaye et al. 2018). Although it's possible that sulfur dioxide and iron chloride are acting as UV absorbers, the albedo, between 330 and 500 nm, and the refraction index between 1.07 and 1.7 (Laven, 2008), and the spatial and temporal changes in contrasts (Markiewicz et al., 2014; Rossi et al., 2015), may be evidence that microbial populations, attached to particles in the lower cloud layers of Venus (Limaye et al. 2018), are waxing and waning in biomass.

Specifically, as based on observations from spacecraft and terrestrial telescopes, it is likely that the clouds of Venus' have a high density of micron-sized particles (Hansen and Hovenier, 1974; Knollenberg et al., 1980). Although a variety of absorbing mechanisms have been proposed (reviewed by Limaye et al. 2018), if these particles originated in space and are from Earth, of if they are comprised of sulfuric acid or other substances elevated from the surface, is unknown (Pérez-Hoyos et al., 2017).

Regardless of their source or composition, it is not likely that these particles, alone, are acting as UV absorbers since their numbers are insufficient and incompatible with Venera 14 observations (Krasnopolsky, 2017; Limaye et al. 2018). The refraction indices also exceed the values expected from sulfuric acid cloud particles except in the presence of FeCl3 (Markiewicz et al. (2014), the latter of which may be contributing to cloud condensation nuclei.

Grinspoon (1997) proposed that the "unknown ultraviolet absorber" may be a photosynthetic pigment and that "terrestrial" photosynthetic, acid-resistant organisms are metabolizing and engaging in phototrophic and chemotrophic activities involving sulfur and iron within the Venusian clouds. Grinspoon and Bullock (2007), have also argued that irregularly shaped cloud particles may be living organisms. Another possibility is that living organisms are attached to and feeding on these particles.

To adequately account for the amount of UV rays absorbed, the biota dwelling in the clouds of Venus would have to be massive in numbers. According to the calculations of Bullock (2018): "If microbes grow to represent one tenth of the aerosol mass, the total biomass in Venus' clouds would be 3.6 billion tonnes, or about the biomass of the Earth's oceans."

Certainly a biomass of this magnitude would absorb a significant amount of UV rays; and this position, and Bullock's calculations, are supported by Limaye et al. (2018) who have argued: "for Venus' lower clouds, the mass loading estimates (0.1–100 mg·m−3) are comparable to the upper biomass value for terrestrial biological aerosols (44 mg·m−3), while the particle size regime (≤8 μm) opens the possibility that the clouds may similarly harbor suspensions of single cells or aggregated microbial communities. In theory, the 2- and 8-μm-sized particles (modes 2′ and 3) could harbor a maximum of 108 and 1010 cells."

The possibility of iron- and sulfur-centered metabolism in these clouds, coupled with particle and biomass density as estimated by Bullock (2018) and Limaye et al. (2018), may account for the spectral refraction anomalies observed in polarized and unpolarized light. UV rays are being absorbed and metabolized by Venusian organisms.

If the cloud layers of Venus harbor biology, then we would expect they would exhibit spectral signatures similar to terrestrial clouds. As pointed out by Limaye et al. (2018) data from the Galileo, Pioneer, Messenger and Akatsuki space probes "are tantalizingly similar to the absorption properties of terrestrial biological molecules."

10. Speculations: Venusian Atmospheric Organisms

Limaye et al. (2018) also propose that microorganisms similar to A. ferro-oxidans may dwell in the clouds of Venus as "this bacterium thrives at extremely low pH values (pH 1 to 2), fixes both CO2 and nitrogen gas from the atmosphere (Valdes et al., 2008), and obtains its energy for growth from the oxidation of hydrogen, ferrous iron, elemental sulfur, or partially oxidized sulfur compounds." Other candidates include chemolithoautotrophic and acidophilic γ-proteobacterium, both of which thrive under low pH and anaerobic conditions and at temperatures of 50–60°C, and oxidize elemental sulfur and using Fe3+ as a terminal electron acceptor. Likewise, green sulfur bacteria and archaeal Stygiolobus oxidize elemental sulfur to yield sulfuric acid under acidic and anaerobic conditions.

Therefore, if the analyses of Bullock (2018) and Limaye et al. (2018) are correct, then species similar to Stygiolobus, green sulfur bacteria, sulfate-reducing bacteria, and A. ferrooxidans, and a biomass equal to that found in the oceans of Earth, may be thriving within the clouds of Venus.

Bullock (2018) and Limaye and colleagues (2018) also point out that cloud-based microbial populations would have to be replenished on relatively fast timescales. According to the calculations of Bullock (2018): "The residence time of mode 3 (r=3.6 μm) aerosols in Venus' lower cloud is about 2 days. The rate equation shows that if the microbial lifetime is shorter than this, a steady state population can exist within the aerosols." That is, as microbes reproduce they replace those which fall to the surface and can transfer between aerosol, otherwise they must be replenished from the surface or from space.

Because of the very strong downdrafts and updrafts (e.g. Blamont et al. 1986), the powerful winds of Venus may be continually transferring and recycling biomass to and from the surface to the atmosphere, and from the upper cloud layers to the lower layers. Conversely, if via galactic winds microbes and microbe-laden dust are expelled from the upper atmosphere of Earth (Joseph & Schild 2010) into the path of Venus, they would continually contaminate the upper atmosphere and descend from the upper to lower cloud layers where they may flourish, reproduce, and perhaps evolve, only to be eventually deposited on the surface. Conversely, those dwelling on or just beneath the surface would be lofted back into the Venusian atmosphere via strong updrafts. Life would continually be recycled from clouds to the surface and from the surface (or subsurface) back to the clouds, and replenished by organisms from Earth.

11. Life Beneath The Subsurface

Venera 13 landed in the Beta Phoebe region at median elevations in the upland rolling plains province, an area described as a "stony desert" with lose soil, compacted sand, pebbles, and rocks similar to terrestrial tholeiitic basalts which make up much of Earth's ocean crust (Surkov et al. 1984). The Venusian basalts were found to contain high levels of potassium and covered with soil and soil patches (Florensky et al. 1983; Surkov et al. 1984). Garvin et al. (1984) likened the area to the bedrock at Snake River Plain, in Idaho, USA, and noted that many large and small rocks were partly buried beneath the sandy soil.

On Earth, endolithic microorganisms flourish in hyper-arid rocky deserts and extreme environmental conditions by colonizing the interior or the undersides of rocks including those just beneath the surface (Weirzchos 2012; Pointing & Belnap 2012); conditions which allow photosynthesis, and within which water molecules may be trapped. However, in the absence of water these organisms are able to "reversibly activate metabolism" until moisture becomes available (Harel et al. 2004). Generally, these hot desert micro-habitats are dominated by fungi, lichens, cyanobacteria, mosses and heterotrophic bacteria (Pointing & Belnap 2012).

The surface temperature of Venus, as determined by Venera 7, was 747 K / 473.85°C for the first 50 seconds after landing and to have a constant temperature of 739 K°/ 465.85°C (Avduesvesky et al. 1971). Basalt has high thermal insulating properties (Eppelbaum et al. 2014; Mastafa et al. 2004) such that temperatures beneath these rocks would be much cooler. Temperatures beneath the Venusian soil are unknown.

In hot and arid climates, temperatures 1 m to 10 m beneath the surface are significantly cooler than surface temperatures (Davis & Schubert 1981; Smerdon et al. 2004; Al-Temeemi & Harris, 2001). At these shallow depths, and due to these more temperate subsurface vs higher surface temperatures, microorganisms are able to engage in considerable metabolic activity (Blume et al. 2002; Buchanan & King, 1992, Kaiser & Heinemyer, 1993). Investigators have found a considerable biomass of microorganisms and fungi up to 15 cm beneath the subsurface (Dobbins et al. 1992, Fierer et al., 2012; Crocker et al. 2000; Krumholz, 2000) even in arid high temperature environments (Weirzchos 2012; Pointing & Belnap 2012; Steven et al., 2013; Mueller et al., 2015).

Because of its density, soil is an excellent modulator of heat such that soil temperatures decrease at a rate exponential to soil depth until reaching 10 m; below which thermal conductivity is fairly uniform (Davis & Schubert 1981; Smerdon et al. 2004; Al-Temeemi & Harris, 2001). In high temperature environments, heat transfer reduction from the surface to the subsurface can be as much as 57% (Al-Temeemi & Harris, 2001); i.e. 43% of surface temperature.

Kuwait is located in a dry desert region and is characterized by high temperature extremes and intense solar radiation with a mean air temperature ranging from 37°C to 45°C during the summer months of June through August (Al-Temeemi & Harris, 2001). There is little precipitation and soil water content is 2% (Al-Sanad & Ismael, 1992; Al-Temeemi & Harris, 2001). However, it is much cooler beneath the surface and temperatures steadily decrease with depth. For example, Al-Temeemi and Harris (2001) established that if the average surface air temperature is 41°C, the subsurface temperature, at a depth of 1 m, reaches a maximum of 36°C, whereas at a depth of 10 m the subsurface temperature is a constant 27°C; i.e. a reduction of 12.2% and 34.2% respectively.

Al-Temeemi and Harris (2001), also determined that when compared to atmospheric temperatures the comparative decline is even more pronounced due to subterranean cooling. They concluded there are "significant temperature drops below-ground when air temperatures are at their highest."

These findings support the likelihood that subsurface temperatures on Venus are much cooler than the surface and may be habitable for a variety of microorganisms. Hypothetically these may include photoautotrophic, lithobiontic, poikilohydric, and fungal hyperthermophiles which have evolved and adapted to the Venusian environment. Specifically, by applying to Venus the 12.2% and 34.2% reductions in subsurface temperatures as determined by Al-Temeemi and Harris (2001) then, at a depth of 1 m temperatures will have fallen by 56.6°C yielding a temperature of 407.4°C, followed by a reduction of 158.7°C and thus a subsurface temperature of 305.3°C at 10 m. However, given the reduction value of 57% (Al-Temeemi & Harris, 2001), temperatures 10 m beneath the surface might fall to 200°C--well within the limit for the hardiest hyper-thermophiles on Earth (Kato & Takei 2000). However, although some species may survive at this temperature, there would likely be little or no biological activity--according to terrestrial standards. On the other hand, because basalt has high thermal insulating properties (Eppelbaum et al. 2014; Mastafa et al. 2004), temperatures beneath these rocks could be much cooler.

The hyper-arid, waterless surface of Venus may also draw moisture and water up from the subterranean depths, just as occurs on Earth and the deserts of Kuwait (Al-Sanad & Ismael, 1992). If so, then any organisms living 10 m below ground may be continually supplied with water as it rises to the surface and completely evaporates.

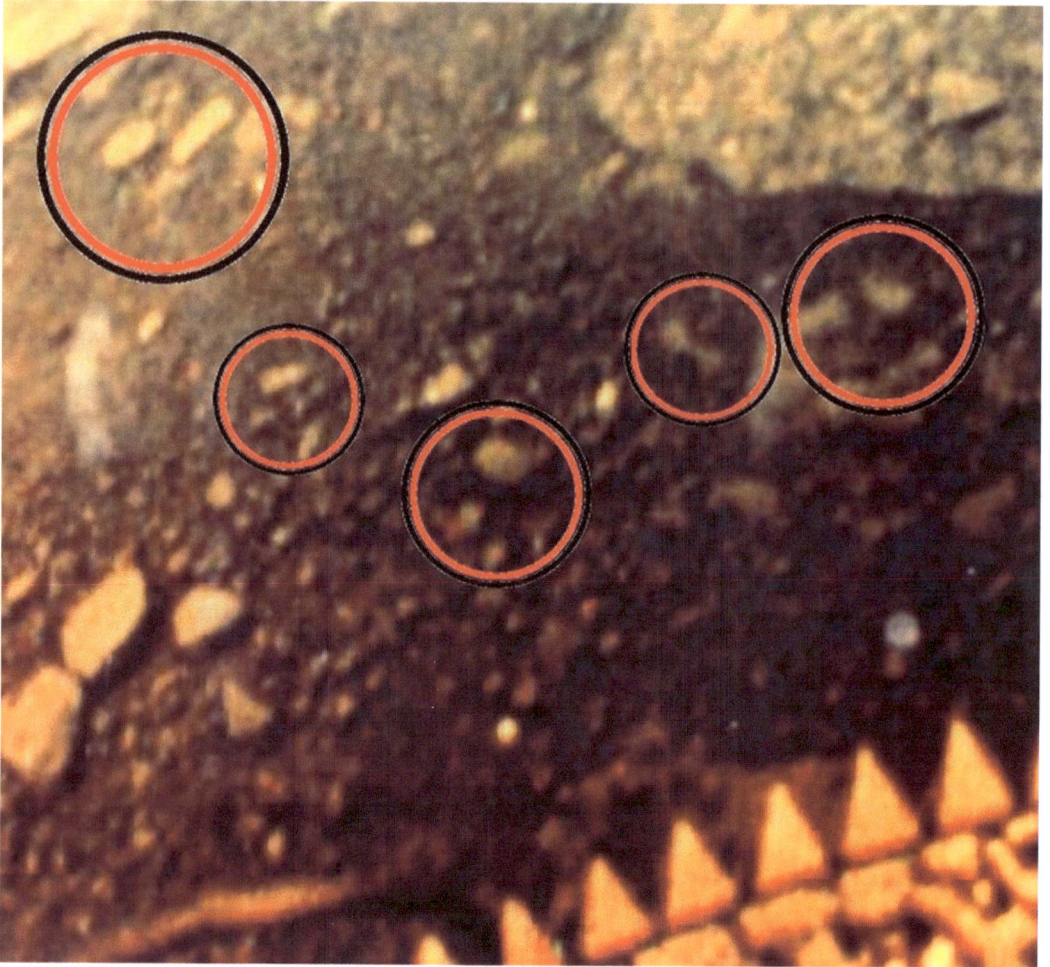

Figure 1. Venera-13. Venus: Fungi-Mushroom-Lichen-shaped specimens.

12. Speculations: Fungi & Mushroom-Shaped Specimens On The Subsurface Of Venus?

Specimens resembling fungi have been observed on the surface of Venus as reported here (Figures 1, 2, 3, 4). Unfortunately, because of the poor quality of the Venera 13 photos, positive identification is impossible, and one can only speculate and propose hypotheticals based on morphological similarities to terrestrial mushrooms.

Specifically, upon examining panoramic color images from the 1982 Soviet Venera 13 mission--reprocessed by NASA-- four well defined specimens with caps and stalks which resemble the classic terrestrial mushroom were observed (center, and bottom left, Figure 1). These specimens protrude approximately 2 to 3 cm from the surface, and the caps are approximately 3 to 4 cm in diameter. Several smaller, less defined specimens with a cap and what may be stalks appear to the far left of center and which are bordered by additional specimens which form a crescent of mushroom-shapes (Figures 1, 2).

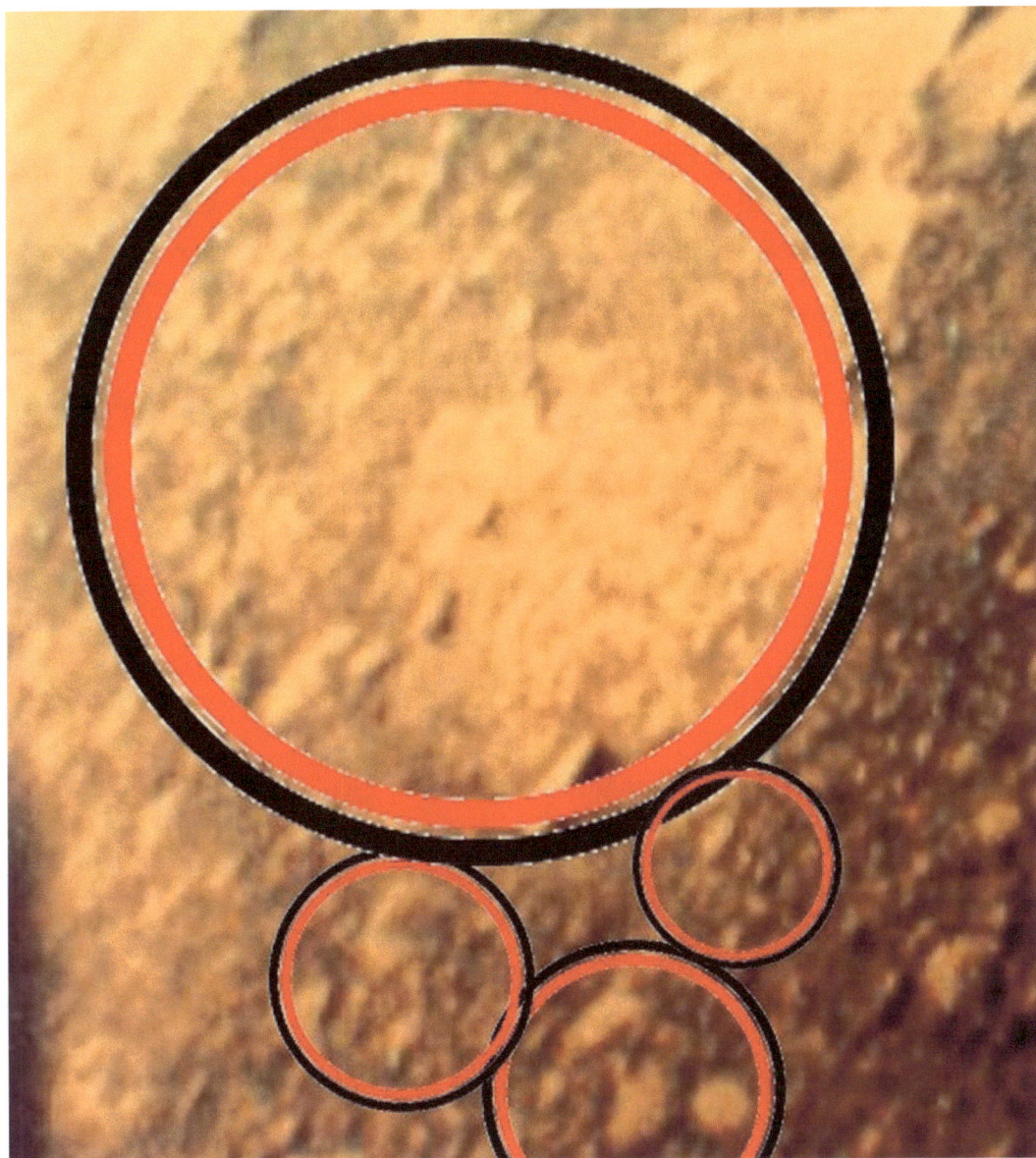

Figure 2. Venera-13. Venus: Fungi-Mushroom-lichen-shaped specimens (from Figure 1).

Examination of the original black and white Venera 13 photos--which provides a 180 degree view--reveals a third well defined specimen at the bottom left (three in total, Figure 3). Clusters of additional poorly defined specimens are observed to the upper left (Figures 4).

Therefore, five well defined specimens and numerous smaller surface structures which resemble classic terrestrial mushrooms have been observed on the surface of Venus. There is no known geological or weathering process on Earth which can sculpt mushroom-like shapes from rocks. However, it must be stressed that similarities in morphology are not proof of life.

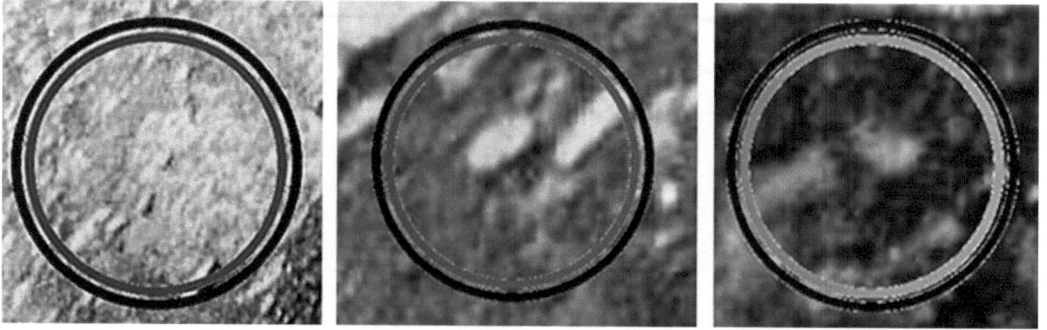

Figure 3. Venera-13. Venus: Enlarged Fungi-Mushroom-lichen-shaped specimens. (From the far lower left of Figure 1.

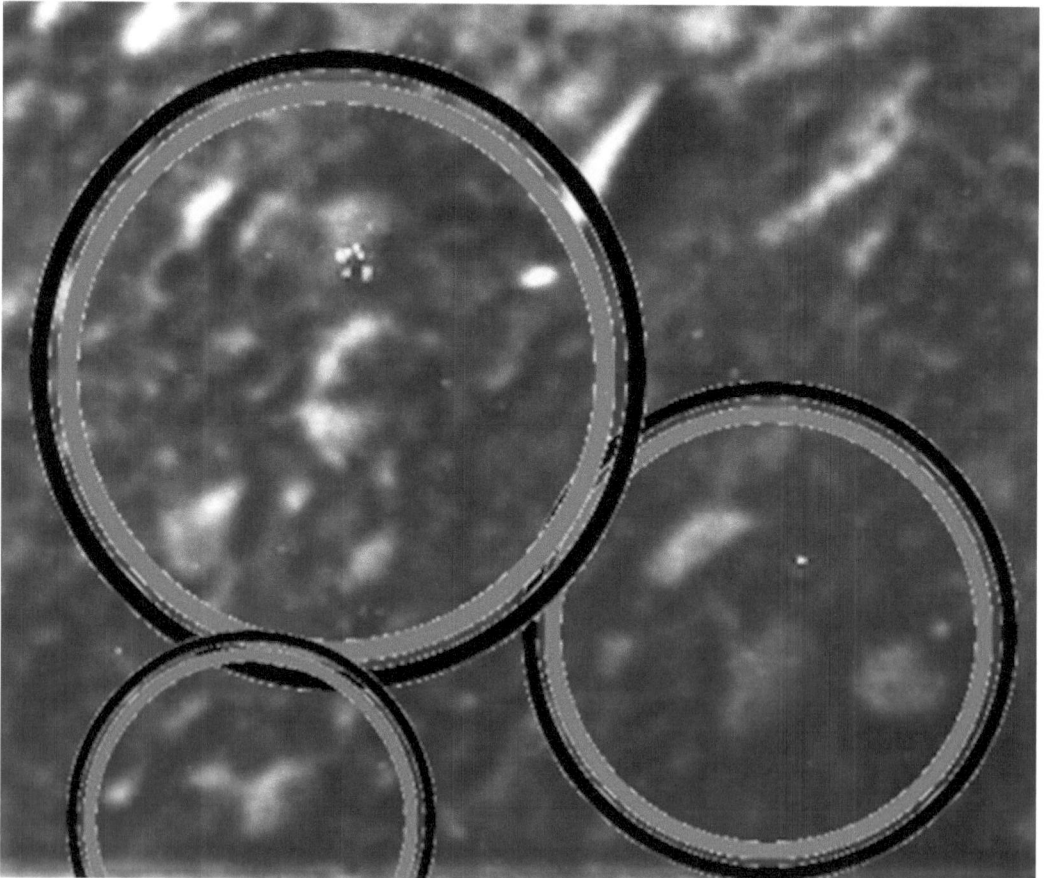

Figure 4. Venera-13. Venus: Fungi-Mushroom-lichen-shaped specimens. (Enlarged and to the far left of Figure 1).

These Venusian specimens may have been uncovered from beneath the surface when Venera 13 landed and blew away top soil, rocks, and dust surrounding the landing site. Garvin (1981) reports that due to surface winds and turbulence, clouds of dust were recorded by the Venera crafts for up to 30 seconds after impact and which persisted for up to 7 minutes.

It is unknown if these are mushroom-shaped rocks, fossils, or living organisms that had evolved and had been growing beneath the surface where temperatures would be cooler and which may provide access to any underground moisture drawn toward the surface. One can only speculate and propose hypotheticals. Based on terrestrial standards, if alive, they would have been obliterated almost instantly upon exposure to surface conditions. However, the number and density of these specimens also increase with increasing distance from the lander (Figures 1, 4), suggesting those closer in proximity were destroyed by the sudden turbulence of the craft's landing. This leaves three explanations: they are mushroom-shaped rocks, they are fossilized organisms, or they have evolved into hyper-extremophiles perfectly adapted to the Venusian surface/subsurface environment.

It must be emphasized this author does not claim to have discovered fossilized or current life on Venus but is speculating and proposing hypotheticals. However, several of these specimens are similar to what may be fungal-lichen organisms growing on Mars (see Joseph et al. 2019) as well as on Earth, which supports the interplanetary transfer hypothesis. Hypothetically, if the specimens observed on Venus are fossils, or organisms which evolved and adapted, this supports the possibility that fungi, in general, may be hyper-extremophiles, capable of colonizing Mars, Venus, and the harshest of alien environments.

The author's observations and interpretation of these mushroom-shaped specimens are supported by the findings of Ksanfomality (2013). Based on examination of NASA-enhanced panoramic images from the 1982 Soviet Venera 13 mission, Ksanfomality (2013) observed what he believed to be two fungal shaped objects at a distance of 15 to 20 cm from the buffer of the landing module. Ksanfomality (2013) has estimated these specimens to be elevated 3 cm above the surface and to have a diameter of approximately 8 cm. These do not have mushroom-shapes and are to the far (90°degrees) right of the specimens observed by this author and cannot be observed in Figures 1-4.

13. Photosynthesis, Water, Hydrocarbons, Anaerobic Respiration

It is unknown if the fungal and mushroom-shaped specimens reported here are fossils, living organism, or mushroom-shaped rocks which may have been uncovered when the Venera-13 landed on the surface. One can only speculate. If fossils, then they may be relics from a time when Venus had oceans, rivers, and a less extreme environment.

It has been suggested that Venus may have had liquid water on the surface and provided a habitable environment for billions of years (Grinspoon and Bullock, 2007; Way et al., 2016. 2019). Way and Del Genio (2019), based on the results from four simulations, determined that Venus "may have had enough condensable water on its surface for a shallow ocean..." and "would have been able to maintain stable temperatures – from a low of 20 °C (68 °F) to a high of 50 °C (122 °F) – for about three billion years"i.e. until about 750 mya. The finding of Donahue and colleagues (Donahue et al., 1982; Donahue and Hodges, 1992)--based on comparisons of atmospheric deuterium/hydrogen ratios with Earth--support those contentions.

On Earth, in this general time frame, a variety of prokaryotes and eukaryotes had evolved. Between 600 to 800 mya, this biota included acritarchs, protozoa, ediacarans, amoebozoans, cercozoans, plankton, algae, coccoid and filamentous cyanobacteria, and fungi (Butterfield 2005a,b; Bottjer et al., 2006; Narbonne 2005; Narbonne and Gehling 2003; Shen et al., 2008). Hypothetically, if Venus was also habitable and wet at this time, then similar species may have also evolved. If this view is correct, then as surface water evaporated and temperatures rose, survivors may have adapted, evolved, migrated beneath the surface, and/or via the powerful winds of Venus (Blamont et al. 1986), were lofted to the clouds (Cockell, 1999; Schulze-Makuch et al., 2004; Grinspoon and Bullock, 2007; Limaye et al. 2018).

It is unknown if there is water beneath the Venusian surface. Budisa and Schulze-Makuch (2014) have hypothesized that in the absence of water and high concentration of carbon dioxide, that life on Venus may be sustained by "supercritical fluids" which "have different properties compared to regular fluids and could play a role as life-sustaining solvents on other worlds... One location where CO_2 should occur in the supercritical state is in the near-subsurface of Venus."

Although the amount of light reaching the surface is only approximately 70% of that of Earth (within 0.5 - 9 klx), Venusian organisms, dwelling in the clouds, or beneath the surface, could derive energy from photosynthesis. Some terrestrial prokaryotes, as a solvent, utilize hydrogen sulfide, H2S as the electron donor, whereas some autotrophic prokaryotes living beneath the surface engage in chemosynthesis instead of photosynthesis (Durvaslua & Rao, 2018; Gerday & Glansdorff, 2007).

The atmosphere of Venus is 96.5% CO_2, 3.5% nitrogen, and due to active volcanism, subject to a high sulfur cycle (Donahue and Hodges, 1992; Barstow et al., 2012). Sebach and Libby (1970) in an experiment designed to mimic the Venusian environment, grew algae in pure CO_2, at elevated temperature and in an acidic medium, and reported these specimens continued to thrive.

A variety of anaerobic species survive and flourish in anoxic, high CO_2, environments (Kelly et al. 2001; Holliger et al. 1998, Emerson et al. 2016) such as Shewanella--a gram-negative, proteobacteria (Fredrickson et al. 2008; Hunt et al. 2010). In the absence of oxygen, anaerobes utilize substances such as sulphate (SO_4^{2-}), nitrate (NO_3), or sulphur (S), as terminal electron acceptors to convert energy into adenosine triphosphate (ATP) which is conserved and utilized when performing energy requiring processes (Grein et al. 2013; Kelly et al. 2001; Richter et al. 2012; Holliger et al. 1998). Some eukaryotic species which normally respire oxygen (e.g. sub-arctic dwarf shrubs), can survive high levels of CO2, under hypoxic conditions, with no indication of damage (Preece & Phoenix, 2013).

The metabolic activity of some eukaryotic fungi and anaerobic archaea is inhibited by oxygen and amplified with increased levels of CO2 (Lenhart, et al. 2012). Given the high sulfur cycle, a variety of sulfate reducing anaerobic organisms (e.g. from the sulfur molybdoenzyme family) could adapt to the Venusian environment as they utilize sulfur compounds for sulfur metabolism to obtain energy (Grein et al. 2013).

So far, hydrocarbons have not been detected in the atmosphere (Plummer 1969); and without this evidence any possibility of life on Venus must be viewed with caution.

On Earth, fungi and microorganisms dwelling in clouds, and on the surface/subsurface, have a major role in organic compound degradation (Amato, et al. 2007; Ariya et al. 2002; Côté, et al. 2008); and they play the same role in regard to hydrocarbons (Aydin et al. 2917; Heider et al. 1998). It is possible that the lack of evidence for Venusian hydrocarbons is due to rapid mineralization and consumption, serving as an energy source for Venusian organisms; similar to what occurs on Earth. A variety of microorganisms, including ferric iron-reducing, sulfate-reducing, and anaerobic hydrocarbon degrading bacteria (Heider et al. 1998) as well as anaerobic fungi (Aydin et al. 2917), catabolize hydrocarbon compounds. These species might be well adapted for life beneath the surface or within the clouds of Venus.

14. Conclusions

Venus may have been contaminated with Earthly life early in its history, and that seeding with life may have continued into the present via microbial-laden debris, dust, and spacecraft. Venus may have had oceans and rivers until as recently as 750 mya, and, hypothetically, species which colonized the planet may have adapted and evolved when those oceans evaporated and temperatures rose. However, Venus may still provide a habitable environment for a variety of hyperextremophiles including fungi which might be continually arriving from Earth.

References

Adey. W. R. (1993). Biological Effects of Electromagnetic Fields. Journal of Cellular Biochemistry 51:410-416.

Adhikari A, Reponen T, Grinshpun SA, Martuzevicius D, LeMasters G. (2006) Correlation of ambient inhalable bioaerosols with particulate matter and ozone: A two-year study. Environ Pollut 140:16–28.

Al-Temeemi, A.A., Harris, D.J (2001). The generation of subsurface temperature profiles for Kuwait, Energy and Buildings, 33, 837-841.

Altobelli, N., S. Kempf, M. Landgraf, R. Srama, V. Dikarev, H. Kru¨ger, G. Moragas-Klostermeyer, and E. Gru¨n (2003), Cassini between Venus and Earth: Detection of interstellar dust, J. Geophys. Res., 108(A10), 8032, doi:10.1029/2003JA009874.

Altobelli,N., S. Kempf,1 H. Kru¨ger, M. Landgraf, M. Roy, and E. Gru¨n (2005) Interstellar dust flux measurements by the Galileo dust instrument between the orbits of Venus and Mars, Journal Of Geophysical Research, 110, A07102, doi:10.1029/2004JA010772, 2005

Altobelli, N., S. Kempf M. Landgraf R. Srama V. Dikarev H. Krüger G. Moragas-Klostermeyer E. Grün (2003). Cassini between Venus and Earth: Detection of interstellar dust, Journal of Geophysical Research, 18, 21-33.

Alvarez, L. W. Alvarez, W., Asaro, F., Michel, H. V. (1980). Extraterritorial cause for the Cretaceous -Tertiary extinction. Science, 208, 1095-1108.

Amato, P.; Demeer, F.; Melaouhi, A.; Fontanella, S.; Martin-Biesse, A.-S.; Sancelme, M.; Laj, P.; Delort, A.-M. A fate for organic acids, formaldehyde and methanol in cloud water: Their biotransformation by microorganisms. Atmos. Chem. Phys. 2007, 7, 4159–4169.

Amato, P., Parazols, M., Sancelme, M., Mailhot, G., Laj, P., and Delort, A.-M. (2007) An important oceanic source of microorganisms for cloud water at the Puy de Do^me (France). Atmos Environ 41:8253–8263.

Al-Sanad, H., Ismael, N. F., (1992) Thermal properties of desert sands in Kuwait, Journal of the University of Kuwait 19, 207-215).

Arimoto, B., B. J. Ray N. F. Lewis U. Tomza R. A. Duce (1997). Mass□particle size distributions of atmospheric dust and the dry deposition of dust to the remote ocean, Atmospheres, 102, 15867-15874.

Ariya, P.A., Nepotchatykh, O., Ignatova, O., Amyot, M. (2002) Microbiological degradation of atmospheric organic compounds. Geophys. Res. Lett. 29, 2077–2080.

Armstrong, R. A. (2019). Could Lichens Survive on Mars? Journal of Astrobiology and Space Science Reviews, 1, 235--241, 2019

Armstrong, R. A. (2019). The Lichen Symbiosis: Lichen "Extremophiles" and Survival on Mars Journal of Astrobiology and Space Science Reviews, 1, 378-397.

Artemieva, N., Ivanov, B. (2004), Launch of Martian Meteroties in Oblique Impacts, Icarus 171, 183

Aydin, S., Karaçay, H. A., Shahia, A., Gökçe S., Ince, B., Incea. O. (2017). Aerobic and anaerobic fungal metabolism and Omics insights for increasing polycyclic aromatic hydrocarbons biodegradation Fungal Biology Reviews, 31, 61-72

Arrhenius, S. (1908). Worlds in the Making. Harper & Brothers, New York.

Arrihenius S. (1918). The Destinies of the Stars. Putnam.

Avduevsky, V. S., MY Marov, Rozhdestvensky, M. K., Borodin, N. F., Kerzhanovich, V. V. (1971) Soft landing of Venera 7 on the Venus surface and preliminary results of investigations of the Venus atmosphere, Journal of Atmsopheric Science, 28, 263-269.

Baross, J.A. & Deming, J. W. (1983). Growth of black smoke bacteria at temperature at least 250 Celsius, Nature 303, 423-426.

Barstow J.K., Tsang C.C.C., Wilson C.F., Irwin P.G.J., Taylor F.W., McGouldrick K., Drossart P., Piccioni G., and Tellmann S. (2012) Models of the global cloud structure on Venus derived from Venus Express observations. Icarus 217:542–560.

Basset C.AL. (1993). Beneficial effects of electromagnetic fields. J Cell Biochem 31:387-393.

Bauer, H.; Kasper-Giebl, A.; Löflund, M.; Giebl, H.; Hitzenberger, R.; Zibuschka, F.; Puxbaum, H. The contribution of bacteria and fungal spores to the organic carbon content of cloud water, precipitation and aerosols. Atmos. Res. 2002, 64, 109–119.

Beckett, R., Kranner, I., & Minibayeva, F. (2008). Stress physiology and the symbiosis. In T. Nash, III (Ed.), Lichen Biology (pp. 134-151). Cambridge: Cambridge University Press. doi:10.1017/CBO9780511790478.009

Beech, M., Comte, M., Coulson. I (2018). Lithopanspermia – The Terrestrial Input During the Past 550 Million YearsAmerican Journal of Astronomy and Astrophysics, 7(1): 81-90.

Beltzer, P. R. KL Carder, RA Duce, JT Merrill, NW Tindale (1988). Long–range transport of giant mineral aerosol particles, Nature, 336, 568–571.

Blamont J.E., Young R.E., Seiff A., Ragent B., Sagdeev R., Linkin V.M., Kerzhanovich V.V., Ingersoll A.P., Crisp D., Elson L.S., Preston R.A., Golitsyn G.S., and Ivanov V.N. (1986) Implications of the VEGA balloon results for Venus atmospheric dynamics. Science 231:1422–1425.

Blume, E. Bischoff, M., Reichert, J.M., .Moorman, T., Konopkab, A., Turcod, R.F. (2002). Surface and subsurface microbial biomass, community structure and metabolic activity as a function of soil depth and season, Applied Soil Ecology Volume 20, Issue 3, June 2002, Pages 171-181.

Boone, D R., Liu, Y., Zhao, Z. J., Balkwill, D. L., Drake, G. R., Stevens, T. O., Aldrich, H. C, (1995). Bacillus infernus sp. nov., an Fe(III)- and Mn(IV)-reducing anaerobe from the deep terrestrial subsurface. International journal of systematic bacteriology. 45(3):441-8.

Boreson, J.; Dillner, A.M.; Peccia, J. (2004). Correlation bioaerosol load with PM2.5 and PM10cf concentrations: A comparison between natural desert and urban fringe aerosols. Atmos. Environ. 38, 6029–6041.

Bottjer, D.J., Clapham, M.E. Xiao, S., Kaufman, A.J. (2006) in Neoproterozoic Geobiology and Paleobiology, Evolutionary paleoecology of Ediacaran benthic marine animals, eds Xiao S, Kaufman AJ (Springer, Dordrecht, Netherlands), pp 91–114.

Bougher, S. W., Hunten, D. M., Phillips, R. J. (1997) Venus II: Geology, Geophysics, Atmosphere, and Solar Wind Environment, University of Arizona Press.

Brandt, A., et al. (2015). Viability of the lichen Xanthoria elegans and its symbionts after 18 months of space exposure and simulated Mars conditions on the ISS-International Journal of Astrobiology, 14, 411-425.

Brodie, E. L., DeSantis, T.Z., Parker, J. P. M., Zubietta, I. X., Piceno, Y. M., Andersen, G. L. (2007). Urban aerosols harbor diverse and dynamic bacterial populations PNAS. 104, 299-304.

Buchanan, M., King, L.D., 1992. Seasonal fluctuations in soil microbial biomass carbon, phosphorus, and activity in no-till and reduced-chemical-input maize agroecosystems. Biol. Fertility Soils 13, 211–217.

Budisa, N., & Schulze-Makuch. D., (2014). Supercritical Carbon Dioxide and Its Potential as a Life-Sustaining Solvent in a Planetary Environment, Life, 4, 331-340.

Bullock, M. (2018). The Total Possible Biomass of Venus' Clouds, American Astronomical Society, DPS meeting #50, id.102.04.

Burchell, M. J., Mann, J., Bunch, A. W., Brandob, P. F. B. (2001). Survivability of bacteria in hypervelocity impact, Icarus. 154, 545-547.

Burchell, M. J. Mann, J., Bunch, A. W. (2004). Survival of bacteria and spores under extreme shock pressures, Monthly Notices of the Royal Astronomical Society, 352, 1273-1278.

Butterfield, N.J (2005a). Probable Proterozoic fungi. Paleobiology. 31, 165–182.

Butterfield, N.J. (2005b). Reconstructing a complex early Neoproterozoic eukaryote, Wynniatt formation, arctic Canada. Lethaia. 38, 155–169.

Calabrese EJ, Baldwin LA. (1999). Tales of two similar hypotheses: the rise and fall of chemical and radiation hormesis. BELLE Newslett 8:47-66.

Calabrese EJ, Baldwin LA. 2000. Radiation hormesis: its historical foundations as a biological hypothesis. Human Exper Toxicol 19:41-75.

Cano, R.J. and Borucki, M.K. (1995) Revival and identification of bacterial spores in 25– to 40-million-year-old Dominican amber. Science 268, 1060–1064.

Carrillo-Sánchez, J.D.,Gómez-Martín, J.C., Bones, D. L., Nesvorný, D., Pokornýde, P., Bennae, M., Flynng, G. J., Planea, J.M.C. (2020). Cosmic dust fluxes in the atmospheres of Earth, Mars, and Venus, Icarus, Volume 335. 33-43.

Chandrasekhar, S. (1989). Radiative Transfer and Negative Ion of Hydrogen, Vol 2) University of Chicago Press.

Chivian, D., Brodie, E. L., Alma, E. J., et al., 2008. Environmental Genomics Reveals a Single-Species Ecosystem Deep Within Earth Science 322, 275-278.

Clarke, A., Morris, G.J., Fonseca, F., Murray, B.J., Acton, E., and Price, H.C. (2013) A low temperature limit for life on Earth. PLoS One 8:e66207.

Cockell C.S. (1999) Life on venus. Planet Space Sci 47:1487–1501.

Colin., J., & Kasting, J. F. (1992). Venus. A search for clues to early biological possibilities. In, Exobiology in the Solar System. NASA Technical Publication 512.

Côté, V.; Kos, G.; Mortazavi, R.; Ariya, P.A. Microbial and "de novo" transformation of dicarboxylic acids by three airborne fungi. Sci. Total Environ. 2008, 390, 530–537.

Covey. C., et al. (1994) Climatic effects of atmospheric dust from an asteroid or comet impact on Earth. Glob al and Planetary Change. 9, 263-273.

Crocker, F.H., Fredrickson, J.K., White, D.C., Ringelberg, D.B., Balkwill, D.L., 2000. Phylogenetic and physiological diversity of Athrobacter strains isolated from unconsolidated subsurface sediments. Microbiology UK 146, 1295–1310.

Dadachova E., Bryan RA, Huang X, Moadel T, Schweitzer AD, Aisen P, et al. (2007) Ionizing Radiation Changes the Electronic Properties of Melanin PLoS One, doi:10.1371/journal.pone.0000457.

Dass, R. S. (2017) The High Probability of Life on Mars: A Brief Review of the Evidence, Cosmology, 27, 62-73.

Davies, P. C. W. (2007) The Transfer of Viable Microorganisms Between Planets, (Editors Gregoy R. Bock Jamie A.) Novartis Foundation Symposia

Davis, A. J. Schubert, R.P. (1981). Alternative Natural Energy Sources inBuilding Design, Van Nostrand Reinhold, New York.

Dehel, T., Lorge, F., and Dickinson, M. (2008) Uplift of microorganisms by electric fields above thunderstorms. J Electrostat 66:463–466.

De la Torre Noetzel, R. et al. (2017). Survival of lichens on the ISS-II: ultrastructural and morphological changes of Circinaria gyrosa after space and Mars-like conditions EANA2017: 17th European Astrobiology Conference, 14-17 August, 2017 in Aarhus, Denmark.

Deleon-Rodriguez, N., Lathem, T.L., Rodriguez, R.L., Barazesh, J.M., Anderson, B.E., Beyersdorf, A.J., Ziemba, L.D., Bergin, M., Nenes, A., and Konstantinidis, K.T. (2013) Microbiome of the upper troposphere: species composition and prevalence, effects of tropical storms, and atmospheric implications. Proc Natl Acad Sci USA 110:2575–2580.

Deming, J. W., Somers, L. K., Straube, W. L., Swartz, D.G., Colin., J., & Kasting, J. F. (1992). Venus. A search for clues to early biological possibilities. In, Exobiology in the Solar System. NASA Technical Publication 512.

De Vera, J. -P. et al. (2014). Results on the survival of cryptobiotic cyanobacteria samples after exposure to Mars-like environmental conditions, International Journal of Astrobiology, 13, 35-44.

De Vera, J.-P., Alawi, M., Backhaus, T. et al. (2019) Limits of life and the habitability of Mars: The ESA space experiment BIOMEX on the ISS. Astrobiology, 18:145-157.

De Vera, J. -P. (2012). Lichens as survivors in space and on Mars. Fungal Ecology, 5, 472-479.

Diehl, R.H. (2013) The airspace is habitat. Trends Ecol Evol 28:377–379.

Dighton, J, Tatyana Tugay, T., ZhdanovaN., (2008) Fungi and ionizing radiation from radionuclides, FEMS Microbiol Lett 281, 109-120.

Dobbins, D.C., Aelion, C.M., Pfaender, F.K., 1992. Subsurface, terrestrial microbial ecology and biodegradation of organic chemicals: a review. Crit. Rev. Environ. Control 22, 67–136.

Doerfert1, S. N. (2009). Methanolobus zinderi sp. nov., a methylotrophic methanogen isolated from a deep subsurface coal seam. Int J Syst Evol Microbiol 59, 1064-1069.

Donahue T.M., Hoffman J.H., Hodges R.R., and Watson A.J. (1982) Venus was wet—A measurement of the ratio of deuterium to hydrogen. Science 216:630–633.

Donahue T.M., Hodges R.R.Jr. (1992) Past and present water budget of Venus. J Geophys Res 97:6083–6091.

Donahue, T. M., Russell, C. T., (1997). The Venus atmosphere and ionosphere and their interaction with the solar wind; an overview. In Bougher, S. W. (Eds0. Venus II. Geology. University of Arizona Press. 3-31.

Durvasula R. V., Rao, D.V.S (2018), Extremophiles: From Biology to Biotechnology, CRC Press. Earth Impact Database (2019). http://www.passc.net/EarthImpactDatabase

Emerson, J.B., Brian C. Thomas Walter Alvarez Jillian F. Banfield (2016) Metagenomic analysis of a high carbon dioxide subsurface microbial community populated by chemolithoautotrophs and bacteria and archaea from candidate phyla, Environmental Microbiology, 18, 1686-1703.

Eppelbaum L., Kutasov I., Pilchin A. (2014) Thermal Properties of Rocks and Density of Fluids. In: Applied Geothermics. Lecture Notes in Earth System Sciences. Springer, Berlin, Heidelberg

Evans, H. (2016. Bacteria: Microbiology and Molecular Genetics, Syrawood, publishing.

Fajardo-Cavazosa, P., Schuerger, A. C, Nicholson W. L (2007). Testing interplanetary transfer of bacteria betw een Earth and Mars as a result of natural impact phenomena and human spaceflight activities, Acta Astronautica, 60, 534-540.

Fierer, N., Leff, J. W., Adams, B. J., Nielsen, U. N., Bates, S. T., Lauber, C. L., et al. (2012). Cross-biome metagenomic analyses of soil microbial communities and their functional attributes. Proc. Natl. Acad. Sci. U.S.A. 109, 21390–21395. doi: 10.1073/pnas.1215210110

Florensky, C. P., A. T. Basilevsky, V. P. Kryuchov, R. O. Kusmin, O. V. Nikolaeva, A. A. Pronin, I. M. Chernaya, Y. S. Tyuflin, A. S. Selivanov, M. K. Naraeva, and L. B.

Ronca, (1983) Venera 13 and 14: Sedimentary rocks on Venus?, Science, 221, 57.

Fritz, J., Artemieva, N.A., Greshake, A. (2005), Ejection of Martian Meteorites, Meteoritics & Planetary Science, 40, 1393

Fröhlich-Nowoisky, J.; Pickersgill, D.A.; Després, V.R.; Pöschl, U. High diversity of fungi in air particulate matter. Proc. Natl. Acad. Sci. USA 2009, 106, 12814–12819.

Gao, P., Zhang, X., Crisp, D., Bardeen, C.G., and Yung, Y.L. (2014) Bimodal distribution of sulfuric acid aerosols in the upper haze of Venus. Icarus 231:83–98.

Garvin, J. B. (1981). Landing Induced Dust Clouds on Venus and Mars, Proceedings of the Lunar and Planetary Society, 12B, 1493-1505.

Garvin, J.B. James W. Head Maria T. Zuber Paul Helfenstein (1984) Venus: The nature of the surface from Venera panoramas, Journal of Geophysical Research, 89, 3381-3399.

Gerday, C. & Glansdorff, N. (2007) Physiology and Biochemistry of Extremophiles, ASM press.

Gibson, C. H. (2000). Laboratory and ocean studies of phytoplankton response to fossil turbulence, Dynamics of Atmospheres and Oceans, 31, 295-306

Gibson,C. H. (1991). Kolmogorov similarity hypotheses for scalar fields: sampling intermittent turbulent mixing in the ocean and galaxy, Proceedings of the Royal Society A. 43

Gibson, C. H., & William H. Thomas, W. T. (1995). Effects of turbulence intermittency on growth inhibition of a red tide dinoflagellate, Gonyaulax polyedra Stein, Journal of Geophysical Research, 100, 24841-24846

Gladman, B. J. Burns, J. A., Duncan, M., Lee, P. C., Levison H. F. (1996), the exchange of impact ejecta between terrestrial planets. Science, 271, 1387-1392.

Gohn, G. S., et al., (2008). Deep Drilling into the Chesapeake Bay Impact Structure Science, 320, 1740-1745.

Grein, F. Ana RaquelRamosSofia S.VenceslauInês A.C.Pereira (2013). Unifying concepts in anaerobic respiration: Insights from dissimilatory sulfur metabolism, Biochimica et Biophysica Acta (BBA) - Bioenergetics, 187, 145-160.

Griffin, D.W. (2004) Terrestrial microorganisms at an altitude of 20,000 m in Earth's atmosphere. Aerobiologia 2004, 20, 135–140.

Griffin, D.W., Kubilay, N.; Kocak, M.; Gray, M.A.; Borden, T.C.; Shinn, E.A. Airborne desert dust and aeromicrobiology over the Turkish Mediterranean coastline. Atmos. Environ. 2007, 41, 4050–4062.

Grinspoon, D.H. (1993) Probing Venus's cloud structure with Galileo NIMS. Planet Space Sci 41:515–542. Grinspoon, D.H. (1997) Venus Revealed: A New Look Below the Clouds of Our Mysterious Twin Planet. Addison Wesley, Reading, MA, USA, 355 pp.

Grinspoon D.H. (1997) Venus Revealed: A New Look Below the Clouds of Our Mysterious Twin Planet. Addison Wesley, Reading, MA.

Grinspoon D.H., Bullock M.A. (2007) Astrobiology and Venus exploration. In: Exploring Venus as a Terrestrial Planet, edited by L.W. Esposito, E.R. Stafan, and T.E. Cravens, American Geophysical Union, pp. 191–206.

Grün, E., M. Baguhl H. Fechtig M. S. Hanner J. Kissel B.□A. Lindblad D. Linkert G. Linkert I. Mann J. A. M. McDonnell G. E. Morfill C. Polanskey R. Riemann G. Schwehm N. Siddique H. A. Zook (1992). Galileo and Ulysses dust measurements: Fz Venus to Jupiter, Geophysical Research Letters, 19, 1311-1314.

Grün, E., Landgraf, M. (1997). Collisional consequences of big interstellar grains, Bull. Am. Astron. Soc.,29, 1045, 1997.

Grün, E., M. Baguhl, H. Svedhem, and A. Zook, (2001). In situ measurements of cosmic dust, inInterplanetary Dust, pp. 295– 346, Springer□Verlag, New York.

Guo, Y., and Ning, K. (2011). Freezing mechanism of supercooled water droplet impinging on metal surfaces, International Journal of Refrigeration, December 2011, Pages 2007-2017

Hansen J.E. and Hovenier J.W. (1974) Interpretation of the polarization of Venus. J Atmos Sci 31:1137–116.

Hara, T., Takagi, K.,m Kajiura, D. (2010). Transfer of Life-Bearing Meteorites from Earth to Other Planets. Journal of Cosmology, 7, 1731-1742.

Harel Y, Ohad I, Kaplan A (2004) Activation of photosynthesis and resistance to photo inhibition in cyanobacteria within biological desert crust. Plant Physiol 136:3070-3079.

Hazael. R., fitzmaurice, B. C., Fogilia, F.,m Appleby-Thomas, G. J., McMilan, P. F. (2017). Bacterial survival following shock compression in the GigaPascal range. Icarus, 293, 1-7.

Heckman, Timothy M.; Thompson, Todd A. (2017). Galactic Winds and the Role Played by Massive Stars". Handbook of Supernovae. Springer, Cham. pp. 2431–2454.

Heider, J., Spormann, A. M., Beller, H.R., Widdel, F. (1998). Anaerobic bacterial metabolism of hydrocarbons. FEMS Microbiology Reviews, 22, 459–473.

Holliger, C., Wohlfarth, G.; Diekert, G. (1998). "Reductive dechlorination in the energy metabolism of anaerobic bacteria" (PDF). FEMS Microbiology Reviews. 22 (5): 383.

Holton, J.R., Haynes, P.H., McIntyre, M.E., Douglass, A.R., Rood, R.B., and Pfister, L. (1995) Stratosphere-troposphere exchange. Rev Geophys 33:403–440.

Horneck, G. (1993). Responses ofBacillus subtilis spores to space environment: Results from experiments in space Origins of Life and Evolution of Biospheres 23, 37-52.

Horneck, G., Bücker, H., Reitz, G. (1994). Long-term survival of bacterial spores in space. Advances in Space Research, Volume 14, 41-45.

Horneck, G., Eschweiler, U., Reitz, G., Wehner, J., Willimek, R., Strauch, G. (1995). Biological responses to space: results of the experiment Exobiological Unit of ERA on EURECA I. Advances in Space Research 16, 105-118

Horneck, G. Mileikowsky, C., Melosh, H. J., Wilson, J. W. Cucinotta F. A., Gladman, B. (2002). Viable Transfer of Microorganisms in the solar system and beyond, In G. Horneck & C. Baumstark-Khan. Astrobiology, Springer.

Horneck, G., Stoffler, D., Ott, S., et al. (2008). Microbial rock inhabitants survive hypervelocity impacts on Mars-like host planets: first phase of lithopanspermia experimentally tested. Astrobiology, 8, 17-44.

Hunt, K.A. Jeffrey M. Flynn, Belén Naranjo, Indraneel D. Shikhare, Jeffrey A. Gralnick (2010) Substrate-Level Phosphorylation Is the Primary Source of Energy Conservation during Anaerobic Respiration of Shewanella oneidensis Strain MR-1, Journal of Bacteriology, 192, 3345-3351

Imshenetsky, A.A., Lysenko, S.V., Kazakov,G.A. (1978). Upper boundary of the biosphere. Applied and Environmental Microbiology, 35, 1-5.

Ito, H, Watanabe H, Takeshia M. & Iizuka H (1983). Isolation and identification of radiation-resistant cocci belonging to the genus Deinococcus from sewage sludges and animal feeds. Agric". Biol. Chem. 47: 1239-47. doi:10.1271/bbb1961.47.1239.

Irwin, L.N. and Schulze-Makuch, D. (2011) Suspended animation: life in the clouds of a dense atmosphere on planets like Venus or the gas giants. In Cosmic Biology: How Life Could Evolve on Other Worlds, Springer, New York, pp 153–172.

Ishibashi, W., Fabian, A. C. (2015). AGN feedback: galactic-scale outflows driven by radiation pressure on dust, Monthly Notices of the Royal Astronomical Society, 451, 93–102,

James, E., Toon, O., and Schubert, G. (1997) A numerical microphysical model of the condensational Venus cloud. Icarus 129:147–171.

Jones, A.M.; Harrison, R.M. The effects of meteorological factors on atmospheric bioaerosol concentrations—A review. Sci. Total Environ. 2004, 326, 151–180

Jones, M.H., Bewsher, D., Brown, D. S. (2013) Imaging of a Circumsolar Dust Ring Near the Orbit of Venus, Science, 342, 960-963

Jones, S.E.; Newton, R.J.; McMahon, K.D. (2008). Potential for atmospheric deposition of bacteria to influence bacterioplankton communities. FEMS Microbiol. Ecol. 64, 388–394.

Joseph, R. (2014) Life on Mars: Lichens, Fungi, Algae, Cosmology, 22, 40-62.

Joseph, R. (2016) A High Probability of Life on Mars, The Consensus of 70 Experts, Cosmology, 25, 1-25.

Joseph, R. (2018). Sterilization Failure and Fungal Contamination of Mars and NASA's Mars Rovers, Journal of Cosmology (Cosmology.com), 30, 51-97, 2018

Joseph, R. & Schild, R. (2010). Biological Cosmology. Journal of Cosmology, 10. 40-75.

Joseph, R. G, Dass, R. S., Rizzo, V., Cantasano, N., Bianciardi, G. (2019), Evidence of Life on Mars? Journal of Astrobiology and Space Science Reviews, 1, 40–81.

Kaiser, E.A., Heinemeyer, O., (1993). Seasonal variations of soil microbial biomass carbon within the plough layer. Soil Biol. Biochem. 25, 1649–1655.

Kasting, J. F. (1988). Runaway and moist greenhouse atmospheres and the evolution of Earth and Venus, Icarus, 74, 195-212.

Kato C. (1999) Molecular Analyses of the Sediment and Isolation of Extreme Barophiles from the Deepest Mariana Trench. In: Horikoshi K., Tsujii K. (eds) Extremophiles in Deep-Sea Environments. Springer, Tokyo.

Kato, C., Qureshi, M.H. (1999). Pressure Response in Deep-sea Piezophilic Bacteria, J. Molec. Microbiol. Biotechnol. (1999) 1(1): 87-92.

Kato, C., Li, L., Nakamura, Y., Nogi, Y., Tamaoka, J., and Horikoshi, K. (1998a). Extremely barophilic bacteria isolated from the Mariana Trench, Challenger Deep, at a depth of 11,000 meters. Appl. Environ. Microbiol. 64: 1510-1513

Kato, C. Li, L., Nogi, Y., Nakamura, Y., Tamoako, H. Horiskoshi, K. (1998b). Extremely barophilic bateria isolated from Mariana trench, Challenger Deep, at a depth of 11,000 Meters. Applied Enviornmental Microbiology, 64, 1510-1513.

Kato, C., & Takai, K. (2000).. Microbial diversity of deep-sea extremophiles-Piezophiles, Hyperthermophiles, and subsurface microorganisms. Biol Sci Space. 2000, 341-52.

Kato, S. Yoshinori Takano, Takeshi Kakegawa, Hironori Oba Kazuhiko Inoue, Chiyori Kobayashi, Motoo Utsumi, Katsumi Marumo, Kensei Kobayashi, Yuki Ito, Jun-ichiro Ishibashi, and Akihiko Yamagishi (2010). Biogeography and Biodiversity in Sulfide Structures of Active and Inactive Vents at Deep-Sea Hydrothermal Fields of the Southern Mariana Trough , Applied And Environmental Microbiology, 76, 2968–2979.

Kelly, D. J. Poole, RK., Hughes, PNJ (2001) Microaerobic physiology: aerobic respiration, anaerobic respiration, and carbon dioxide metabolism, (edited by Mobley HLT, Mendz GL, Hazell SL,) In Helicobacter pylori: Physiology and Genetics. ASM press.

Kidron, G. J. (2019. Cyanobacteria and Lichens May Not Survive on Mars. The Negev Desert Analogue Journal of Astrobiology and Space Science Reviews, 1, 369-377, 2019

Knollenberg R.G. and Hunten D.M. (1980) The microphysics of the clouds of Venus—Results of the Pioneer Venus particle size spectrometer experiment. J Geophys Res 85:8039–8058

Konesky, G. (2009) Can Venus shed microorganisms? Proc SPIE 7441, doi:10.1117/12.828643.

Krasnopolsky V.A. (2017) On the iron chloride aerosol in the clouds of Venus. Icarus 286:134–137

Krasnopolsky V.A., Krysko, A.A. (1979). Venera 9, 10: Is there a dust ring around Venus? Planetary and Space Science, 27, 951-957.

Krumholz, L.R., 2000. Microbial communities in the deep subsurface. Hydrogeol. J. 8, 4–10.

Krupa, T. A. (2017). Flowing water with a photosynthetic life form in Gusav Crater on Mars, Lunar and Planetary Society, XLVIII.

Ksanfomality, L. W., (2013). An Object of Assumed Venusian Flora Doklady Physics, 2013, Vol. 58, No. 5, pp. 204–206.

Kuchner, M. J., Reach, W. T. & Brown, M. E., (2000) A search for resonant structures in the Zodiacal Cloud with COBE DIRBE: The Mars ... The role of secular resonances in the history of Trojans, Icarus 146, 232-239.

Kuchner, M. J. & Holman, M. J., (2003), The geometry of resonant signatures in debris disks with planets, Astrophysical Journal, 588, 1110

Kuhlman KR, Venkat P, La Duc MT, Kuhlman GM, McKay CP (2008) Evidence of a microbial community associated with rock varnish at Yungay, Atacama Desert, Chile. J Geophys Res Biogeosci 113, G04022.

Kurr M; Huber R; Konig H; Jannasch HW; et al. (1991). "Methanopyrus kandleri, gen. and sp. nov. represents a novel group of hyperthermophilic methanogens, growing at 110°C". Arch. Microbiol. 156 (4): 239–247. doi:10.1007/BF00262992

Lacey, M.E.; West, J.S. The Air Spora: A Manual for Catching and Identifying Airborne Biological Particles; Springer: Dordrecht, The Netherlands, 2006.

La Duc, M. T., K. Venkateswaran, C. A. Conley (2014) A genetic inventory of spacecraft and associated surfaces. Astrobiology. 2014;14:15- 23.

Landgraf, M. (2000), Modeling the motion and distribution of interstellar dust inside the heliosphere, J. Geophys. Res., 105, 10,303 – 10,316.

Landgraf, M., K. Augustsson, E. Grün, and B. A. S. Gustafson, Deflection of the local interstellar dust flow by solar radiation pressure, Science, 286, 2319– 2322, 1999.

Landgraf, M., W. J. Baggaley, E. Grün, H. Krüger, and G. Linkert, Aspects of the mass distribution of interstellar dust grains in the solar system from in situ measurements, J. Geophys. Res., 105,10,343– 10,352, 2000.

Landgraf, M., H. Krüger, N. Altobelli, and E. Grün, (2003) Penetration of the heliosphere by the interstellar dust stream during solar maximum, J. Geophys. Res., 108, 8030

Laven P. (2008) Effects of refractive index on glories. Appl Opt 47:H133.

Lenhart K. et al. (2012) Evidence for methane production by saprotrophic fungi. Nat Commun. 2012;3:1046. doi: 10.1038/ncomms2049.

Leinert, C., Moster, B. (2007). Evidence for dust accumulation just outside the orbit of Venus. Astron. Astrophys. 472, 335-340.

Levin, G.V. and Straat, P.A. (2016) The Case for Extant Life on Mars and its Possible Detection by the Viking Labeled Release Experiment. Astrobiology 16 (10): 798-810,

Lewis, T. J., Wang, K. (1992) Influence of terrain on bedrock temperatures Global and Planetary Change, 6, 87-100

Limaye, S. J., Rakesh Mogul, David J. Smith, Arif H. Ansari, Grzegorz P. Słowik, and Parag Vaishampayan (2018). Venus' Spectral Signatures and the Potential for Life in the Clouds, Astrobiology, 18.

Macdonell, M. T. (1988). Isolation of an obligately basophilic bacterium and description of a new genus. Colwellia. gen. nov. Syst. Applied Microbiology, 10, 152-160.

Maffei, M. E. (2014). Magnetic field effects on plant growth, development, and evolution (2014). Front. Plant Sci., 04.

Maring, H., D. L. Savoie, M. A. Izaguirre, C. McCormick, R. Arimoto, J. M. Prospero, and C. Pilinis, (2000) Aerosol physical and optical properties and their relationship to aerosol composition in the free troposphere at Izaña, Tenerife, Canary Islands during July 1995. J. Geophys. Res., 105, 14,677–14,700,

Markiewicz W.J., Petrova E., Shalygina O., Almeida M., Titov D.V., Limaye S.S., Ignatiev N., Roatsch T., and Matz K.D. (2014) Glory on Venus cloud tops and the unknown UV absorber. Icarus 234:200–203.

Marov, M.H. David H. Grinspoon Tobias Owen Natasha Levchenko Ronald Mastaler (1998),The Planet Venus, Yale University Press.

Marquis, R. E., Shin, S. Y. (2006). Mineralization and responses of bacterial spores to heat and oxidative agents FEMS Microbiology Reviews 14375 - 379.

Mastrapaa, R.M.E., Glanzbergb, H ., Headc, J.N., Melosha, H.J, Nicholsonb, W.L. (2001). Survival of bacteria exposed to extreme acceleration: implications for panspermia, Earth and Planetary Science Letters 189, 30 1-8.

Mautner, M. N. (1997b) Biological potential of extraterrestrial materials. 1. Nutrients in carbonaceous meteorites and effects on biological growth. Planet. Space Sci. 45, 653-664.

Mautner, M. N. (2002) Planetary bioresources and astroecology. 1. Planetary microcosm bioassays of Martian and carbonaceous chondrite materials: Nutrients, electrolyte solutions, and algal and plant responses. Icarus, 158, 72-86.

McLean, R.J.C., Welsh, A.K., Casasanto, V.A., (2006). Microbial survival in space shuttle crash. Icarus 181, 323-325.

McLean, R.J.C., McLean, M. A. C. (2010). Microbial survival mechanisms and the interplanetary transfer of life through space. Journal of Cosmology, 7, 1802-1820.

Melosh, H. J., (1989). Impact Cratering – a geological process. Oxford University Press, Oxford.

Melosh, H. J. (2003). Exchange of Meteorites (and Life?) Between Stellar Systems. Astrobiology, 3, 207-215.

Meteoritical Bulletin Database, (2019)

Mileikowskya. C., Cucinotta, F.A., Wilson, J.W., Gladman, B., Horneck, G., Lindegren, L., Melosh, J., Rickman, H., Valtonen, M., Zheng, J.Q. (2000), Risks threatening viable transfer of microbes between bodies in our solar system Planetary and Space Science, 48, 1107-1115.

Mileikowsky, C., Cucinotta, F.A., Wilson, J.W., Gladman, B., Horneck, G., Lindegren, L., Melosh, J., Rickman, H., Valtonen, M., Zheng, J.Q. (2000), Natural transfer of viable microbes in space. Part 1: From Mars to Earth and Earth to Mars. Icarus, 145, 391-427.

Mitchell, F. J., & Ellis, W. L. (1971). Surveyor III: Bacterium isolated from lunar retrieved TV camera. In A.A. Levinson (ed), Proceedings of the second lunar science Conference, MIT press, Cambridge.

Misconi, N. Y., Weinberg, J. L (1978) is Venus Concentrating Interplanetary Dust Toward Its Orbital Planet Science, 200, 1484-1485. DOI: 10.1126/science.200.4349.1484

Moeller, R., et al. (2012). Protective Role of Spore Structural Components in Determining Bacillus subtilis Spore Resistance to Simulated Mars Surface Conditions. Applied and Environmental Microbiology, DOI: 10.1128/AEM.02527-12.

Möhler, O., DeMott, P.J., Vali, G., Levin, Z. Microbiology and atmospheric processes: The role of biological particles in cloud physics. Biogeosci. Discuss. 2007, 4, 2559–2591.

Moment GB. 1949. On the relation between growth in length, the formation of new segments, and electric potential in an earthworm. J Exp Zool 112:1-12. M

Moseley BE, & Mattingly A (1971). Repair of irradiated transforming deoxyribonucleic acid in wild type and a radiation- sensitive mutant of Micrococcus radiodurans". J. Bacteriol. 105 (3): 976-83. PMC 248526 Freely accessible. PMID 4929286.

Moser, D. P. Gihring, T. M., Brockman, F. J., et al., *2005). Desulfotomaculum and Methanobacterium spp. Dominate a 4- to 5-Kilometer-Deep Fault. Applied and Environmental Microbiology, 71, 8773-8783.

Mostafa M. , Afify N. , Gaber A. , Abu Zaid E. (2004) Investigation of thermal properties of some basalt samples in Egypt, Journal of Thermal Analysis and Calorimetry 75, 2004, https://doi.org/10.1023/B:JTAN.000001734.

Mueller, R. C., Belnap, J., and Kuske, C. R. (2015). Soil bacterial and fungal community responses to nitrogen addition are constrained by microhabitat in an arid shrubland. Front. Microbiol. 10:e0117026. doi: 10.1371/journal.pone.0117026

Murray, N., Brice Ménard1, and Todd A. Thompson, 2011. Radiation Pressure From Massive Star Clusters As A Launching Mechanism For Super-Galactic Winds, The Astrophysical Journal, 735, 66-71.

Narbonne GM (2005) The Ediacara biota: Neoproterozoic origin of animals and their ecosystems. Annu Rev Earth Planet Sci 33:421–442.

Narbonne, G. M., & Gehling, J. G. (2003). Life after snowball: The oldest complex Ediacaran fossils. Geology, 31, 27-30.

Nicholson, W. L., Munakata, N., Horneck, G., Melosh, H. J., Setlow, P. (2000). Resistance of Bacillus Endospores to Extreme Terrestrial and Extraterrestrial Environments, Microbiology and Molecular Biology Reviews 64, 548-572.

Nicholson, W.L., et al. (2012). Growth of Carnobacterium spp. from permafrost under low pressure, temperature, and anoxic atmosphere has implications for Earth microbes on Mars. PNAS. https://doi.org/10.1073/pnas.1209793110.

Noffke, N. (2015). Ancient Sedimentary Structures in the < 3.7b Ga Gillespie Lake Member, Mars, That Compare in macroscopic Morphology, Spatial associations, and Temporal Succession with Terrestrial Microbialites. Astrobiology 15(2): 1-24.

Novikova, N (2009) Mirobiological research on board the ISS, Planetary Protection. The Microbiological Factor of Space Flight. Institute for Biomedical Problems, Moscow, Russia.

Novikova, N et al. (2016) Long-term spaceflight and microbiological safety issues. Space Journal, https://roomeu.com/article/long-term-spaceflight-and-microbiological-safety-issues.

Onofri, S., R. de la Torre, J-P de Vera, E., et al. (2012) Survival of rock-colonizing organisms after 1.5 years in outer space." Astrobiology. 2012;12:508-516.

Onofri, S., Selbman, L., Pacelli, C. et al. (2019) Survival, DNA and ultrastructural integrity of a cryptoendolithic Antarctic fungus on Mars and lunar rock analogues exposed outside the International Space Station. Astrobiology, 19, 2, DOI:10.1089/ast.2017.1728. <P>

Osman, S., Peeters, Z., La Duc, M.T., Mancinelli, R., Ehrenfreund, P., Venkateswaran, K., (2008). Effect of shadowing on survival of bacteria under conditions simulating the Martian atmosphere and UV radiation. Applied and Environmental Microbiology 74, 959-970.

Pacelli, C., L. Selbmann, L. Zucconi, J. P. P. De Vera, E. Rabbow, G. Horneck, R. de la Torre, S. Onofri (2016) "BIOMEX experiment: Ultrastructural alterations, molecular damage and survival of the fungus Cryomyces antarcticus after the Experiment Verification Tests." (Origin of Life and Evolution of Biospheres 47(2):187-202.

Peccia, J.; Hernandez, M. (2006). Incorporating polymerase chain reaction-based identification, population characterization, and quantification of microorganisms into aerosol science: A review. Atmos. Environ. 2006, 40, 3941–3961.

Pérez-Hoyos S., Sánchez-Lavega A., García-Muñoz A., Irwin P.G.J., Peralta J., Holsclaw G., McClintock W.M., and Sanz-Requena J.F. (2017) Venus upper clouds and the UV-absorber from MESSENGER/MASCS observations. J Geophys Res Planets 123:43

Pirt, M. W., & Pirt, S. J. (1980). The influence of carbon dioxide and oxygen partial pressures on Chlorella growth in photosynthetic stead-state cultures. J. General Microbiology, 119, 321-326.

Plummer, W.T. (1969) Venus clouds: test for hydrocarbons. Science 163:1191–1192.

Pointing, S.B., & Belnap, J., (2012) Microbial colonization and controls in dry-land systems, Nature Reviews Microbiology volume 10, pages 551–562.

Polymenakou, P. P. (2012). Atmosphere: A Source of Pathogenic or Beneficial Microbes? Atmosphere 2012, 3(1), 87-102.

Preece, C., Phoenix, G. K. (2013). Responses of sub-arctic dwarf shrubs to low oxygen and high carbon dioxide conditions, Environmental and Experimental Botany, 85, 7-15.

Prospero, J.M.; Blades, E.; Mathison, G.; Naidu, R. Interhemispheric transport of viable fungi and bacteria from Africa to the Caribbean with soil dust. Aerobiologia 2005, 21, 1–19.

Puleo JR, Fields ND, Bergstrom SL, Oxborrow GS, Stabekis PD, Koukol R. (1977) Microbiological profiles of the Viking spacecraft. Appl Environ Microbiol. 33:379-384.

Rabb, H. (2018). Life on Mars, Astrobiology Society, SoCIA, University of Nevada, Reno, USA. April 14, 2018.

Raggio J, Pintado A, Ascaso C, De La Torre R, De Los Ríos A, Wierzchos J, Horneck G, Sancho LG (2011). Whole lichen thalli survive exposure to space conditions: results of Lithopanspermia experiment with Aspicilia fruticulosa. Astrobiology. 2011 May;11(4):281-92. doi: 10.1089/ast.2010.0588.

Randel, W. J. Russell, J.M., Rochie A., Waters, J. W. (1998). Seasonal Cycles and QBO Variations in Stratospheric CH4 and H2O Observed in UARS HALOE Data. Journal of the Atmospheric Sciences, 55. 163-185.

Randel, W.J., Park, M., Emmons, L., Kinnison, D., Bernath, P., Walker, K.A., Boone, C., and Pumphrey, H. (2010) Asian monsoon transport of pollution to the stratosphere. Science 328:611–613.

Reponen, T., S. A. Grinshpun , K. L. Conwell , J. Wiest & M. Anderson (2001) Aerodynamic versus physical size of spores: Measurement and implication for respiratory deposition, Grana, 40:3, 119-125.

Richter, Katrin; Schicklberger, Marcus; Gescher, Johannes (2012-02-01). "Dissimilatory reduction of extracellular electron acceptors in anaerobic respiration". Applied and Environmental Microbiology. 78 (4): 913–921.

Rizzo, V., & Cantasano, N. (2009) Possible organosedimentary structures on Mars. International Journal of Astrobiology 8 (4): 267-280.

Rizzo, V. & Cantasano, N. (2016). Structural parallels between terrestrial microbialites and Martian sediments. International Journal of Astrobiology, doi:10.1017/S1473550416000355

Robb, F., Antranikian, G., Grogan, D., Driessen, A. (2007). Thermophiles: Biology and Technology at High Temperatures, CRC Press.

Roffman, D. A. (2019) Meteorological Implications: Evidence of Life on Mars? Journal of Astrobiology and Space Science Reviews, 1, 329-337, 2019

Rohatschek, H. (1996) Levitation of stratospheric and mesospheric aerosols by gravito-photophoresis. J Aerosol Sci 27:467–475.

Rossi L., Marcq E., Montmessin F., Fedorova A., Stam D., Bertaux J.-L., and Korablev O. (2015) Preliminary study of Venus cloud layers with polarimetric data from SPICAV/VEx Planet. Space Sci 113-114 (2015):159–168.

Ruffi, W. & Farmer, J.D., (2016). Silica deposits on Mars with features resemblinghot spring biosignatures at El Tatio in Chile. Nature Communications, 7, 13554

Russell C.T., Vaisberg O. (1983) The Interaction Of The Solar Wind With Venus, University Of Arizona Press.

Sagan, C. & Morowtz, H. (1967), Life in the clouds of Venus. Nature 215, 1259-

1260.

Saleh, Y. G., M. S. Mayo, and D. G, Ahearn (1988) Resistance of some common fungi to gamma irradiation." Appl. Environm. Microbiol. 1988, 54: 2134-2135.

Sanchez, F. J., Mateo-Martí. E., Raggio, J., Meeßen, J., Martínez-Frías, J., .SanchoL.G, Ott,. S..de la Torrea, R. (2012) The resistance of the lichen Circinaria gyrosa (nom. provis.) towards simulated Mars conditions-a model test for the survival capacity of an eukaryotic extremophile." Planetary and Space Science, 2012, 72(1), 102-110.

Sancho L. G., de la Torre, R., Horneck, G., Ascaso, C. , de los Rios, A. Pintado,A., Wierzchos, J.,Schuster, M. (2007). Lichens Survive in Space: Results from the 2005 LICHENS Experiment Astrobiology. 7, 443-454.

Satoh, K, Y, Nishiyama, T. Yamazaki, T. Sugita, Y. Tsukii, K. Takatori, Y. Benno, and K. Makimura. "Microbe-I (2011) Fungal biota analyses of the Japanese experimental module KIBO of the International Space Station before launch and after being in orbit for about 460 days." Microbiol Immunol. 2011 Dec;55(12):823-9. doi: 10.1111/j.1348-0421.2011.00386.x.

Sattler, B., Puxbaum, H.; Psenner, R. (2001). Bacterial growth in supercooled cloud droplets. Geophys. Res. Lett. 28, 239–242.

Schaber, G. G., Kirk, L., Strom, R. G. (1992). Geology and distribution of impact craters on Venus: what are they telling us? JGR Planets, 97

Schroder, K-P, Smith, R. C. (2008). Distant future of the Sun and Earth revisited. Mon. Not. R. Astron. Soc. 000, 1-10.

Schuerger, A.C., Ulrich, R., Berry, B.J., and Nicholson, W.L. (2013) Growth of Serratia liquefaciens under 7 mbar, 0C, and CO2-enriched anoxic atmospheres. Astrobiology 13:115–131.

Schulze-Makuch, D. and Irwin, L.N. (2002) Reassessing the possibility of life on Venus: proposal for an astrobiology mission. Astrobiology 2:197–202.

Schulze-Makuch, D., Grinspoon D.H., Abbas O., Irwin L.N., and Bullock M.A. (2004) A sulfur-based survival strategy for putative phototrophic life in the Venusian atmosphere. Astrobiology 4:11–18.

Schulze☐Makuch, D. Irwin, L.N., Lips, J.H., LeMone, D., Dohm, J.M., Farien, A. G. (2005) Scenarios for the evolution of life on Mars, Journal of Geophysical Research: Planets, 110, E12.

Sebach, J., & Libby, W. F. (1970) Vegetative life on Venus? Investigations with algae which grow under pure CO2 in hot acid media at elevated pressures. Space Life Sciences, 2, 121-143.

Seckbach, J., Baker, F. A., Shugarman, P. M., (1970). Algac thrive under pure CO2. Nature 227, 744-745.

Selbman, L, Zucconi, D. Isola, and D. Onofri (2015) Rock black fungi: excellence in the extremes. From the Antarctic to Space." 2015. Current Genetics 61: 335-345. DOI 10.1007/s00294-014-0457-7

Setlow, P. (2006). Spores of Bacillus subtilis: their resistance to and killing by radiation, heat and chemicals. Journal of Applied Microbiology 101, 514-525.

Setlow, B., Setlow, P. (1995). Small, acid-soluble proteins bound to DNA protect Bacillus subtilis spores from killing by dry heat. Appl Environ Microbiol. 61, 2787-2790.

Shen, Y., Zhang, T., & Hoffman, P. F. (2008). On the coevolution of Ediacaran oceans and animals --PNAS, 105, 7376-7381.

Smerdon, J..E. Henry N. Pollack Vladimir Cermak John W. Enz Milan Kresl Jan Safanda John F. Wehmiller(2004) Journal of Geophysical Research, 109,. 44-54.

Smith, D.J., Griffin, D.W., McPeters, R.D., Ward, P.D., and Schuerger, A.C. (2011) Microbial survival in the stratosphere and implications for global dispersal. Aerobiologia 27:319–332.

Soffen, G.A. (1965). NASA Technical Report, N65-23980.

Stetter, K. O. (2006). Hyperthermophiles in the history of life, Philos Trans R Soc Lond B Biol Sci. 361(1474): 1837–1843.

Steven, B., Gallegos-Graves, L. V., Belnap, J., and Kuske, C. R. (2013). Dryland soil bacterial communities display spatial biogeographic patterns associated with soil depth and soil parent material. FEMS Microbiol. Ecol. 86, 101–113. doi: 10.1111/1574-6941.12143

Surkov, Y. A., L. P. Moskalyeva, O. P. Shcheglov, V. P. Kharyukova, O. S. Manvelyan, V. S. Kirichenko, and A.D. Dudin, (1983). Determination of the elemental composition of rocks on Venus by Venera 13 and Venera 14 (preliminary results), Proc. Lunar

Planet. Sci. Conf. 13th, Part 2, J. Geophys. Res., 88, suppl. A481.

Sunde, E.P., Setlow, P., Hederstedt, L., Halle, B. (2009). The physical state of water in bacterial spores. Proceedings of the National Academy of Sciences of the United States of America 106, 19334-19339.

Szewczyk, N.J., Mancinelli, R.L., McLamb, W., Reed, D., Blumberg, B.S., Conley, C.A., (2005). Caenorhabditis elegans survives atmospheric breakup of STS-107, Space Shuttle Columbia. Astrobiology 5, 690-705.

Tegen, I. Andrew A. Lacis (2012). Modeling of particle size distribution and its influence on the radiative properties of mineral dust aerosol, Aerosol Atmospheric Optics, 21, 19237-19244.

Thomas, W. H., & . Gibson, C. H. (1990) Effects of small-scale turbulence on microalgae, Journal of Applied Phycology, 2, 71–77.

Thomas, W. H., Vernet, M., Gibson, C. H. (1995). Effects Of Small☐Scale Turbulence On Photosynthesis, Pigmentation, Cell Division, And Cell Size In The Marine Dinoflagellate Gomaulax Polyedra (Dinophyceae)1 Journal of Phycology, 1995, 31, 50-59.

Tugay, T. Zhdanova, N.N., Zheltonozhsky, V., Sadovnikov, L., Dighton, J. (2006). The influence of ionizing radiation on spore germination and emergent hyphal growth response reactions of microfungi, Mycologia, 98(4), 521-527.

Vaishampayan P, Rabbow E, Horneck G, Venkateswaran K .(2012) Survival of Bacillus pumilus spores for a prolonged period of time in real space conditions. Astrobiology. 2012;12:487-497.

Valdes J., Pedroso I., Quatrini R., Dodson R.J., Tettelin H., Blake R.2nd, Eisen J.A., and Holmes D.S. (2008) Acidithiobacillus ferrooxidans metabolism: from genome sequence to industrial applications. BMC Genomics 9, 597-601.

Van Den Bergh, S., (1989) Life and Death in the Inner Solar System, Publications of the Astronomical Society of the Pacific, 101, 500-509.

Van Eaton, A.R., Harper, M.A., and Wilson, C.J.N. (2013) Highflying diatoms: widespread dispersal of microorganisms in an explosive volcanic eruption. Geology doi:10.1130/G34829.1.

Venkateswaran, K., M. T. La Duc, and MT Vaishampay (2012) Genetic Inventory Task: Final Report," JPL Publication 12-12. California Institute of Technology; Pasadena.

Vesper, S.J., W. Wong, C.M. Kuo and D.L. Pierson. (2008) Mold species in dust from the ISS identified and quantified by mold-specific quantitative PCR. Research in Microbiology. 159: 432-435.

Vinogradov, A.P., Yu, P., Surkov, A., Kirnozov, F. F. 1973) The content of uranium, thorium and potassium in the rock of Venus as measured by Venera 8, Icarus, 20, 253-259.

Vreeland, R.N., Rosenzweig, W.D., and Powers, D.W. (2000) Isolation of a 250 million-year-old halotolerant bacterium from a primary salt crystal. Nature 407, 897–900.

Wainwright, M., Fawaz Alshammari, F., Alabri, K. (2010). Are microbes currently arriving to Earth from space? Journal of Cosmology, 2010, Vol 7, 1692-1702.

Way M.J., Del Genio A.D., Kiang N.Y., Sohl L.E., Grinspoon D.H., Aleinov I., Kelley M., Clune T.(2016) Was Venus the first habitable world of our solar system? Geophys Res Lett43:8376–8383.

Way M.J., Del Genio A.D., (2019). A view to the possible habitability of ancient Venus over three billion years, EPSC Abstracts Vol. 13, EPSC-DPS2019-1846-1, 2019 EPSC-DPS Joint Meeting 2019

Weirzchos, J. (2012). Microorganisms in desert rocks: the edge of life on Earth, Int Microbiol (2012); 15 171-181.

Wember, V. V., Zhdanova, N. N. (2001) Peculiarities of linear growth of the melanin-containing fungi Cladosporium sphaerospermum Penz. and Alternaria alternata (Fr.) Keissler. Mikrobiol. Z. 63: 3-12.

White, O. Eisen, J.A., Heidelberg, J. F., et al. (1999) Genome Sequence of the Radioresistant Bacterium Deinococcus radiodurans R1, Science, 286, 1571-1577.

Wibking, B..D., Thompson, T. A., Krumholz, M. R. (2018) Radiation pressure in galactic discs: stability, turbulence, and winds in the single-scattering limit, MNRAS 477, 4665–4684.

Worth, R. J., Steinn Sigurdsson, and Christopher H. House (2013) Seeding Life on the Moons of the Outer Planets via Lithopanspermia, Astrobiology, 13

Zakharova,K., Marzban, G., de Vera, J-P., Lorek, A., Sterflinger, K. (2014). Protein patterns of black fungi under simulated Mars-like conditions. Scientific Reports, 4, 5114.

Zhdanova NN, Lashko TN, Vasiliveskaya AI, Bosisyuk LG, Sinyavskaya OI, Gavrilyuk VI, Muzalev PN. (1991). Interaction of soil micromycetes with 'hot' particles in the model system. Microbiol J 53:9-17.

Zhdanova, N. N., T. Tugay, J. Dighton, V. Zheltonozhsky and P McDermott, (2004) Ionizing radiation attracts soil fungi." Mycol Res. 2004, 108: 1089-1096.

Zhuravskaya AN, Kershengoltz BM, Kuriluk TT, Shcherbakova TT. (1995). En-zymological mechanisms of plant adaptation to the conditions of higher natural radiation background. Rad Biol Radioecol 35:249-355.

II. EVIDENCE OF LIFE ON MARS

R. Gabriel Joseph1, Regina Dass2, V. Rizzo3,4, N. Cantasano5,
G. Bianciardi6

1Astrobiology Associates, Northern California, USA
2Molecular Fungal Genetics and Mycotoxicology Laboratory, Department of
Microbiology, School of Life Sciences, Pondicherry University, Kalapet, India
3Emeritus, Universit`a di Firenze, Via G. La Pira, 4, Firenze
4Emeritus, Consiglio Nazionale delle Ricerche, I.S.A.FO.M. U.O.S.,
Cosenza, Italy
5Consiglio Nazionale delle Ricerche, I.S.A.FO.M. U.O.S., Cosenza, Italy
6Department of Medical Biotechnology, Siena University, Italy

Abstract

Evidence is reviewed which supports the hypothesis that prokaryotes and eukaryotes may have colonized Mars. One source of Martian life, is Earth. A variety of species remain viable after long term exposure to the radiation intense environment of space, and may survive ejection from Earth following meteor strikes, or ejection from the stratosphere and mesosphere via solar winds, whereas simulations studies have demonstrated that prokaryotes, fungi and lichens can survive in simulated Martian environments--findings which support the hypothesis life may have been repeatedly transferred to Mars from Earth. Four independent investigators have reported what appears to be fungi and lichens on the Martian surface, whereas a fifth investigator reported what may be cyanobacteria. In another study, a statistically significant majority of 70 experts, after examining Martian specimens photographed by NASA, identified and agreed fungi, basidiomycota ("puffballs"), and lichens may have colonized Mars. Fifteen specimens resembling and identified as "puffballs" were photographed emerging from the ground over a three day period. It is possible these latter specimens are hematite and what appears to be "growth" is due to a strong wind which uncovered these specimens--an explanation which cannot account for before and after photos of what appears to be masses of fungi growing atop and within the Mars rovers. Terrestrial hematite is in part fashioned and cemented together by prokaryotes and fungi, and thus Martian hematite is also evidence of biology. Three independent research teams have identified sediments on Mars resembling stromatolites and outcroppings having micro meso and macro characteristics typical of terrestrial microbialites constructed by cyanobacteria. Quantitative morphological analysis determined these latter specimens are statistically and physically similar to terrestrial stromatolites. Reports of water, biological residue discovered in Martian meteor ALH84001, the seasonal waning and waxing of atmospheric and ground level Martian methane which on Earth is 90% due to biology and plant growth and decay, and results from the 1976 Mars Viking labeled Release Experiments indicating biological reproduction, also strongly support the hypothesis that Mars was, and is, a living planet.

1. Overview: The Evidence

Presented here is a body of evidence which does not prove but supports the hypothesis Mars was and is a living planet, hosting prokaryotes, lichens, and fungi. This evidence includes: 1) Results from simulation studies demonstrating a variety of species can survive in a Mars-like environment (Cockell et al. 2005; Osman et al. 2008; Mahaney & Dohm, 2010; Pacelli et al. 2016; Sanchez et al. 2012; Schuerger et al., 2017; Selbman et al. 2015), particularly if shielded by soil and stone and provided water for which there is now evidence (Malin &Edgett 1999, 2000; Perron et al. 2007; Renno et al. 2009; Villanueva et al. 2015); 2) NASA's Mars Viking Labeled Release experiments (Levin 1976a,b) which detected evidence which met the criteria established by pre-mission field-tests for biological activity; 3) The observation of specimens which resemble fungi, "puffballs," algae and lichens (Dass 2017; Joseph 2014; Kupa 2017; Rabb 2015, 2018; Small 2015); 4) Seasonal waxing, waning, and continual replenishment of Martian methane (Formisano et al. 2004; Mumma et al. 2009; Webster et al. 2018) and which has no obvious purely geological source and 90% of which on Earth is due to biological activity including seasonal plant growth; 5) Observations of digitate silica structures that closely resemble complex sedimentary formations produced by a combination of abiotic and biotic processes (Ruffi & Farmer 2016) as well what appears to be microbial mats (stromatolites) which may have been built by water-dwelling cyanobacteria, possibly between 3.2 to 3.7 billion years ago (Bianciardi et al. 2014, 2015; Noffke 2015; Rizzo & Cantasano 2009, 2011, 2016); 6) Specimens identified as "hematite" which were likely produced in thermal ("hot") springs (Squyres et al. 2004) and which, on Earth, are fashioned, in part, via water-dwelling prokaryotes and fungi (Ayupova et al 2016; Bosea et al. 2009; Claeys 2006; Lower 2009; Fredrickson et al., 2008; Gralnick and Hau 2007; Owocki et al. 2916); 7) Detection of carbonates and polycyclic aromatic hydrocarbons in Martian meteorite ALH84001 which has been dated to 4 billion years ago and which were also generated in the presence of water (Clement et al. 1998; McKay et al. 2009; Thomas-Keprta et al. 2009).

2. The Transfer of Life from Earth To Mars

One obvious source of life on Mars would be Earth. It is probable that solar winds striking and propelling microbe-laden dust and debris in the stratosphere and mesosphere (cf Arrhenius, 1908) and microbes dwelling in rock and stone and ricocheted into space from Earth by meteor strikes have repeatedly contaminated Mars and other planets (Davies, 2007; Fajardo-Cavazosa et al. 2007; Hara et al. 2010; Melosh 2003; Mileikowsky et al. 2000; Schulze-Makuch, et al. 2005) and vice-versa.

3. Solar Winds vs Microbes in the Stratosphere and Mesosphere

Over 1,8000 different types of bacteria as well as fungi and algae thrive and flourish within the troposphere, the first layer of Earth's atmosphere (Brodie et al. 2007). Air is an ideal transport mechanism and serves as a major pathway for the dispersal of bacteria, virus particles, algae, protozoa, lichens, and fungi including those which dwell in soil and water. Viable microorganisms and spores have been recovered at heights of 40 km (Soffen 1965), 61 km (Wainwright et al., 2010) and up to 77 km (Imshenetsky, 1978) within the mesosphere. These include Mycobacterium, Micrococcus, and fungi Aspergillus niger, Circinella muscae, and Penicillium notatumm 77 km above Earth (Imshenetsky, 1978).

In one study designed to disprove the possibility NASA might contaminate Mars, samples of Bacillus pumilus were launched via a high-altitude NASA balloon to an altitude of 31 km above sea level (Khodad et al. 2017). Nevertheless, a large number of Bacillus pumilus remained viable; and it only takes one bacterium to produce billions of bacterial offspring.

Moreover, due to tropical storms, monsoons, and even seasonal upwellings of columns of air (Randel et al., 1998), microbes, spores, fungi, along with water, methane, and other gases may be transported to the stratosphere and mesosphere where they may remain viable (Imshenetsky, 1978; Soffen 1965; Wainwright et al., 2010). As first formally proposed by Nobel Laureate Dr. Svante Arrhenius (1908) solar winds and photons could disperse space-borne organisms throughout the cosmos.

Hence, it can be readily assumed that microbes not only flourish in the troposphere, but when lofted into the stratosphere and mesosphere many remain viable and may then be blown into space by powerful solar winds (Arrhenius 1908; Joseph & Schild, 2010) where, as shown experimentally, they can easily survive (Horneck, et al. 1994, 2002, Nicholson et al. 2000; Novikova et al. 2016; Onofri et al. 2012; Raggio et al. 2011; Sancho et al. 2007; Setlow 2006).

Between September 22-25, 1998, as measured by NASA's Ultraviolet Imager aboard the Polar spacecraft, a series of coronal mass ejections (CME) and a powerful solar wind created a shock wave which struck the magnetosphere and the polar regions with sufficient force to cause oxygen, helium, hydrogen, and other gases (Moore & Horwitz, 1998), as well as water molecules and surface dust (Schroder & Smith, 2008), to gush from Earth's upper atmosphere into space. Normally the pressure is around two or three nanopascals. However, when the CME struck on September 24, 1998, the pressure jumped to ten nanopascals. Such events may have occurred repeatedly throughout Earth's history.

Thus it could be predicted that some airborne microbes, fungi, lichens, and algae, as well as water and dust, have been repeatedly lofted into the upper atmosphere; that a significant number remained viable, and were then swept into space and propelled by solar winds throughout the solar system (Arrhenius,1908), some of which may have landed on Mars only to go forth and multiply.

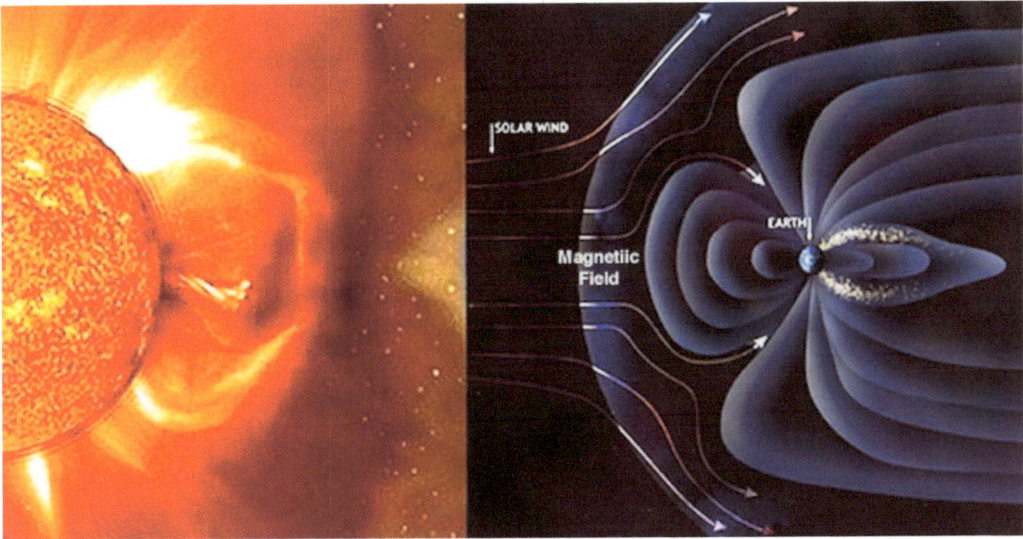

4. Meteorites, Microbes and Ejecta from Earth to Mars

Although innumerable meteorites disintegrate upon striking Earth's upper atmosphere, those at least ten kilometers across will punch a hole in the atmosphere and continue their descent (Van Den Bergh, 1989). When meteors this size or larger strike the surface, tons of dust, rocks, and other debris may be jettisoned over 100 km above the planet (Covey et al. 1994; Hara et al. 2010) and ejected into space, some passing through that atmospheric hole before air can rush back in to fill the gap (Van Den Bergh, 1989).

Asteroids and meteors striking Earth may have repeatedly sheared away masses of earth and rock, and blasted this material (and presumably any adhering microbes, fungi, algae, and lichens) into space (Davies, 2007; Fajardo-Cavazosa et al. 2007; Hara et al. 2010; Melosh 2003; Schulze☐Makuch, et al. 2005), where they can easily survive (Horneck, et al. 1994, 2002, Nicholson et al. 2000; Novikova et al. 2016; Onofri et al. 2012; Raggio et al. 2011; Sancho et al. 2007; Setlow 2006). Some of this microbe-laden debris may have later crashed on Mars (Davies, 2007; Fajardo-Cavazosa et al. 2007; Hara et al. 2010; Schulze☐Makuch, et al. 2005) where, as demonstrated by simulation studies, a variety of organisms can also survive (Cockell et al. 2005; Osman et al. 2008; Mahaney & Dohm, 2010; Pacelli et al. 2016; Sanchez et al. 2012; Selbman et al. 2015).

Experiments have shown that microbes can survive the shock of a violent impact casting them into space (Mastrapaa et al. 2001; Burchell et al. 2004; Burchella et al. 2001). Further, a substantial number could easily survive the descent to the surface of a planet (Burchella et al. 2001; Horneck et al. 2002), even following high atmospheric explosions, i.e. the Columbia space shuttle explosion (Szewczyk et al., 2005), and despite reentry speeds of up 9700 km h-1 (McLean et al., 2006).

When meteors strike Earth's atmosphere, they are subjected to extremely high temperatures for only a few seconds. If of sufficient size, the interior of the meteor will stay relatively cool, with the surface material acting as a heat shield. Heat does not affect the material uniformly. The interior may never be heated above 100 C whereas spores can survive post shock temperatures of over 250 C (Horneck et al. 2002). Mars has a very thin atmosphere.

Thus, many species of microbe have evolved the ability to survive a violent hypervelocity impact and extreme acceleration and ejection into space, including extreme shock pressures of 100 GPa; the frigid temperatures and vacuum of an interstellar environment; the UV rays, cosmic rays, gamma rays, and ionizing radiation they would encounter; and the descent through the atmosphere and the crash landing onto the surface of a planet.

Certainly, surviving organisms dwelling within ejecta from Earth might land on Mars (Davies, 2007; Fajardo-Cavazosa et al. 2007; Hara et al. 2010; Melosh 2003; Mileikowsky et al. 2000; Schulze-Makuch, et al. 2005). And those which can adapt, would likely go forth and multiply.

5. Simulation Studies of Life on Mars

An increasing number of independent investigators have found that a variety of species, including bacteria, algae, fungi and lichens, can survive a simulated Mars-like environment, and that survival rates dramatically increase if provided water or shielded by rock, sand, or soil (Cockell et al. 2005; Osman et al. 2008; Mahaney & Dohm, 2010; Pacelli et al. 2016; Sanchez et al. 2012; Selbman et al. 2015; Villanueva et al. 2015). These simulated environments have included those which imitate Martian radiation, temperature extremes and variations, the low surface pressures, atmospheric gas pressures, the distance between Mars and the sun, the Martian summer/winter solstices and spring/fall equinoxes, environmental parameters analogues to the 24 hours 39 minutes circadian cycle of the Red Planet, effects of shielding and aqueous vs desert vs arctic vs subsurface conditions, and in a CO_2-enriched anoxic atmosphere.

For example, Moeller et al. (2010) found that spores of Bacillus subtilis survived simulated Martian atmospheric and UV irradiation conditions, whereas de Vera and colleagues (2014) reported that cyanobacteria collected from cold and hot deserts on Earth were found to survive "Mars-like conditions such as atmospheric composition, pressure, variable humidity (saturated and dry conditions) and strong UV irradiation." Nicholson et al. (2012) reported that six subspecies of the genus Carnobacterium

collected from a permafrost borehole in northeastern Siberia—considered to be analogs of the subsurface environment of Mars—and nine additional species of Carnobacterium were all capable of flourishing and growing under Mars-like conditions. In yet another study, four methanogen species (Methanosarcina barkeri, Methanococcus maripaludis, Methanothermobacter wolfeii, Methanobacterium formicicum) survived exposure to low pressure conditions similar to Martian surface pressure (Tarasashvili et al., 2013).

Cyanobacteria also tolerate Mars-like conditions. Olsson-Francis and colleagues (2009) exposed akinetes (dormant cells formed by filamentous cyanobacteria) to extraterrestrial conditions, including periods of desiccation, temperature extremes (−80 to 80°C), and UV radiation (325–400 nm), and which displayed high levels of viability under these environments similar to Mars.

Eukaryotes (fungi, lichens) are also survivors (de Vera 2012; Sanchez et al. 2012; Zakharova et al. 2014). Zakharova et al. (2014) report that microcolonial fungi, Knufia perforans and Cryomyces antarcticus, as well as Exophiala jeanselmei (a species of black yeasts), not only survived but adapted and showed no evidence of stress after long term exposure to thermo-physical Mars-like conditions. Likewise, Onofri et al. (2019), after growing dried colonies of the Antarctic cryptoendolithic black fungus Cryomyces antarcticus and exposing them for 16 months to simulated Mars-like conditions on the International Space Station, found that "C. antarcticus was able to tolerate the combined stress of different extraterrestrial substrates, space, and simulated Mars-like conditions in terms of survival, DNA, and ultrastructural stability."

Lichens are a symbiotic organism which have been classified as both a prokaryote and eukaryote and are comprised of cyanobacteria (algae) and fungi (Armstrong 2017; Techler & Wedin, 2008)—two species which remain viable when exposed to Mars-like stimulated environments (Olsson-Francis et al 2009; Zakharova et al. 2014). Likewise, lichens easily survive environmental extremes, lack of water, desiccation, temperatures as low as -196°C (Armstrong 2017; Becket et al. 2009), as well as high levels of UV radiation and direct exposure to the radiation intense environment of space (Raggio et al. 2011; Sancho, et al. 2007). Hence, perhaps not surprisingly, studies have demonstrated that lichens remain viable and maintain photosynthetic activity when exposed to simulated Martian temperatures, atmosphere, humidity, and UV radiation (de Vera 2012; De la Torre Noetzel, 2017; Sanchez et al. 2012).

For example, Noetzel et al. (2017) exposed lichens to real space outside the ISS and to a Mars simulated environment for 18 months. The samples remained viable and these investigators reported normal metabolic activity of those exposed to the Mars-like environment.

Simulation studies performed by numerous teams of independent investigators have thus demonstrated that a variety of prokaryotes and eukaryotes, including cyanobacteria, methanogens, fungi and lichens, could survive and even flourish on Mars, especially if dwelling beneath the soil or rock shelters and provided water—for which there is now evidence (Malin & Edgett 1999, 2000; Perron et al. 2007; Renno et al. 2009; Villanueva et al. 2015). Although controversial, the results from NASA's Mars Viking Labeled Release experiments also suggests that prokaryotes and simple eukaryotes not only survive on Mars but are engaging in biological activity (Levin 1976a,b; Levin & Straat 2009).

6. Radiation and Martian Life

Martian ground level radiation has been estimated to equal "0.67 millisieverts per day" (Hassler et al. 2013). This is significantly and profoundly below the radiation tolerance levels of a variety of prokaryotes (Moseley and Mattingly 1971; Ito et al. 1983) and simple eukaryotes, including fungi which can withstand radiation doses up to 1.7×104 Gy (Saleh, et al. 1988).

Moreover, fungi, lichens and numerous species of microbe are attracted to and thrive in highly radioactive environments (Becket et al. 2008; Dadachova et al. 2007; Tugay et al. 2006; Wember & Zhdanova 2001), even in space. Novikova et al. (2016) and Vesper et al. (2008) reported that fungi are invigorated and grow rapidly within the International Space Stations as a consequence of the heightened radiation levels. Moreover, these scientists discovered that these space-fungi are impossible to eradicate. Moreover, fungi flourish on the outskirts and along the walls of the damaged and highly radioactive Chernobyl nuclear power plant (Dighton et al. 2008; Zhdanova et al. 2004). Fungi, lichens, and prokaryotes also survive long-term direct exposure to space, gamma, and solar UV radiation and remain viable (Horneck et al. 1994, 2002; Nicholson et al. 2000; Novikova 2009; Novikova et al 2016; Onofri et al. 2012; Sato et al. 2011; Tugay et al. 2006; Sancho et al. 2007; Raggio et al. 2011).

Moreover, fungi (Wember & Zhdanova 2001; Zhdanova et al. 2004) and radiation-tolerant bacteria (Moseley and Mattingly 1971; Ito et al. 1983) will seek out and grow towards sources of radiation which serve as an energy source which they metabolize (Dighton et al. 2008; Tugay et al. 2006). Even if their DNA is damaged by radiation levels above their tolerance levels, they can easily repair these genes due to a redundancy of genes with repair functions (White et al. 1999).

These and other species may also develop adaptive features—a property described as "radiostimulation," "radiation hormesis," and "adiotropism" (Tugay et al. 2006; Zhdanova et al 2004)—and which also occurs in animals and plants living with increasing levels of background radiation (Alshits et al 1981; Calabrese & Baldwin 2000; Zhuravskaya et al 1995). These radiation-induced adaptations include tissue and cellular regeneration and growth (Becker 1984; Becker & Sparado 1972; Occhipinti et al. 2014; Levin 2003; Maffei 2014; Moment, 1949).

Tugay and colleagues (2006; Zhdanova et al. 1991, 2004) exposed microfungi and fungi to pure or mixed radiation (137 Cs, 123 Te, 109 Cd, 121 Sn), gamma irradiation (121 Sn) 200–400 Gy, and mixed gamma and beta radiation (137 Cs) (100–150) Gy (equivalent to an electron dose of 300–500 Gy), and found that 60% of fungal strains exhibited positive radiotropism, significant growth, and enhanced spoor production.

The varying levels of radiation on Mars would not be a hinderance to fungi, lichens, and numerous other species.

7. Evidence of Lichens and Fungi on Mars

Four independent investigators, after examining photos taken by NASA's Mars Rovers Opportunity and Curiosity, have observed hundreds of specimens resembling fungi, mushrooms, and lichens on the surface (Dass 2017; Joseph 2014; Rabb 2018; Smith 2015). In 2016, Joseph devised a computerized system coded and programmed to quantify, validate, and statistically analyze expert judgements and developed a research-study website which was quantitatively coded and programmed to enable experts to link their computers to that website and examine and rate 25 separate photos of Martian specimens, type in the names of the specimens, and determine the probability these were living organisms. This methodology has been repeatedly demonstrated to yield scientifically valid and reliable results (Dommeyer et al. 2004; Hewson & Stewart, 2016; Richardson 2005; Watt et al. 2002). The study website was also programmed to link all 25 ratings and responses to that expert's computer IP address.

Next, Joseph and his research assistants searched the faculty rosters of every university in the English-speaking world and located over 1000 scientists who had been identified by their universities as experts in fungi, algae and lichens, and over 1000 experts identified as experts in geomorphology and mineralogy, all of whom were invited to participate.

Therefore, the "life on Mars" study was based on the judgments of two homogeneous "closed populations" of exerts in subfields of biology and geology (Joseph 2016). Samples from "closed populations" have a high degree of reliability and validity and accurately represent the views of other scientists belonging to those homogenous populations (Dommeyer et al. 2004; Hewson & Stewart, 2016; Richardson 2005; Watt et al. 2002).

Seventy scientists—40 biologists and 30 geologists—completed the invitation-only, computerized-study which enabled each expert to view; examine; type in the name of each specimen; and to judge, utilizing a computerized four-point probability scale, the likelihood each of these specimens photographed by NASA on Mars, are living organisms:

1 (0% Probability) - 2 (33% Probability) - 3 (66% Probability) -
4 (100% Probability).

Each of the participants were informed these specimens were photographed on Mars. Examination of the raw data indicated geologists and biologists agreed on five of their top seven choices and this data was analyzed. Chi-square analyses indicated a significant difference between scientists choosing "1" vs "2" but no difference between those choosing "2" vs "3" and "4," meaning that a significant majority of experts believe there is a high probability these are living organisms.

A Fisher's exact statistical test was performed and demonstrated that a majority geologists and biologists agreed there is a high probability (vs no probability) of life on Mars, as based on the comparisons of the top five specimens chosen by biologists (p = <0.0008) and geologists (p = <0.0004); and the same is true of the top seven specimens, biologists (p = <0.0001) and geologists (p = <0.0001). Dozens of experts also identified these living specimens as "puff balls," "Basidiomycota," "mushrooms," and "lichens." Therefore, a statistically significant majority of experts agree there is a high probability fungi and lichens may have colonized the Red Planet (Joseph 2016).

Figure 1. Sol 88, photographed by Opportunity's "left" Panoramic Camera "eye." A significant majority of experts in fungi, lichens, geomorphology, and minerology agreed these may be lichens (Joseph 2016). These lichen-like specimens are estimated to be approximately 2 mm to 7 mm in size/length (based on bore hole specs) and are similar to terrestrial lichens (see text for details).

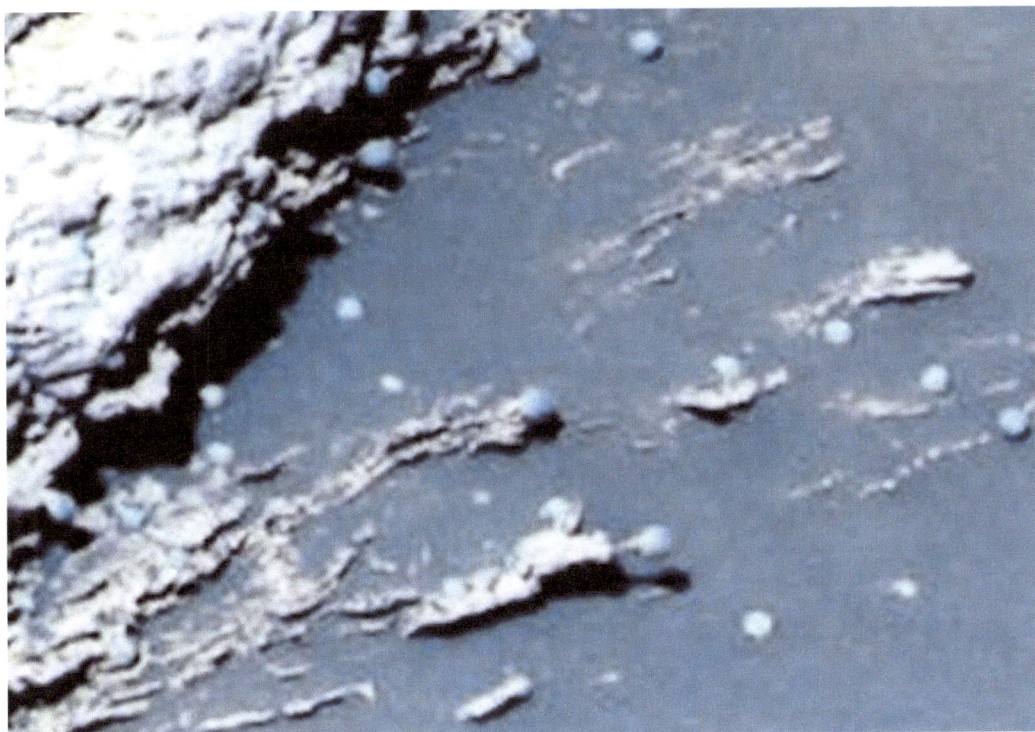

Figure 2. Sol 37, photographed by Opportunity's "left" Panoramic Camera "eye." A significant majority of experts in fungi, lichens, geomorphology, and minerology agreed these may be lichens (Joseph 2016). The average size of these lichen-like specimens are estimated to be approximately 2 mm to 7 mm, and are similar to terrestrial lichens (see Figure 3).

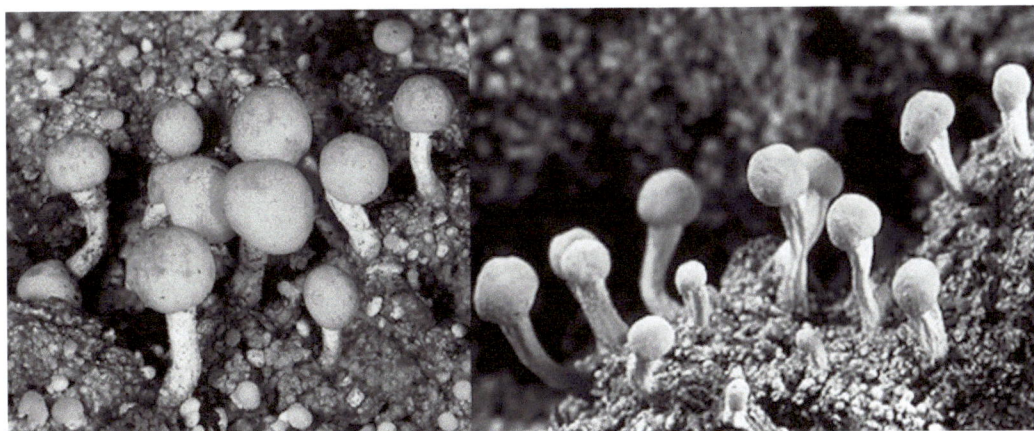

Figure 3. Terrestrial Lichens / Dibaeis baeomyces. Ranging from 2 mm to 7 mm in size. Photos reproduced by permission: Courtesy of Dragisa Savic (left) and Stephen and Sylvia Sharnoff (right).

Figure 4. Hematite concretions the size of "pebbles" "marbles" and "golf balls" (the largest five cm) collected from Utah's national parks. Reproduced with permission, courtesy of Fantasia Mining and Ashley Rouech.

8. Evidence of Lichens on Mars

Lichens are composite life forms comprised and consisting of a symbiotic relationship involving algae/cyanobacteria (photobiont) and fungi (mycobiont), the latter of which is largely responsible for the lichens' thallus, mushroom shape, and fruiting bodies (Armstrong 2017). The specimens observed on Mars and identified by experts as lichens (Dass 2017; Joseph 2016) closely resemble Dibaeis baeomyces, a fruticose lichen belonging to the Icmadophilaceae family—characterized by stalks which may grow to 6 mm topped by a bulbous apothecia, 1–4 mm in diameter. Dibaeis baeomyces have been found growing on rocks, in desert sand, dry clay, and in the arctic (Jonsson et al. 2008; Ryan et al. 2002; Seminara et al. 2018; Thompson, 1997; U.S. Department of the Interior 2010). The U.S. National Park Service, Department of the Interior (2010), which published a Lichen Inventory of lichen growth in five different areas of the Arctic—considered by many scientists to be a Mars-like analog—determined the average height of a mature lichen from each area, in cm, is 7.29, 3.72, 4.83, 5.32, 5.42, or when considered as a whole 5.32 centimeters (53 millimeters).

The Martian specimens depicted in Figures 1 and 2 clearly resemble and are in most respects identical to Dibaeis baeomyces in morphology, shape, and size. All were photographed by Opportunity's Panoramic Camera and which, according to NASA's website, have a 1024 x 1024 pixel array and the following specs: The camera's "right eye" specializes in infrared wavelengths and the "left eye" in visible colors, thereby enabling NASA to colorize images. Both "eyes" are mounted at a height of about five feet (1.5 meters), with 11.8 inches (30 centimeters) between them. Image resolution is ~0.04 inch (1 millimeter) per pixel at a distance of 9.8 feet (3 meters), whereas focal length is capable of close ups at ~1.5 inches (39 millimeters), with optimal focus from five feet (1.5 meters).

According to parameters provided by NASA, drill holes are 45 mm in diameter, five mm deep. Hence, using drill hole comparative parameters, specimens resembling lichens are approximately two mm to seven mm in size/length. Therefore, all the Opportunity images of what appear to be lichens (Figures 1, 2) are nearly identical to terrestrial lichens (Dibaeis baeomyces) in size, shape, growth patterns, and morphology as depicted in Figure 3. By contrast, these specimens (Figures 1, 2) have absolutely no resemblance to terrestrial hematite (Figure 4), NASA's favored hypothesis. Hematite does not have a thallus or a fruiting-body mushroom shape, or a stalk/stem attached to rocks, or jut-out from rocks at varying angles. These are characteristics of lichens (Figure 3), not hematite (Figure 4).

9. Evidence of Fungi on Mars

Lichens are comprised of algae and fungi, and four different investigators (Dass 2017; Joseph 2014; Rabb 2015, 2018; Small 2015) and a significant majority of experts in fungi, lichens, geomorphology, and mineralogy (Joseph 2016) have identified what appears to be fungi on the Martian surface and beneath Martian rock shelters (Figures 5, 6, 13). Fifteen specimens were photographed by the Rover Opportunity increasing in size and emerging from the ground over a three day period (Figure 8).

Puffballs (phylum Basidiomycota) are round-shaped fruiting bodies that contain trillions of spores which are released as dry powdery "puffs" and which can resemble flakes of dry paint. They sit directly on and are usually attached by short stalks to the ground (Petersen 2013; Roberts & Evans 2011). Thus, the specimens in Figure 5 and 6 clearly resemble puffballs (Figure 7). What appears to be spores appear to litter the surrounding Martian surface (Figure 6). By contrast, NASA's favored hypothesis is these specimens are hematite produced in hot-springs (NASA 2009; Squyres et al. 2004).

The specimens depicted in Figures 5 and 6 were photographed by the "microscopic imager" attached to the rover Opportunity. According to specs provided on NASA's website, the Microscopic imager has a focal length ranging from 0.8 inch (21 millimeters) an optimal focus distance of 2.67 inches (68 millimeters) and is able to resolve features as small at 0.004 inches (0.1 mm). The original image's size of Figures 6 and 7 was 1024 x 1024 pixels (0.001 inch or 0.031 millimeter per pixel). Based on these stats, the approximate size of the two specimens in Figures 5, 6, and 7 (sol 221/right) are approximately 30 to 50 mm (3 to 5 cm). Mature terrestrial puffballs, on average, are approximately 1.68 inches (4.267 cm) in size (Petersen 2013; Roberts & Evans 2011).

The specimens depicted in Figures 5, 6, 8, also resemble hematite (Figure 4) but are nearly identical to terrestrial puffballs (Figure 7). However, hematite does not shed white fluffy spore-like material, which litters the surface in Figures 5, 6.

Evidence favoring the fungal/puffball hypothesis is what appears to be the growth and emergence of 15 specimens, over a three day period (Figure 8). Specifically, five appear to increase in size whereas ten emerge from the ground. If they are not growing, and are in fact hematite, then the only other reasonable explanation is that a powerful wind uncovered these specimens by blowing away dust, dirt, and sand.

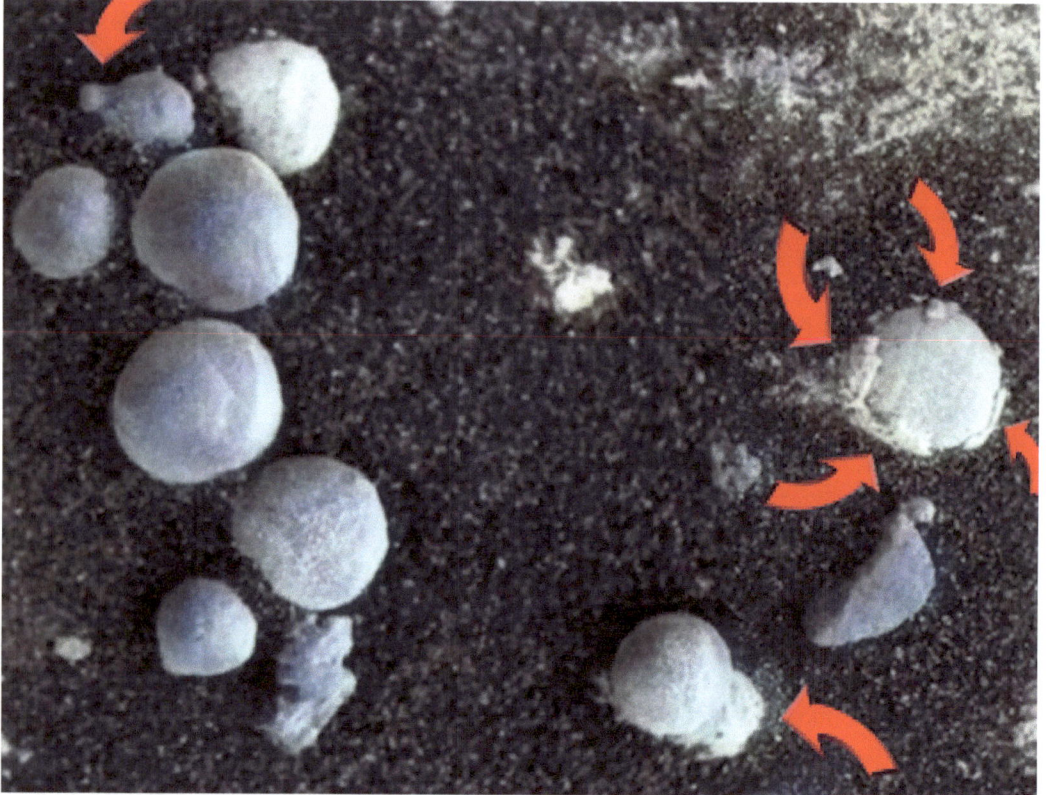

Figure 5. Sol 257 photographed by NASA's Mars Rover Opportunity. Martian specimens resembling Puffballs (Basidiomycota), some with stalks and shedding what appears to be spores and the outer cap, lower cup, and universal veil that covers embryonic fungi.

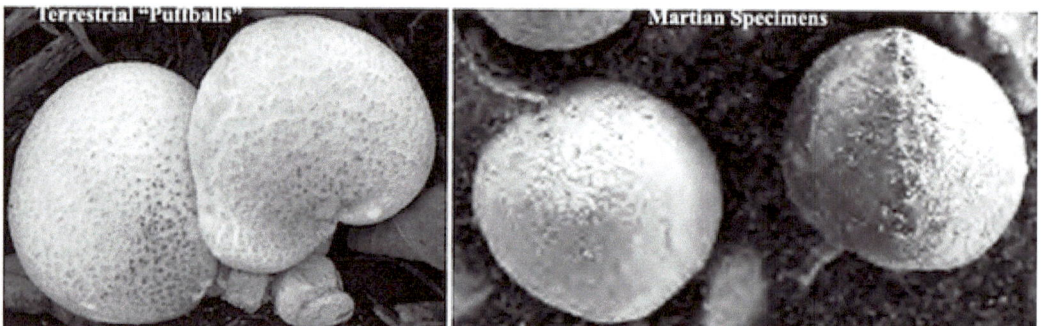

Figure 6. Comparing terrestrial fungi (left) with Martian specimens (right, Sol 221) photographed by the Rover Opportunity at Meridian Planum, Mars.

Figure 7. Sol 182 photographed by NASA Rover Opportunity. A majority of experts identified these specimens as "fungi" and "puffballs" (Joseph 2016). Note what appears to be spores littering the surface. NASA favors a hematite hypothesis.

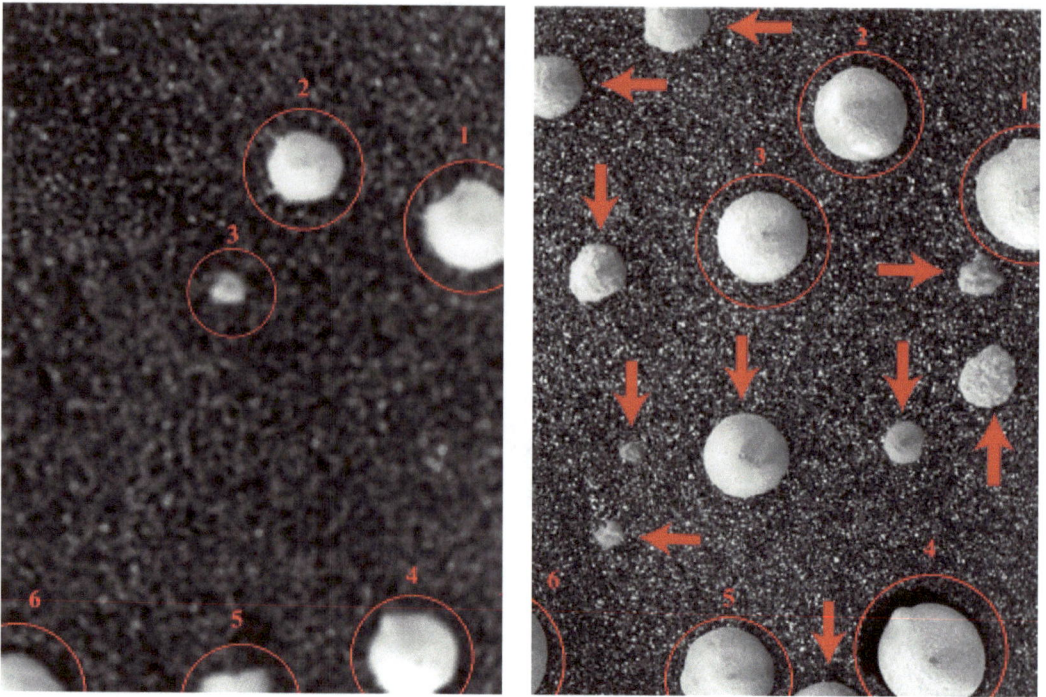

Figure 8. Sol 1145-left v Sol 1148-right). Comparing Sol 1145-left vs Sol 1148-right. Growth of fifteen Martian specimens over three days. Specimens labeled 1-5 and marked with red circles have increased in size. Those specified by arrows—Sol 1148-right—demarcate the emergence of ten new specimens which were not visible in Sol 1145-left photographed three days earlier by NASA/JPL. Differences in photo quality are presumably secondary to changes in camera-closeup-focus by NASA. The majority of experts in fungi, lichens, geomorphology, and mineralogy agreed these are likely living specimens, i.e. fungi, puffballs. An alternate explanation is a strong wind uncovered hematite which had been beneath sand and dust.

10. Wind or Fungal Growth?

The Opportunity was not equipped to measure wind. However, Opportunity has been subject to extremely dusty conditions. For example, in December of 2013 the average dust factor was estimated by NASA to be .467 (relatively dusty); 0.964 (mildly dusty) in May of 2014; and 0.725 (extremely dusty) in June of 2016 (NASA 2018). In fact, rather than strong Martian winds blowing away dust, sand, and dirt, they have instead blanketed the Opportunity and its solar panels with so much debris that Opportunity has been subject to repeated episodes of reduced power (e.g. from 700 watt hours to 400), thereby severely limiting its activities. Furthermore, Opportunity twice stopped functioning for long periods, including in July of 2007 when solar-panel output dropped to 128 watt hours (NASA 2007) and in June of 2018 when Opportunity finally ceased to function (NASA 2018) and has yet to recover as of February 14, 2019.

Given these dusty conditions, what is the likelihood that a strong wind would have uncovered the specimens in Figure 8, and not covered them up (and the Opportunity's solar panels) with dust, sand, and dirt? The answer is unknown and a cleansing wind remains a distinct possibility.

The wind explanation cannot explain why before and after photos, taken by NASA, depict what appears to be large masses of bacteria and fungi growing on the rovers Opportunity and Curiosity (Figures 9, 10, 11, 12). Moreover, what experts identified as fungi growing beneath a Martian rock shelter (Figure 14) is very similar to what appears to be fungi growing within the shelter of the rover Curiosity's upper deck (Figures 12, 14). Hence, wind is not a likely explanation for what appears to growth. Instead the evidence supports the hypothesis that fungi (and lichens) have colonized Mars.

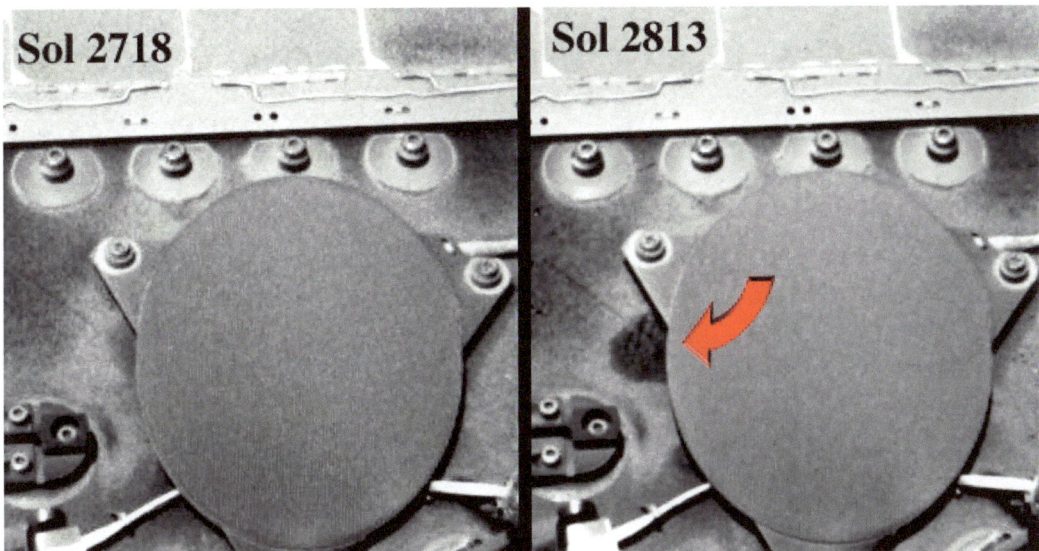

Figure 9. Mars Sol 2718 vs Sol 2813—Growth of what appears to be a mass of bacteria and fungi on the Mars Rover, Opportunity, after 95 (Martian) days. Photo, NASA/JPL.

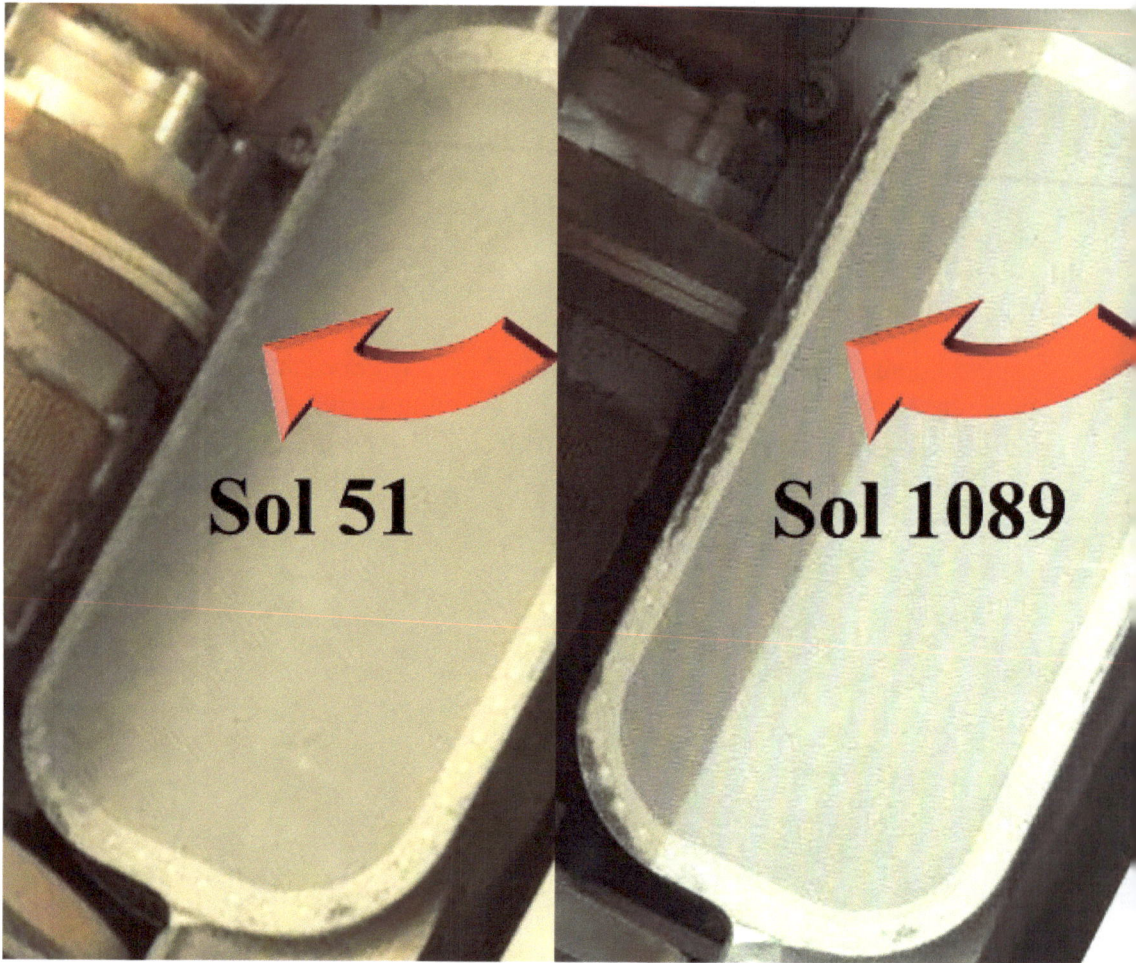

Figure 10. (opposite page) Mars Sol 51 vs Sol 1089—Growth of what appears to be a mass of bacteria and fungi on the Mars Rover Curiosity after 1038 Martian days. Photo NASA/JPL.

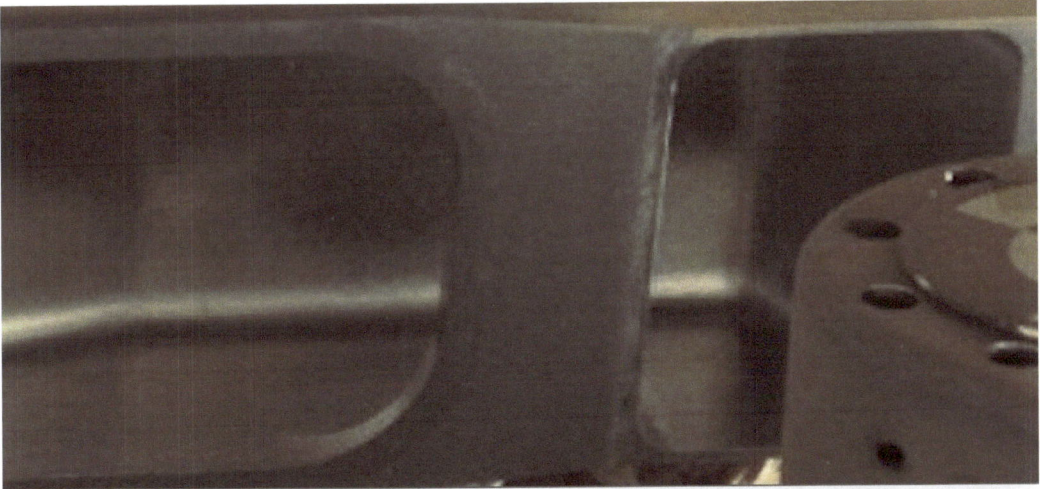

Figure 11. (Top) Sol 51. Mastcam photo of the interior and shelter of the rover Curiosity's chem cam deck after 51 Martian days. Photo by NASA/JPL.

Figure 12. (Bottom) Sol 1089. Fungal contamination of the interior and shelter of the rover Curiosity's chem cam deck after 1089 Martian days (compare with Figure 11). Photo Mastcam, by NASA/JPL.

Figure 13. Sol 1162 Photographed by NASA Rover Curiosity. A majority of experts identified the white specimens as fungi (Joseph 2016).

Figure 14. Comparison of Curiosity's Chem Cam Deck (Sol 1089 / Left, Figure 12) with Sol 1162 (Right, Figure 13).

11. The Biology of Hematite

Four independent investigators and a majority of experts in fungi, lichens, geomorphology, and mineralogy favor a biological explanation for Martian specimens resembling fungi/puffballs as depicted in Figures 5, 6, and 8. NASA (2009) and Squyres and colleagues (2004) argue in favor of hematite. Hematite, however, is not evidence against biology, but would be further proof of biology. It is well established that prokaryotes and fungi play an important role in the formation of this mineral (Ayupova et al. 2016; Claeys 2006; Owocki et al. 2016).

NASA has identified what may be "hematite" on the Martian surface as based on photographs utilizing color filters taken from space by NASA's Mars Global Surveyor spacecraft's infrared spectrometer (NASA 2009). Chemical (but non-biological) studies at ground level—via the Rovers "Opportunity" and "Spirit"—also detected the presence of hematite (Squyres et al. 2004).

Hematite is a mineralized iron oxide which, over thousands of years, slowly forms in hot springs (Anthony et al. 2005; Morel 2013), as well as in volcanoes when temperatures rise above 950 C (1740 F). NASA (2009) and Squyres et al. (2004) have argued that Martian hematite was most likely created in boiling hot springs and hydrothermal vents thousands, millions, or billions of years ago.

Likewise, numerous species of bacteria and archaea flourish in hot springs and hydrothermal vents including anaerobic hyperthermophiles, sulfate reducing bacteria (Desulfovibrio desulfuricans), and microbes such as thermophilic archaebacteria Thermus aquaticus (Gerday & Glansdorff 2007; Durvasula & Rao 2018; Robb et al. 2007). And, as noted, hematite is often fashioned in association with biological activity (Ayupova et al 2016; Bosea et al. 2009; Claeys 2006; Lower 2009; Fredrickson et al., 2008; Gralnick & Hau 2007; Owocki et al. 2016).

On Earth, hematite filaments and tubes are similar to structures produced by iron-oxidizing bacteria which suggests the former are fashioned by the latter (Ayupova et al 2016; Claeys 2006; Rajendrana et al. 2017). Moreover, a variety of bacteria help form (by cementing together) or feed upon hematite by extracting energy from iron which precipitates hematite formation (Bosea et al. 2009; Fredrickson et al. 2008; Gralnick and Hau 2007). Oolitic hematite, for example, is fashioned when sediment particles (ooids) on the seafloor accumulate thin layers of lime. Bacteria replace the lime thus assisting in the fashioning of hematite (Claeys 2006; Lowy et al. 2006). Likewise, facultative anaerobes such as Shewanella—a gram-negative, proteobacteria—grows and feeds on hematite and respires on a variety of organic electron acceptors found in hematite (Bosea et al. 2009; Fredrickson et al. 2008, Gralnick and Hau 2007; Lowy et al. 2006).

Fungi also play a major role in hematite formation (Ayupova et al. 2016, Claeys 2006; Owocki et al. 2016), the mineral substrates of which have been found "attached to fungal filaments, embedded in the fungal mycelium" (Claeys 2006). As determined by Clayes (2006) fungal interactions with hematite also produce "significant biomechanical and biochemical bioweathering features:

strong pitting of the mineral surfaces, exfoliation, tunnelling, dissolution, honey-comb-alveolar structures, perforations, fragmentation, and cementation." There is a strong attachment of fungal hyphae to these minerals, such that "fungi engulf whole blocks of minerals in the hyphal network, irrespective of mineral surface topography" (Claeys 2006). Spherical hematite contains numerous filaments with structures similar to fungal hyphae (Ayupova et al. 2016; Claeys 2006;).

Hematite fashioned in hot springs or large bodies of water is therefore cemented together and then shaped and fashioned via the assistance of fungal and bacterial activity (Ayupova et al. 2016; Claeys 2006, Owocki et al. 2016; Lowy et al. 2006; Morel 2013). Therefore, like the hematite of Earth, the Martian hematite identified by NASA (2009) and Squyres et al. (2004) is likely, at least in part, a byproduct of biological activity and is further evidence that Mars was (and may still be) a living planet which was long ago colonized by fungi and prokaryotes including those typically dwelling in water.

12. Martian Meteorite ALH 84001

Claims of "nanobacteria" in Martian meteorite ALH 84001 have been vigorously disputed and are not an issue here. As summed up by Martel and colleagues (2012), "...structures resembling terrestrial life forms known as nanobacteria—can be deemed ambiguous at best."

Likewise, there is controversy over evidence of biological residue, carbonates, and fossilized polycyclic aromatic hydrocarbons (PAHs)—a byproduct of cellular decay—discovered in ALH 84001, and other Martian meteorites (see Treiman, 2003 vs Thomas-Keprta et al. 2009). For example, Steele et al. (2012), after examining 11 Martian meteorites, report that 10 contain a ubiquitous distribution of carbon found in association with oxide grains and magmatic minerals and which indicate an abiotic origin, perhaps secondary to Martian magmas. Treiman and Essen (2011) also favor an abiogenic explanation.

Thomas-Keprta et al. (2009) agrees that much of what appears to be biological residue is probably abiogenic in origin. However, her team also determined that at least 25% is biological (McKay et al. 2009; Thomas-Keprta et al. 2009). Martel et al (2012), who argues against nanobacteria, also admits that "the presence of polycyclic aromatic hydrocarbons, magnetite crystals, carbonate globules... are compatible with living processes."

In fact, the highest concentration of Martian PAHs was embedded in or found alongside those areas of the meteor rich in carbonates (Clemett et al. 1998). Moreover, the magnetotactic residue is not random, but has the characteristic chain-like organization associated with biological activity (Clemett et al. 1998; McKay et al. 1996, 2009). Hence, as based on evidence marshalled by Thomas-Keprta and colleagues (McKay et al. 1996, 2009; Thomas-Keprta et al. 2002), at least 25% of the Martian PAHs found within this ejected sediment were most likely produced by carbonate and iron-eating bacteria, magnetotactic bacteria, algae, or fungi around 4 billion years ago (Thomas-Keprta et al. 2009).

Moreover, the carbonates and biological residue appear to have been produced in an aqueous environment (Halevy et al. 2011; McKay et al. 1996, 2009; Thomas-Keprta et al. 2009). For example, Shaheen et al. (2015), upon measuring carbonate phases and distinct oxygen isotope compositions with ALH 84001, found several episodes of aqueous activity. Halevy et al (2011) have come to similar conclusions which they attribute to "a gradually evaporating, subsurface water body—likely a shallow aquifer (meters to tens of meters below the surface)."

13. Biology and Martian Stromatolites?

A detailed analysis of Martian meteorite ALH 84001 revealed high concentrations of carbon compounds, including elongated spheroids and rounded carbonate globs and which had been recycled through water (McKay et al. 1996, 2009; Thomas-Keprta et al. 2002, 2009). Carbonates are typically found in fossil beds of dead sea life. The Martian ALH 84001 carbonate globules also contain cores rich in calcium coupled with dissolved carbonates and magnetite and iron-sulfides which were most likely produced biologically (McKay et al. 1996, 2009; Thomas-Keprta et al. 2002, 2009); i.e. carbonate- and iron-metabolizing bacteria, including cyanobacteria. The outer rims were also oxidized in a pattern typically associated with biological activity; that is reducing and rusting, secondary, perhaps, to photosynthesis, which is also a characteristic of cyanobacteria.

In 2009, Rizzo and Cantasano reported observations of microbial induced sedimentary structures which they interpreted to be evidence of stromatolites, i.e. microbialites (Rizzo & Cantasano 2009, 2011, 2016), most likely created by photosynthesizing, carbonate metabolizing cyanobacteria living in water (Noffke 2015).

In 2014, Bianciardi, Rizzo and Cantasano, conducted an extensive observational analysis of sedimentary microstructures coupled with a quantitative, objective, image analysis and compared what appears to be Martian stromatolites photographed by the Mars rover Opportunity, with terrestrial stromatolites and microbialites, so as to evaluate the geometric textural complexity vs "randomness."

Bianciardi and colleagues (2014) reported that "the morphometric analysis reveals that both textures, from microbialites (Earth) and from selected MI images (Mars), present a multifractal aspect" and that "Martian and terrestrial textures were extremely similar to each other." Moreover, they found a "textural pattern that is also present in living microbialites as well in recent and fossil stromatolites... characteristic of microbial communities of cyanobacteria." Other investigators have reached similar conclusions (Ruffi & Farmer, 2016; Noffke 2015).

Ruffi and Farmer (2016), upon examining sediments photographed by the rover Spirit, detected silica structures which closely resemble "microbially mediated micro-stromatolites" as well what could be biofilms with filaments and sheaths. In addition, morphological analyses of sedimentary specimens—photographed by the rover Spirit, at Gusev crater—revealed microstructures organized as intertwined microspherule filaments. These textures are also observed in Earthly stromatolites and microbialites (Bianciardi et al. 2014; Noffke 2015).

Figure 15. Sol 820. Green algae, stromatolites, microbial mats? Photographed by the Rover Curiosity Mars Science Laboratory Mars Hand Lens Imager which takes color images of features as small as 12.5 micrometers and at distances between 20 mm and infinity and having a depth of field of 1.6 mm to 2 mm. Note: specimen "A" at bottom center (also depicted in Figure 16). Photograph: NASA/JPL.

Noffke (2015) reported that rocks in the < 3.7 Ga Gillespie Lake region of Mars display sedimentary characteristics which mirror ecological changes over time typical of microbiological mats on Earth and which are produced by biological interactions with the environment in regressive bodies of water. Based on an analysis of photographs made by Curiosity's Mastcam camera of Martian outcrops, Noffke (2015) observed macroscopic morphologies and spatial relationships associated with a temporal change in the stratigraphic succession which are typically produced secondary to colonization by microbial mats; i.e. "centimeter- to meter-scale structures similar in macroscopic morphology"… that include "'erosional remnants and pockets,' 'mat chips,' 'roll-ups,' 'desiccation cracks,' and 'gas domes' which do not have a random distribution but were arranged in spatial associations and temporal successions similar to the "growth of a microbially dominated ecosystem that thrived in pools that later dried completely."

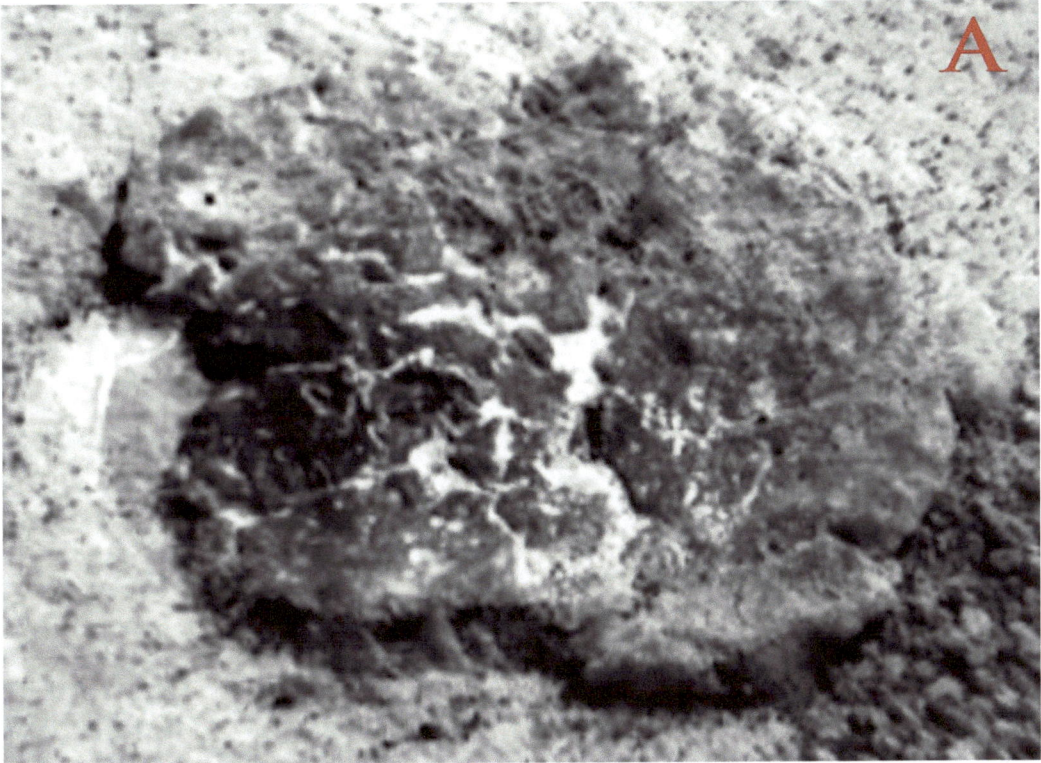

Figure 16. Sol 820. Specimen A (from Figure 15). Evidence of bacterial mat? Photographed by the Rover Curiosity Mars Science Laboratory Mars Hand Lens Imager.

In a further examination of bio-mineralization processes it was found that Martian sediments are characterized by highly organized microspherules similar to terrestrial stromatolites which consist of voids, gas domes, and layer deformations (Rizzo and Cantasano 2016). Moreover, a quantitative image statistical analysis comparing 45 microbialites samplings with 50 photographed by the rover also determined the specimens from Mars are statistically and morphologically similar to terrestrial samples with a probability of this occurring by chance being less than 1/28, $p<0.004$ (Bianciardi et al. 2014).

The evidence, therefore, supports the hypothesis that algae (cyanobacteria) may have colonized Mars over 3 billion years ago in the presence of water.

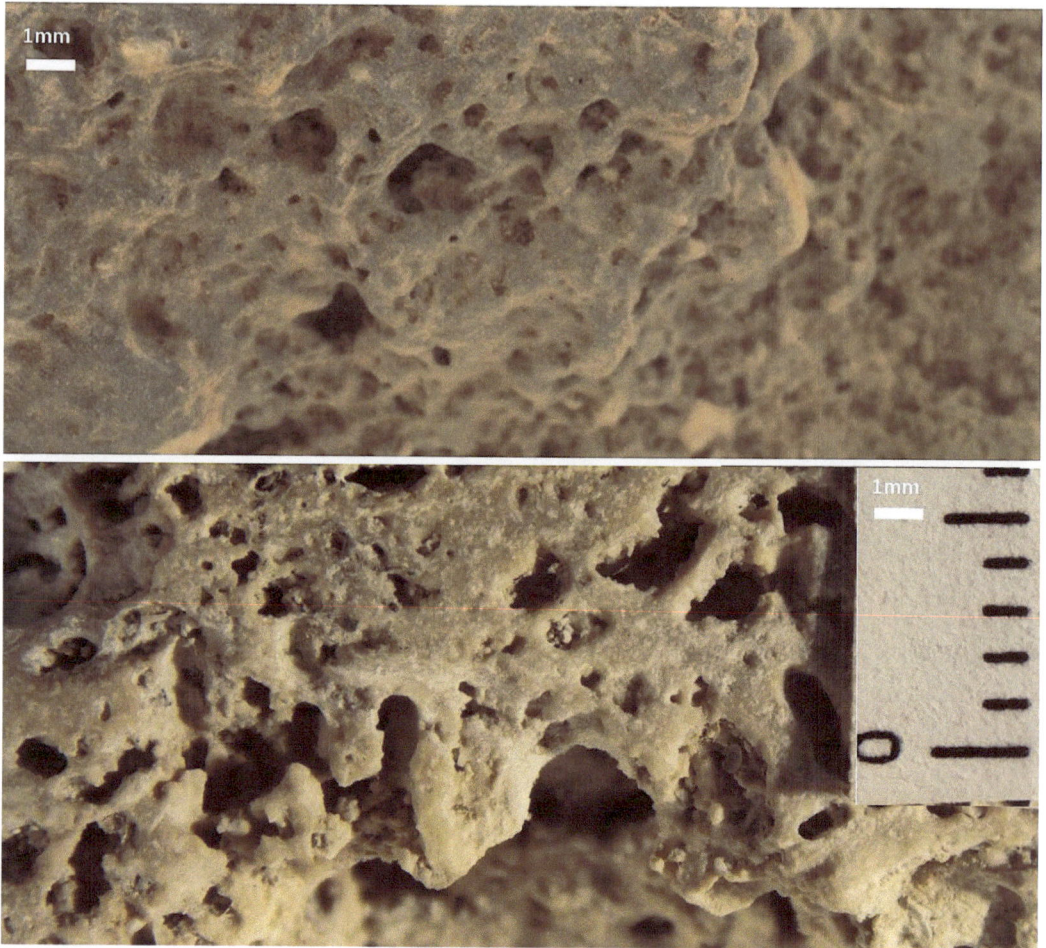

Figure 17. Microanalyses of a Martian stromatolite (top) photographed by the Rover Curiosity compared with a terrestrial stromatolite from Lagoa Salgada, Brazil (bottom). Highly organized microspherules and thrombolytic microfacies are common to both. Earthly Cyanobacteria typically form voids, intertwined filaments, and layer deformation within stromatolites.

14. Evidence of Martian Algae/Cyanobacteria?

Several investigators, based on an examination of photos of Martian specimens, have suggested that algae may be present on Mars (Joseph 2014, Rabb 2018; Smart 2015), perhaps adjacent to areas which may occasionally fill with water (Krupa 2018) secondary to runoff from Martian surface or pooling from subsurface reservoirs—the likely presence of which has been endorsed by a number of investigators (e.g. Renno et al., 2009; Webster et al 2018).

In 2017, T. A. Krupa presented evidence of what may be algae at the Lunar and Planetary Society. Krupa performed a detailed analysis of low albedo, "anomalous" images from Columbia Hills in Gusev Crater, and those photographed by the Spirit Rover's Pan Cam in an area of Mars dubbed "Larry's Outcrop/Larry's Lookout." These photos were initially analyzed by employing "red," "blue," and "green" filters, which were "radiometrically corrected with the radiance correction parameters found in the image file header." These images were analyzed next with a computer Geographic Information System, SAGA.

Krupa (2017) reported the analysis revealed what appears to be water pathways which may intermittently fill with water and that "the hillside at Larry's outcrop is covered by a very thin layer of green material" and "green spherules" which clearly resembles algae in the soil. Krupa (2017) concluded "that these spherules are a life form supported by that water.... and their green color suggests that the spherules contain a photosynthetic compound similar to green chlorophyll...The distribution of these spherules in a single layer...is also consistent with the hypothesis that the spherules are photosynthetic life forms," e.g., algae/cyanobactera.

Krupa's (2017) computerized analyses which revealed water pathways that may intermittently fill with water and which may support the growth of photosynthetic algae is therefore consistent with the findings of Martian water by Renno et al., (2009) and other investigators. Specifically, in 2004 European Space Agencies' Mars orbiter found evidence of water ice and detected vapors of water molecules via an infrared camera aboard the Mars Express spacecraft which was circling the Red Planet's south pole.

Recently, Webster and colleagues (2018) have also argued in favor of surface and subsurface reservoirs of water on Mars, but as an explanation against life on Mars.

15. Biology of Seasonal Fluctuations in Martian Methane

If prokaryotes and fungi contributed to the formation of Martian sedimentary structures billions of years in the past, the accumulation of these decaying organisms should also be a source of sedimentary and atmospheric methane. As based on simulation studies (Tarasashvili et al., 2013), methanogens can flourish in a Mars-like environment. If methanogens have colonized Mars, they too would be a source of methane. In fact, high levels of methane have been detected at ground level and in the atmosphere of Mars and which varies in concentration depending on the season and which is continually replenished (Formisano et al. 2004; Mumma et al. 2004, 2009; Webster et al. 2013, 2015, 2018).

Specifically, it has been determined that Martian atmospheric methane levels are punctuated by major spikes in concentration which later decline, only to later increase again (Formisano et al. 2004; Mumma et al. 2004, 2009; Webster et al. 2015, 2018). Three separate methane plumes consisting of 19,000 metric tons of methane gas were detected in the Martian atmosphere by Europe's Mars Express spacecraft in 2003 (Formisano et al. 2004). Employing infrared spectrometers on three Earth-based telescopes several possible methane emission sources were found in the vicinity of Syrtis Major, Arabia Terra, and Nili Fossae in the southern and northern hemispheres (Formisano et al. 2004; Mumma et al. 2004).

In July of 2013, "an upper limit of 2.7 parts per billion of methane" was detected in the general vicinity of the Gale Crater fluctuating between a value of 0.18 ppbv to 1.3 ppbv as measured on September of 2013 (Webster et al. 2013). A "tenfold spike" in methane levels followed with increases in late 2013 and early 2014, averaging "7 parts of methane per billion in the atmosphere" (Webster et al. 2015).

In 2018, Webster and colleagues reported that "in situ measurements at Gale crater made over a five-year period by the Tunable Laser Spectrometer on the Curiosity rover "revealed a strong, repeatable seasonal variation...which is greater than that predicted from either ultraviolet degradation of impact-delivered organics on the surface or from the annual surface pressure cycle." According to Webster et al (2018), the findings "are consistent with small localized sources of methane released from Martian surface or subsurface reservoirs."

On Earth, 90% of methane is produced biologically by living and decaying organisms (U.S. Department of Agriculture 2017; U.S. Department of Energy, 2017) and released as a waste product by prokaryotes (Bruhn et al. 2012; Kepler et al. 2006) and certain species of fungi (Lenhart et al. 2012; Liu et al 2015; Mukhin & Voronin, 2007). Terrestrial atmospheric methane levels also vary with the seasons and are directly attributed to biological activity.

For example, Rasmussen and Khalil (1981) found "stable seasonal cycles with peak concentrations in October and minimum concentrations in July." In 1983 these investigators reported that in the southern hemisphere the lowest concentrations are found during the late Australian summer and fall; likewise, there is less atmospheric methane in the northern hemisphere during summer. These seasonal variations have as their source biological activity in wetlands (Chen et al. 2008; Whalen 2005) and on farms and in rice paddies, just prior to harvest (Chen et al. 2008; Cicerone et al. 1983).

Chen et al. (2008) report a direct correlation between atmospheric methane and the growing seasons (14.45 mg CH_4 $m-2$ $h-1$ [0.17 to 86.78 mg CH_4 $m-2$ $h-1$]) vs non-growing seasons (0.556 mg CH_4 $m-2$ $h-1$ (0.002 to 6.722 mg CH_4 $m-2$ $h-1$). Major contributing factors include surface temperatures, standing water depths, and the degree of plant growth, whereas in anerobic environments the absence of oxygen and the amount of degradable material are controlling influences.

Based on a thorough review of the evidence, Whalen (2005) determined that "emission from wetlands is also a significant component of the atmospheric CH_4 budget...about 25% of total emissions from all anthropogenic and natural sources." Much of "this methane is produced by subsurface, anaerobic methanogenic bacteria and surficial oxidation by methanotrophic bacteria."

As documented in this report, methanogens, cyanobacteria, fungi, and lichens can easily survive in a Mars-like environment. Further there is evidence that these same species of prokaryote and eukaryotes have colonized Mars and that there is water beneath and above the surface.

Although considered controversial, NASA's 1976 Viking Labeled Release (VLR) studies, at two landing sites 4,000 miles apart on Mars, detected evidence of methane and subsurface biological activity that could be attributed to fungi, lichens, algae, and a variety of prokaryotes (Levin 1976a,b; Levin and Straat, 2009).

It is also well established that fungi and stromatolite-building cyanobacteria (algae) produce and are sources of methane (Hansson 1983; Klassen et al. 2017; Lenhart et al. 2012). Fungi and other eukaryotes generate methane via interactions with methanogenic archaea (see Olsson et al. 2017).

Algae and Saprotrophic fungi also produce methane independently of archaea (Hansson 1983; Klassen et al. 2017; Lenhart, et al. 2012) whereas fungal (and archaea) methane production is inhibited by the presence of oxygen and increases with increased levels of carbon dioxide (Lenhart, et al. 2012)—a finding which is true for most methane-producing species. Therefore, Mars is an ideal habitat for methanogens as there are minimal levels of free oxygen and the atmosphere is 96% carbon dioxide (Mahaffy et al. 2013), whereas the electron acceptor in methanogenesis is carbon dioxide.

Martian radiation may also promote the biological production of methane. It has been established that saprotrophic micro-fungi biologically decompose carbon-based radioactive debris from the damaged Chernobyl nuclear reactor (Zhdanova et al., 1991). Saprotrophic fungi are even adapted for accumulation and uptake of radiocesium fallout (Dighton et al. 1991) and, as noted, are a source of methane (Lenhart, et al. 2012). As indicated by Figures 8, 9, 10, 12, there is evidence of Martian fungal growth. Hence, it would be expected that the growth and biological activity of any Martian organisms might wax and wane, thereby resulting in a waxing and waning of Martian methane and contributing to seasonal variations.

Hence, using Earth as an example, the most probable contributors to seasonal variations in Martian methane emissions are variations in water availability, temperature, and degradable and methanogen biomass and the growth and decay of various organisms. As there is no evidence of abiotic or geological methane production on Mars (Khayat et al. 2017; Roos-Serote et al. 2016; Webster et al. 2018), and given that 90% of terrestrial methane is biological in origin, then it is reasonable to assume biological activity is the primary source of the fluctuating levels of and seasonal variations in Martian methane.

16. Geology of Martian Methane?

Certainly, and as most investigators insist, it is possible that Martian methane is produced geologically and through abiogenic processes. Perhaps Martian methane is vented periodically and naturally released via gas permeable fissures; faults and fractures in rocks; sandstone and sediment; and the leakage of deep gas reservoirs through geothermal activity, and especially through vents leading deep beneath the surface from inactive mud-volcanoes. as these are the source of abiogenic methane on Earth (U.S. Department of Energy, 2017; Etiope & Klusman 2002; Vanneste et al. 2001).

Khayat and colleagues (2017), however, searching for geological sources found none. Khayat et al. (2017) examined two volcanic districts via a high resolution spectrometer at NASA's Infrared Telescope Facility and using the high resolution heterodyne receiver at the James Clerk Maxwell Telescope facility, and "no active release of such gasses was detected."

Organic molecules are also a source of terrestrial methane (U.S. Department of Energy, 2017) and several investigators have reported what may be trace amounts of organic molecules on the surface of Mars (Ming et al. 2009; Sutter et al. 2016) which has been attributed to material deposited by meteors and dust drifting down from space (Frantseva et al. 2018; Moores & Schuerberg 2012; Schuerger et al. 2012). Therefore, it's been argued, Martian methane is produced by meteors and via the UV photolysis of the minimal amounts of organic carbon drifting down upon the surface (Fries et al. 2016; Keppler et al. 2012).

However, as determined by Webster et al. (2018) these scenarios are not probable and the levels and variations in methane are "greater than that predicted from either ultraviolet degradation of impact-delivered organics on the surface or from the annual surface pressure cycle." Furthermore and as summed by Roos-Serote, Atreva, Webster and colleagues (2016), "We find no compelling evidence for any correlation between atmospheric methane and predicted meteor events."

Moreover, UV photolysis of Martian organic carbon is carbon-limited and constrained by the accretion rate of IDP organics (Moores & Schuerberg 2012). There is insufficient carbon on the surface (Bieman et al. 1976; Ming et al. 2009) to account for the varying and large concentrations of methane which are periodically pumped into the Martian atmosphere (Moores & Schuerberg 2012). And, organics buried one to two mm below ground would not be subject to UV photolysis, and any methane could only be liberated biologically (U.S. Department of Energy, 2017; U.S. Department of Agriculture 2017).

Thus, other than speculation, there is absolutely no evidence for the abiogenic production of methane on Mars. By contrast, there is evidence of biological activity, including the growth of what appears to be fungi, algae and lichens, on the Martian surface. Thus the most logical, rational, scientific explanation for the replenishment of and seasonal fluctuations in Martian methane is biological activity—the first evidence of which was documented by the 1976 Mars Viking Labeled Release experiments (Levin 1976a,b).

17. The Viking Labeled Release Experiment Detects Biological Activity on Mars

As summarized by Levin (2010), the Viking Labeled Release (LR) experiments were designed to detect biological activity on Mars. Thousands of field tests were performed and it was proved that the LR experiment was capable of accurately detecting a very wide range of microorganisms including aerobic, anaerobic, and facultative bacteria, as well as lichens, fungi, and algae.

Once on Mars a nutrient containing radioactive carbon was added to a Martian soil sample and the presence of radioactivity in the gasses released served as evidence of active metabolism. A control experiment heat-treated a second sample to kill microorganisms. The control (heat-treated) sample yielded negative results in every experiment conducted, whereas positive results including evidence of biological reproduction were obtained from the raw sample.

As described by Levin (2010), "The LR instruments operated flawlessly on Mars. Both Viking landing sites, some 4,000 miles apart, produced strong responses and met the pre-mission criteria for the detection of life including biological reproduction."

To distinguish between non-biological and biological agents, additional experiments were executed via commands from Earth. Each such ad hoc series of tests again demonstrated on-going Martian metabolism. Four different LR experiments were conducted, each of which yielded positive results, and five controls, all of which supported the positive results as biological.

Levin concluded that the "amplitudes and kinetics of the Mars LR results were similar to those of terrestrial results, especially close to those of soils in, or from, frigid areas," and that the LR experiment had found evidence of biological activity on Mars (Levin 1976a,b, 2010).

The results, however, were rejected by NASA administrators who argued that since the addition of more nutrients into the soil—during the ad hoc experiments—temporarily decreased the level of biological activity "the LR therefore had not detected life on Mars, but had detected a chemical or physical agent that had produced false positive results" (Levin 2010). NASA's arguments (detailed on the NASA/Mars website), though interesting, are not based on factual evidence, but post-hoc theorizing and the interested reader is encouraged to review NASA's claims to arrive at their own conclusions. In fact, according to Levin (2010) a temporary decline was exactly as predicted and that "NASA-bonded Antarctic soil 664 had reacted to its second injection as had the Martian soils" and "the decline in gas level was caused by re-adsorption of the evolved gas into the dampened soil." The fact that only trace amounts of carbon and organic molecules have been detected on the Martian surface (Bieman et al. 1976, 1977; Ming et al. 2009; Sutter et al. 2016) also does not support NASA's physical-chemical-false-positive hypothesis.

Subsequently, Bianciardi, Miller, Straat, and Levin (2012) performed a mathematical complexity deep analysis of the Viking LR data, employing seven complexity variables. It was determined that the Viking LR positive responses demonstrated a different pattern from control responses which resembled near-random noise. By contrast, the active experiments exhibited highly organized responses typical of biology thereby further demonstrating that extant microbial life had been detected on Mars in 1976.

18. Conclusions

We have presented a body of observations and evidence which supports the hypothesis that Mars may have been and may still be a living planet inhabited by methanogens, cyanobacteria/algae, lichens, fungi, and other species. It is possible the ancestors of these and other species originated on Earth, survived a journey through space, and were deposited on Mars by solar winds and terrestrial ejecta. If life has also been transferred from Mars to Earth is unknown, but possible.

Specifically, we have reviewed evidence indicating that specimens identical to lichens have been photographed on Mars. Fungi also appear to be growing on the Martian surface and on the rovers Curiosity and Opportunity.

Hematite has also been identified on the Martian surface. As documented in this review, hematite is in part fashioned and cemented together by fungi and prokaryotes.

There is also evidence of what may be algae/cyanobacteria growing on the surface. There are numerous sedimentary structures on Mars resembling stromatolites which may have been constructed via a combination of abiotic and biotic processes and outcroppings having micro meso and macro structures typical of terrestrial microbialites, some of which may have been constructed over three billion years ago by cyanobacteria.

Fluctuations and seasonable variation in Martian atmospheric and ground level methane may be a direct consequence of biological activity including growth, decay, and microbial reproduction which was first detected by the Viking LR experiments.

There is no factual, scientific evidence supporting a purely abiotic explanation for the data presented here. Thus, in conclusion: evidence that the Martian environment may be habitable for some prokaryotes and simple eukaryotes; the presence of specimens which resemble lichens and fungi; evidence of fungal growth on the Rovers; biological contributions to hematite and Martian sediments; seasonal fluctuations in methane, 90% of which on Earth is related to biological activity; biological residue in Martian meteorite ALH84001; and the Mars Viking LR results are mutually supportive and collectively serve as evidence of life on Mars.

References

Adey. W. R. (1993). Biological Effects of Electromagnetic Fields. Journal of Cellular Biochemistry 51:410-416.

Alshits LK, Kulikov NV, Shevchenko VA, Yushkov PI. 1981. Changes in radio-sensitivity of pea seeds affected by lowlevel radiation. Radiobiol 21:459–463.

Anthony, J. W., Bideaux, R. A., Bladh, K. W., Nichols, M. C. (2005.). "Hematite". Handbook of Mineralogy. Mineralogical Society of America. Chantilly, VA, ISBN 0962209724.

Armstrong R.A. (2017) Adaptation of Lichens to Extreme Conditions. In: Shukla V., Kumar S., Kumar N. (eds) Plant Adaptation Strategies in Changing Environment. Springer, Singapore.

Arrhenius, S. (1908). Worlds in the Making. Harper & Brothers, New York.

Ayupova, N., Maslennikov, V. V., Tessalina, S., Statsenko, E. O. (2016). Tube fossils from gossanites of the Urals VHMS deposits, Russia: Authigenic mineral assemblages and trace element distributions. Ore Geology Reviews 85, DOI: 10.1016/j.oregeorev.2016.08.003

Basset CAL. 1993. Beneficial effects of electromagnetic fields. J Cell Biochem 31:387–393.

Becker RO. 1984. Electromagnetic controls over biological growth processes. Journal of Bioelectricity 3:105–118.

Becker RO, Sparado JA. 1972. Electrical stimulation of partial limb regeneration in mammals. Bull NY Acad Med 48:627– 641.

Becket, Kranner, & Minibayeva 2008. Stress Tolerance in Lichens. In Lichen Biology (T, H. Nash III Ed) Cambridge University Press.

Bianciardi, G., Miller, J. D., Straat, P.N., Levin, G. V. (2012). Complexity Analysis of the Viking Labeled Release Experiments, Int'l J. of Aeronautical & Space Sci. 13(1), 14–26.

Bianciardi, G., Rizzo, V., Cantasano, N. (2014). Opportunity Rover's image analysis: Microbialites on Mars? International Journal of Aeronautical and Space Sciences, 15 (4) 419-433.

Bianciardi, G., Rizzo, V., Farias, M. E., & Cantasano (2015). Microbialites at Gusev Craters, Mars. Astrobiology Outreach, 2,5.

Bosea, S., HochellaJr., M. F., .Gorby, Y.A. Kennedy, D. W., McCready, D. E., Madden, A. S., Lower, B. H. (2009) Bioreduction of hematite nanoparticles by the dissimilatory iron reducing bacterium Shewanella oneidensis MR-1, Geochimica et Cosmochimica Acta, 73, Issue 4, 962-976.

Brandt, A., et al. (2015). Viability of the lichen Xanthoria elegans and its symbionts after 18 months of space exposure and simulated Mars conditions on the ISS—International Journal of Astrobiology, 14, 411-425.

Brodie, E. L., DeSantis, T.Z., Parker, J. P. M., Zubietta, I. X., Piceno, Y. M., Andersen, G. L. (2007). Urban aerosols harbor diverse and dynamic bacterial populations PNAS. 104, 299-304.<

Bruhn, D., et al. (2012) Terrestrial plant methane production and emission. Physiol. Plant 144, 201-209.

Burchella, M. J., Manna, J., Bunch, A. W., Brandob, P. F. B. (2001). Survivability of bacteria in hypervelocity impact, Icarus. 154, 545-547.

Burchell, J. R. Mann, J., Bunch, A. W. (2004_. Survival of bacteria and spores under extreme shock pressures, Monthly Notices of the Royal Astronomical Society, 352, 1273-1278.

Calabrese EJ, Baldwin LA. 1999. Tales of two similar hypotheses: the rise and fall of chemical and radiation hormesis. BELLE Newslett 8:47–66.

Calabrese EJ, Baldwin LA. 2000. Radiation hormesis: its historical foundations as a biological hypothesis. Human Exper Toxicol 19:41–75.

Chen, H., et al. (2008). Determinants influencing seasonal variations of methane emissions from alpine wetlands in Zoige Plateau and their implications, JGR Atmospheres 113,

Cicerone, R. J., Shetter, J. D., Delwiche, C. C. (1983). Seasonal variation of methane flux from a California rice paddy, JGR Oceans, 1983, 88, 11022-11024.

Claeys. P. (2006). Experimental Observations of the Patterns of Fungi-Mineral Surfaces Interactions with Muscovite, Biotite, Bauxite, Chromite, Hematite, Galena, Malachite, Manganite and Carbonate Substrates. SAO/NASA ADS Physics Abstract Service. http://adsabs.harvard.edu/abs/2006AGUFM.B11A1005C

Clemett, S. J., M. T. Dulay, J. S. Gillette, X. D. Chillier, T. B. Mahajan, and R. N. Zare. (1998). Evidence for the extraterrestrial origin of polycyclic aromatic hydrocarbons in the Martian meteorite ALH84001. Faraday Discuss. 109:417-436.

Cockell, C.S., Schuerger, A.C., Billi, D., Friedmann, E.I. & Panitz, C. (2005). Astrobiology 5, 127–140.

Covey. C., et al. (1994) Climatic effects of atmospheric dust from an asteroid or comet impact on Earth. Global and Planetary Change. 9, 263-273.

Dadachova E., Bryan RA, Huang X, Moadel T, Schweitzer AD, Aisen P, et al. (2007) Ionizing Radiation Changes the Electronic Properties of Melanin PLoS One, doi:10.1371/journal.pone.0000457.

Dass, R. S. (2017) The High Probability of Life on Mars: A Brief Review of the Evidence, Cosmology, Vol 27, April 15, 2017.

Davies, P. C. W. (2007) The Transfer of Viable Microorganisms Between Planets P. C. W. Davies (Editors Gregoy R. Bock Jamie A.) Novartis Foundation Symposia.

De la Torre Noetzel, R. et al. (2017). Survival of lichens on the ISS-II: ultrastructural and morphological changes of Circinaria gyrosa after space and Mars-like conditions EANA2017: 17th European Astrobiology Conference, 14-17 August, 2017 in Aarhus, Denmark

De Vera, J. -P. et al. (2014). Results on the survival of cryptobiotic cyanobacteria samples after exposure to Mars-like environmental conditions, International Journal of Astrobiology, 13, 35-44.

De Vera, J. -P. (2012). Lichens as survivors in space and on Mars. Fungal Ecology , 5, 472-479.

Dighton, J, Tatyana Tugay, T. ZhdanovaN., (2008) Fungi and ionizing radiation-fromradionuclides, FEMS Microbiol Lett 281, 109–120.

Dohm, J.M., Anderson, R.C., Barlow, N.G., et al. (2008) Recent geological and hydrological activity on Mars: The Tharsis/Elysium Corridor. Planet. Space Sci. 56, 985–1013.

Dohm, J.M., Barlow, N.G., Anderson, R.C., et al. (2007) Possible ancient giant basin and related water enrichment in the Arabia Terra province, Mars. Icarus, doi:

10.1016/j.icarus.2007.03.006.

Dommeyer, C.J., P. Baum, R.W. Hanna, and K.S. Chapman. (2004). Gathering faculty teaching evaluations by in-class and online surveys: their effects on Response Rates and Evaluations. Assessment & Evaluation in Higher Education, Vol. 29, No. 5, October 2004

Drewnowska, J. M., et al. (2015) "Melanin-Like Pigment Synthesis by Soil Bacillus weihenstephanensis Isolates from Northeastern Poland, Plos One. 2015, dx.doi. org/10.1371/journal.pone.

Durvasula R. V., Rao, D.V.S (2018), Extremophiles: From Biology to Biotechnology, CRC Press.

Ettwig, K. F. et al. (2010) Nitrite-driven anaerobic methaneoxidation by oxygenic bacteria. Nature 464, 543-548.

Fajardo-Cavazosa, P., Schuerger, A. C, Nicholson W. L (2007). Testing interplanetary transfer of bacteria between Earth and Mars as a result of natural impact phenomena and human spaceflight activities, Acta Astronautica, 60, 534-540.

Formisano V, Atreya S, Encrenaz T, Ignatiev N, Giuranna M. (2004). Detection of methane in the atmosphere of Mars. Science. 2004 Dec 3;306(5702):1758-61. Epub 2004 Oct 28

Frantseva, K., Mueller. M., Kate, I. L., van der Tak, F.F.S., Greenstreetde, S. (2018). Delivery of organics to Mars through asteroid and comet impacts. Icarus, 309, 125-133.

Fredrickson, J, et al. (2008). Towards environmental systems biology of Shewanella." Nature Reviews in Microbiology. Volume 6:592-603.

Gerday, C. & Glansdorff, N. (2007) Physiology and Biochemistry of Extremophiles, ASM press.

Gostincar, C., Grube, M., de Hoog, S., Zalar, P., and Cimerman, N. G. (2010) Extremotolerance in fungi evolution on the edge. FEMS Microbiology Ecology, 71: 2-11.

Gralnick, R. and Hau, S. (2007). Ecology and biotechnology of genus Shewanella." Annu Rev Microbiol. 61:237-58.

Gupta, V. K., R. L. Mach, and S. Sreenivasaprasad (2015) Fungal Biomolecules: Sources, Applications and Recent Developments," Wily.

Halevy, I., Fischer, W. W., Eiler. J. M. (2011).Carbonates in the Martian meteorite Allan Hills 84001 formed at 18 ± 4 °C in a near-surface aqueous environment. PNAS, 108 (41) 16895-16899.

Hansson, G. (1983) Methane production from marine, green macro-algae. Resources and Conservation. 8, Issue 3, 185-194. https://doi.org/10.1016/0166-3097(83)90024-X

Hassler, D, M., et al. (2013) Mars' Surface Radiation Environment Measured with the Mars Rover. Science Science, 2013; doi: 10.1126/science.1244797

Hewson, C., Stewart, D. W. (2016) Internet Research Methods. John Wiley & Sons. DOI: 10.1002/9781118445112.stat06720.pub2

Horneck, G., Bücker, H., Reitz, G. (1994). Long-term survival of bacterial spores in space. Advances in Space Research, Volume 14, 41-45.

Horneck, G. Mileikowsky, C., Melosh, H. J., Wilson, J. W. Cucinotta F. A., Gladman, B. (2002). Viable Transfer of Microorganisms in the solar system and beyond, In G. Horneck & C. Baumstark-Khan. Astrobiology, Springer.

Imshenetsky, A.A., Lysenko, S.V. and Kazakov,G.A. (1978). Upper boundary of the biosphere. Applied and Environmental Microbiology, 35, 1-5.

Ito, H, Watanabe H, Takeshia M. & Iizuka H (1983). Isolation and identification of radiation-resistant cocci belonging to the genus Deinococcus from sewage sludges and animal feeds. Agric". Biol. Chem. 47: 1239–47. doi:10.1271/bbb1961.47.1239.

Jonsson, A.V., Moen, J., Palmqvist, K. 2008. Predicting lichen hydration using biophysical models. Journal of the Royal Society Interface, Oecologia, 156:259-273.

Joseph, R. (2014) Life on Mars: Lichens, Fungi, Algae, Cosmology, 22, 40-62.

Joseph, R. (2016) A High Probability of Life on Mars, The Consensus of 70 Experts Cosmology, 25, 1-25.

Joseph, R. & Schild, R. (2010). Biological Cosmology. Journal of Cosmology, Vol 10. 40-75.

Kepler, F., et al., (2006) Methane emissions from terrestrial plants under aerobic conditions. Nature 439, 187-191.

Keppler, F. et al. (2012). Ultraviolet-radiation-induced methane emissions from meteorites and the Martian atmosphere. Nature, 486, 93-96.

Khalil, M.A.K., Rasmussen, R. A., (1983). Sources, sinks, and seasonal cycles of atmospheric methane, JGR Oceans, 1983, 88, 5131-5144

Khayatabg. A.S.J., villanuevaam, G.L., Mumma, M. J., Tokunagac, A.T. (2017). A deep search for the release of volcanic gases on Mars using ground-based high-resolution infrared and submillimeter spectroscopy: Sensitive upper limits for OCS and SO2. Icarus, 296, 1-14.

Khodadad, C.L., et al. (2017). Stratosphere Conditions Inactivate Bacterial Endospores from a Mars Spacecraft Assembly Facility, Astrobiology, 17,

Klassen, V., Blifernez-Klassen, O ., Wibberg, D., Winkler, A., Kalinowski J., Posten, C., Kruse, O (2017) Highly efficient methane generation from untreated microalgae biomass, Biotechnology for Biofuels201710:186, https://doi.org/10.1186/s13068-017-0871-4

Kupa, T. A. (2017). Flowing water with a photosynthetic life form in Gusav Crater on Mars, Lunar and Planetary Society, XLVIII.

Lau, M.C.Y. et al. (2015) An active atmospheric methane sink in high Arctic mineral cryosols. The ISME Journal, DOI:10.1038/ismej.2015.13

Lenhart K. et al. (2012) Evidence for methane production by saprotrophic fungi. Nat Commun. 2012;3:1046. doi: 10.1038/ncomms2049.

Levin, G. (1976a) The Viking Biological Investigation: Preliminary Results, Science, 194, 4260, 99-105.

Levin, G. (1976b) Viking Labeled Release Biology Experiment: Interim Results, Science, 194, 1322-1329.

Levin, G. (2010).Extant Life on Mars: Resolving the Issues, Journal of Cosmology, 5, 920-929.

Levin, G. V. and P. A. Straat (1976) Labeled Release - An Experiment in Radiorespirometry" Origins of Life, 7, 293-311.

Levin, G. V. and Straat, P. A. (1977) Life on Mars? The viking labeled release experiment, Biosystems 9 :2-3, pp. 165-174.

Levin, G. V. and P. A. Straat (1979) Completion of the Viking Labeled Release Experiment on Mars, J. Mol. Evol., 14, 167-183.

Levin, G. V. and P. A. Straat, (2009). The Likelihood of Methane-producing Microbes on Mars," Instruments, Methods, and Missions for Astrobiology XII, SPIE Proc., vol. 7441, invited paper 744110D.

Levin, M. (2003). Review: Bioelectromagnetics in Morphogenesis. Bioelectromagnetics 24:295-315.

Liu, J., Chen, H., Zhu, Q., Shen, Y., Wang, X., Wang, M., Peng, C. (2015) A novel pathway of direct methane production and emission by eukaryotes including plants, animals and fungi: An overview. Atmospheric Environment, Volume 115, August 2015, Pages 26-35

Lowy, D.A. et al. (2006) "Harvesting energy from the marine sediment- water interface II – kinetic activity of anode materials." Biosens. Bioelectron. 21, 2058–2063

Maffei, M. E. (2014). Magnetic field effects on plant growth, development, and evolution (2014). Front. Plant Sci., 04.

Mahaffy, P. R. et al., (2012) The Sample Analysis at Mars Investigation and Instrument Suite. Space Sci. Rev. 170, 40-478.

Mahaffy P. R. et al. (2013) Abundance and isotopic composition of gases in the martian atmosphere from the Curiosity rover. Science 341, 263-266.

Mahaney, W. C. & Dohm, J. (2010) Life on Mars? Microbes in Mars-like Antarctic Environments, Journal of Cosmology, 5, 951-958.

Malin, M. C., Edgett, K. S., (1999). Oceans or Seas in the Martian Northern Lowlands: High Resolution Imaging Tests of Proposed Coastlines, Geophys. Res. Letters, V. 26, No. 19, p. 3049-3052.

Malin MC, Edgett KS. (2000b). Evidence for recent groundwater seepage and surface runoff on Mars. Science 288(5475):2330-2335.

Martel, J., Young,D., Peng, H-H., Wu,C-W., Young, J. D. (2012). Biomimetic Properties of Minerals and the Search for Life in the Martian Meteorite ALH84001-042711-10540. Annual Review of Earth and Planetary Sciences Volume 40, 167-193.

Mastrapaa, R.M.E., Glanzbergb, H ., Headc, J.N., Melosha, H.J, Nicholsonb, W.L. (2001). Survival of bacteria exposed to extreme acceleration: implications for panspermia, Earth and Planetary Science Letters 189, 30 1-8.

McKay, D.S., Gibson, E.K., Thomas-Keprta, K.L., Hojatollah, V., Romanek, C.S., Clemmett, S.J., Chillier, X.D.F., Maechling, C.R., and Zare, R.N. (1996) Search for past life on Mars: possible relic biogenic activity in Martian meteorite ALH84001. Science 273: 924-930.

McKay, D.S., Thomas-Keprta, K.L., Clemett, S.J., Gibson Jr, E.K., Spencer, L. and Wentworth, S.J. (2009) Life on Mars: new evidence from martian meteorites. In, Instruments and Methods for Astrobiology and Planetary Missions, 7441, 744102.

McLean, R.J.C., Welsh, A.K., Casasanto, V.A., (2006). Microbial survival in space shuttle crash. Icarus 181, 323-325.

McLean, R.J.C., McLean, M. A. C. (2010). Microbial survival mechanisms and the interplanetary transfer of life through space. Journal of Cosmology, 7, 1802-1820.

Mickol, R.L., Kral, T. A. (2017). Low Pressure Tolerance by Methanogens in an Aqueous Environment: Implications for Subsurface Life on Mars. Origins of Life and Evolution of Biospheres, 47, 511–532.

Mileikowskya. C., et al. (2000), Risks threatening viable transfer of microbes between bodies in our solar system. Planetary and Space Science, 48, 1107-1115.

\Mileikowsky, C., Cucinotta, F.A., Wilson, J.W., Gladman, B., Horneck, G.,

Lindegren, L., Melosh, J., Rickman, H., Valtonen, M., Zheng, J.Q. 2000, Icarus, 145, 391-427.

Ming, D. W., H. V. Lauer Jr., P. D. Archer Jr., B. Sutter, D. C. Goldern,R. V. Morris, P. B. Niles and W. V. Boynton (2009), Combustion of organic molecules by the thermal decomposition of perchlorate salts:Implications for organics at the Mars Phoenix Scout landing site,Proc.Lunar Planet. Sci. Conf.,40, Abstract 2241.

Moeller, R., et al. (2012). Protective Role of Spore Structural Components in Determining Bacillus subtilis Spore Resistance to Simulated Mars Surface Conditions. Applied and Environmental Microbiology, DOI: 10.1128/AEM.02527-12.

Moore, T. E., Horwitz, J. L. (1998). Thirty Years of Ionospheric Outflow: Causes and Consequences. American Geophysical Union. San Francisco, December

Moores, J. E., Schuerger, A. C. (2012). UV degradation of accreted organics on Mars: IDP longevity, surface reservoir of organics, and relevance to the detectionof methane in the atmosphere, Journal of Geophysical Research, Vol. 117, E08008, doi:10.1029/2012JE004060.

Morel, D. (2013). Hematite: Sources, Properties and Applications , Nova Biomedical

Moment GB. 1949. On the relation between growth in length, the formation of new segments, and electric potential in an earthworm. J Exp Zool 112:1–12.

Moseley BE, & Mattingly A (1971). Repair of irradiated transforming deoxyribonu- cleic acid in wild type and a radiation- sensitive mutant of Micrococcus radiodurans". J. Bacteriol. 105 (3): 976–83. PMC 248526 Freely accessible. PMID 4929286.

Mukhin, V. & Voronin, P. (2007) Methane emission during wood fungal decomposition. Doklady Biol. Sci. 413, 159-160.

Mumma, M.J., Novak, R.E., DiSanti, M.A., Bonev, B.P., (2003) A sensitive search for methane on Mars. Bull. Am. Astron. Soc. 35, 937.

Mumma, M.J., Villanueva, G.L., Novak, R.E., Hewagama, T., Bonev, B.P., DiSanti, M.A., Mandell, A.M., and Smith, M.D. (2009) Strong release of methane on Mars in northern summer 2003. Science. doi:10.1126/science.11,65,243.

Navarro-Gonzalez, R., Rainey, F.A., Molina, P., Bagaley, D.R., Hollen, B.J., de la Rosa, J., Small, A.M., Quinn, R.C., Grunthaner, F.J., Caceres, L., Gomez-Silva, B., McKay, C.P. 2003, Science (New York, N.Y, 302(5647), pp. 1018-1021.

NASA (2007). (https://science.nasa.gov/science-news/science-at-nasa/2007/20jul_duststorm)

NASA (2018). (https://mars.nasa.gov/mer/mission/status_opportunityAll.html#sol4413)

Nicholson, W. L., Munakata, N., Horneck, G., Melosh, H. J., Setlow, P. (2000). Resistance of Bacillus Endospores to Extreme Terrestrial and Extraterrestrial Environments, Microbiology and Molecular Biology Reviews 64, 548-572.

Nicholson, W.L., et al. (2012). Growth of Carnobacterium spp. from permafrost under low pressure, temperature, and anoxic atmosphere has implications for Earth microbes on Mars. PNAS. https://doi.org/10.1073/pnas.1209793110.

Noffke, N. 2015. Ancient Sedimentary Structures in the < 3.7b Ga Gillespie Lake Member, Mars, That Compare in macroscopic Morphology, Spatial associations, and Temporal Succession with Terrestrial Microbialites. Astrobiology 15(2): 1-24.

Novikova, N (2009) Mirobiological research on board the ISS, Planetary Protection. The Microbiological Factor of Space Flight. Institute for Biomedical Problems, Moscow, Russia.

Novikova, N et al. (2016) Long-term spaceflight and microbiological safety issues. Space Journal, https://room.eu.com/article/long-term-spaceflight-and-microbiological-safety-issues.

Occhipinti, A., De Santis, A., and Maffei, M. E. (2014). Magnetoreception: an unavoidable step for plant evolution? Trends Plant Sci. 19, 1–4. doi: 10.1016/j.tplants.2013.10.007

Olsson, S., Bonfante, P, Pwlowska, T. E. (2017) Ecology and Evolution of Fungal-Bacterial Interactions. In: Dighton, J. & White J. F. (Eds) The Fungal Community: Its Organization and Role in the Ecosystem, CRC Taylor & Francis

Olsson-Francis, K., et al. (2009). Survival of Akinetes (Resting-State Cells of Cyanobacteria) in Low Earth Orbit and Simulated Extraterrestrial Conditions. Origins of Life and Evolution of Biospheres, 39, 565.

Onofri, S., R. de la Torre, J-P de Vera, et al. (2012) Survival of rock-colonizing organisms after 1.5 years in outer space." Astrobiology. 2012;12:508–516.

Onofri, S., et al (2018). Survival, DNA, and Ultrastructural Integrity of a Cryptoendolithic Antarctic Fungus in Mars and Lunar Rock Analogues Exposed Outside the International Space. Astrobiology, 19, 2.

Osman, S., Peeters, Z., La Duc, M.T., Mancinelli, R., Ehrenfreund, P., Venkateswaran, K., (2008). Effect of shadowing on survival of bacteria under conditions simulating the Martian atmosphere and UV radiation. Applied and Environmental Microbiology 74, 959-970.

Owocki. K., Kremer, B., Wrzosek, B., Królikowska, A., Kaźmierczak J. (2016). Fungal Ferromanganese Mineralisation in Cretaceous Dinosaur Bones from the Gobi Desert, Mongolia. Plos One. https://doi.org/10.1371/journal.pone.0146293

Pacelli, C., L. Selbmann, L. Zucconi, J. P. P. De Vera, E. Rabbow, G. Horneck, R. de la Torre, S. Onofri "BIOMEX experiment: Ultrastructural alterations, molecular damage and survival of the fungus Cryomyces antarcticus after the Experiment Verification Tests." (2016) Origin of Life and Evolution of Biospheres 2016, 47(2):187-202.

Perron, J. Taylor; Jerry X. Mitrovica; Michael Manga; Isamu Matsuyama & Mark A. Richards (2007). Evidence for an ancient Martian ocean in the topography of deformed shorelines Nature. 447: 840-843.

Petersen, J. H. (2013). The Kingdom of Fungi. Princeton University Press.

Rabb, H. (2018). Life on Mars - Visual Investigation, Astrobiology Society, SoCIA, University of Nevada, Reno, USA. April 14, 2018.

Rajendrana, K, Shampa, S. .Sujac, S. Lakshmana S.T. VinothKumarc. V. (2017) Evaluation of cytotoxicity of hematite nanoparticles in bacteria and human cell lines, Colloids and Surfaces B: Biointerfaces Volume 157, 101-109.

Raggio J, Pintado A, Ascaso C, De La Torre R, De Los Ríos A, Wierzchos J, Horneck G, Sancho LG (2011). Whole lichen thalli survive exposure to space conditions: results of Lithopanspermia experiment with Aspicilia fruticulosa. Astrobiology. 2011 May;11(4):281-92. doi: 10.1089/ast.2010.0588.

Rasmussen, R. A., Khalil, M.A.K. (1981). Atmospheric methane (CH4): Trends and seasonal cycles, JGR Oceans, 86, 9826-9832.

Renno, N. O., and 22 colleagues (2009) Physical and Thermodynamic Evidence for Liquid Water on Mars, Lunar and Planetary Science Conference, Houston, March 23-27.

Richardson, J.T.E. (2005). Instruments for obtaining student feedback: a review of the literature. Assessment & Evaluation in Higher Education 30, no. 4: 387–415.

Rizzo, V., & Cantasano, N. (2009) Possible organosedimentary structures on Mars. International Journal of Astrobiology 8 (4): 267-280.

Rizzo, V., & Cantasano, N. (2011), Cyanobacteria on Terrestrial Meteorites and Stromatolites on Mars, Journal of Cosmology, 13, 15.

Rizzo, V. & Cantasano, N. (2016). Structural parallels between terrestrial microbialites and Martian sediments. International Journal of Astrobiology, doi:10.1017/S1473550416000355

Robb, F., Antranikian, G., Grogan, D., Driessen, A. (2007). Thermophiles: Biology and Technology at High Temperatures, CRC Press.

Roos-Serote, M., Atreya, S. K., Webster, C. R. Mahaffy, P. R. (2016). Cometary origin of atmospheric methane variations on Mars unlikely. JGR Planets, 121, 2108-2119.

Roberts, P., & Evans S. (2011). The book of Fungi. University of Chicago Press.

Ryan, BD, Bungartz F, Nash TH (2002). Morphology and anatomy of the lichen thallus. In Lichen Flora of the greater Sonoran Desert region (eds Nash TH, Ryan BD, Gries C, Bungartz F), pp. 8–23. Tempe, AZ.

Ruffi, W. & Farmer, J.D., (2016). Silica deposits on Mars with features resemblinghot spring biosignatures at El Tatio in Chile. Nature Communications, 7: 13554, DOI: 10.1038/Ncomms13554

Saleh, Y. G., M. S. Mayo, and D. G, Ahearn (1988) Resistance of some common fungi to gamma irradiation." Appl. Environm. Microbiol. 1988, 54: 2134–2135

Sanchez, F. J., E. et al. (2012) The resistance of the lichen Circinaria gyrosa (nom. provis.) towards simulated Mars conditions-a model test for the survival capacity of an eukaryotic extremophile." Planetary and Space Science, 2012, 72(1), 102-110.

Sancho L. G., de la Torre, R., Horneck, G., Ascaso, C. , de los Rios, A. Pintado,A., Wierzchos, J.,Schuster, M. (2007). Lichens Survive in Space: Results from the 2005 LICHENS Experiment Astrobiology. 7, 443-454.

Satoh, K, Y, Nishiyama, T. Yamazaki, T. Sugita, Y. Tsukii, K. Takatori, Y. Benno, and K. Makimura. "Microbe-I (2011) Fungal biota analyses of the Japanese experimental module KIBO of the International Space Station before launch and after being in orbit for about 460 days." Microbiol Immunol. 2011 Dec;55(12):823-9. doi: 10.1111/j.1348-0421.2011.00386.x.

Schulze-Makuch, D. et al. (2005) Scenarios for the evolution of life on Mars, Journal of Geophysical Research: Planets, 110, E12.

Schroder, K-P, Smith, R. C. (2008). Distant future of the Sun and Earth revisited. Mon. Not. R. Astron. Soc. 000, 1-10.

Selbmann L, Zucconi, D. Isola, and D. Onofri (2015) Rock black fungi: excellence in the extremes. From the Antarctic to Space." 2015. Current Genetics 61: 335-345. DOI 10.1007/s00294-014-0457-7

Setlow, P. 2006. Spores of Bacillus subtilis: their resistance to and killing by radiation, heat and chemicals. Journal of Applied Microbiology 101, 514-525.

Setlow, B., Setlow, P. 1995. Small, acid-soluble proteins bound to DNA protect Bacillus subtilis spores from killing by dry heat. Appl Environ Microbiol. 61, 2787-2790.

Schuerger, A. C., J. E. Moores, C. A. Clausen, N. G. Barlow, D. T. Britt

(2012), Methane from UV-irradiated carbonaceous chondritesunder simulated Martian conditions,J. Geophys. Res., doi:10.1029/2011JE004023

Schuerger, A. C., Ming, D. W., Golden, D.C. (2017). Biotoxicity of Mars soils: 2. Survival of Bacillus subtilisand Enterococcus faecalis in aqueous extracts derived from six Mars analog soils, Icarus 290, 215-223.

Seminara, A. et al (2018). A universal growth limit for circular lichens.

Shaheen, R. Et al. (2015). Carbonate formation events in ALH 84001 trace the evolution of the Martian atmosphere, PNAS, 112 (2) 336-341.

Small, L. W, (2015]) On Debris Flows and Mineral Veins - Where surface life resides on Mars. https://www.scribd.com/doc/284247475/On-Debris-Flows-eBook

Squyres, S. W. and 18 colleagues (2004) In Situ Evidence for an Ancient Aqueous Environment at Meridiani Planum, Mars, Science 306 (5702), 1709-1714. 2004 Dec 03.

Soffen, G.A. 1965. NASA Technical Report, N65-23980.

Steele, A., McCubbin, F.M., Fries, M. (2012). A Reduced Organic Carbon Component in Martian Basalts, Science, 337, 212-215.

Sutter, B.; Eigenbrode, J. L.; Steele, A.; McAdam, A.; Ming, D. W.; Archer, D., Jr.; Mahaffy, P. R. (2016). The Sample Analysis at Mars (SAM) Detections of CO_2 and CO in Sedimentary Material from Gale Crater, Mars: Implications for the Presence of Organic Carbon and Microbial Habitability on Mars. American Geophysical Union, Fall General Assembly 2016, abstract id.P21D-07.

Szewczyk, N.J., Mancinelli, R.L., McLamb, W., Reed, D., Blumberg, B.S., Conley, C.A., (2005). Caenorhabditis elegans survives atmospheric breakup of STS-107, Space Shuttle Columbia. Astrobiology 5, 690-705.

Tarasashvili,M. V., et al. (2013). New model of Mars surface irradiation for the climate simulation chamber 'Artificial Mars. International Journal of Astrobiology, 12, 161-170

Tehler, A. & Wedin, M. (2008). Systematics of Lichenized Fungi. In Nash, T. H. (editor) Lichen Biology, Cambridge University Press.

Thomas-Keprta K.L., Clemett S.J., Bazylinski D.A., Kirschvink J.L., McKay D.S., Wentworth S.J., Vali H., and Gibson E.K., Romanek C.S. (2002) "Magnetofossils from Ancient Mars: A Robust Biosignature in the Martian Meteorite ALH84001." Applied and Environmental Microbiology 68, 3663-3672.

Thomas-Keprta, K. L., et al., (2009). Origins of magnetite nanocrystals in Martian meteorite ALH84001. Geochimica et Cosmochimica Acta, 73, 6631-6677.

Treiman A.H. (2003) The Nakhla martian meteorite is a cumulate igneous rock: Comment on Varela et al. (2001). Mineralogy and Petrology 77 , 271-277.

Treiman, A. H., & Essen, E. J. (2011). Chemical composition of magnetite in Martian meteorite ALH 84001: Revised appraisal from thermochemistry of phases in Fe–Mg–C–O. Geochimica et Cosmochimica Acta, 75, 5324-5335.

Tugay, T. Zhdanova, N.N., Zheltonozhsky, V., Sadovnikov, L., Dighton, J. (2006)The influence of ionizing radiation on spore germination and emergent hyphal growth response reactions of microfungi, Mycologia, 98(4), 521–527.

U.S. Department of Agriculture (2017). Complete Guide to Biogas and Methane: Agricultural Recovery, Manure Digesters, AgSTAR, Landfill Methane, Greenhouse Gas Emission Reduction and Global Methane Initiative. U.S. Government Printing Office, WDC

U.S. Department of Energy, (2017) Complete Guide to Methane Hydrate Energy:

Ice that Burns, Natural Gas Production Potential, Effect on Climate Change, Safety, and the Environment.U.S. Government Printing Office, WDC

U.S. Department of the Interior (2010) Lichen Inventory Synthesis Western Arctic National Parklands and Arctic Network, Alaska. Natural Resource Technical Report NPS/AKR/ARCN/NRTR—2010/385.

Van Den Bergh, S., (1989) Life and Death in the Inner Solar System, Publications of the Astronomical Society of the Pacific, 101, 500-509.

Vesper, S.J., W. Wong, C.M. Kuo and D.L. Pierson. (2008) Mold species in dust from the International Space Station identified and quantified by mold-specific quantitative PCR. Research in Microbiology. 159: 432-435.

Villanueva, G. Mumma, M. Novak, R. Kufl, H. Hartogh, P., Encrenaz, T., Tokunaga, A., Khayat, A., Smith, M. (2015) Strong water isotopic anomalies in the martian atmosphere: Probing current and ancient reservoirs". Science. 348: 218-21.

Watt, S., C. Simpson, C. McKillop, and V. Nunn. (2002). Electronic course surveys: does automating feedback and reporting give better results? Assessment & Evaluation in Higher Education 27, no. 4: 325–337.

Webster, G. et al. (2013) Isotope ratios of H, C, and O in CO2 and H2O of the Martian atmosphere. Science 341, 260–263.

Webster, C. R. et al. (2015) Mars methane detection and variability at Gale crater, Science, 347, 415-417.

Webster, C.R. et al. (2018). Background levels of methane in Mars' atmosphere show strong seasonal variations Science 360,1093-1096.

Wember, V. V. and Zhdanova, N. N. (2001) Peculiarities of linear growth of the melanin-containing fungi Cladosporium sphaerospermum Penz. and Alternaria alternata (Fr.) Keissler. Mikrobiol. Z. 2001, 63: 3–12.

Whalen, S.C. (2005). Biogeochemistry of Methane Exchange between Natural Wetlands and the Atmosphere, Environmental and Atmospheric Science, 22, 1093-1096.

White, O. et al. (1999) Science, 286, 1571.

Zhakharova,K., et al. (2014). Protein patterns of black fungi under simulated Mars-like conditions. Scientific Reports, 4, 5114.

Zhdanova NN, Lashko TN, Vasiliveskaya AI, Bosisyuk LG, Sinyavskaya OI, Gavrilyuk VI, Muzalev PN. 1991. Interaction of soil micromycetes with 'hot' particles in the model system. Microbiol J 53:9–17.

Zhdanova, N. N., T. Tugay , J. Dighton, V. Zheltonozhsky and P McDermott, (2004) Ionizing radiation attracts soil fungi." Mycol Res. 2004, 108: 1089–1096.

Zhuravskaya AN, Kershengoltz BM, Kuriluk TT, Shcherbakova TT. (1995). Enzymological mechanisms of plant adaptation to the conditions of higher natural radiation background. Rad Biol Radioecol 35:249–355.

III. MARS: ALGAE, LICHENS, FOSSILS, MINERALS, MICROBIAL MATS, AND STROMATOLITES IN GALE CRATER

R. G. Joseph1*, L. Graham2, Burkhard Büdel3, Patrick Jung4,
G. J. Kidron5, K. Latif6, R. A. Armstrong7,
H. A. Mansour8, J.G. Ray9, G. J. P. Ramos10, L. Consorti11, V. Rizzo12, C. H. Gibson13,14, R. Schild15

1. Astrobiology Associates of California,
2. Dept. of Botany, University of Wisconsin-Madison, USA
3. Plant Ecology and Systematics, Biology Institute, University of Kaiserslautern, Germany
4. University of Applied Sciences Kaiserslautern,
Applied Logistics and Polymer Sciences, Germany.
5. Institute of Earth Sciences, The Hebrew University of Jerusalem, Israel
6. National Centre of Excellence in Geology, University of Peshawar,
Khyber Pakhtunkhwa, Pakistan
7. Aston University, Birmingham, UK.
8. Dept. of Botany, Ain Shams University, Cairo, Egypt
9. School of Biosciences, Mahatma Gandhi University, Kerala, India
10. Ficology Laboratory, State University of Feira de Santana, Brazil
11. Department of Mathematics and Geosciences,
University of Trieste, Trieste, Italy.
12. National Research Council (Emeritus), I.S.A.FO.M. U.O.S., Cosenza, Italy
13. Dept. of Mechanical and Aerospace Engineering, University of California,
14. Scripps Center for Astrophysics and Space Sciences, San Diego.
15. Dept. Astrophysics, Harvard-Smithsonian (Emeritus), Cambridge

ABSTRACT

Gale Crater was an ancient Martian lake that has periodically filled with water and which may still provide a watery environment conducive to the proliferation and fossilization of a wide range of organisms, especially algae. To test this hypothesis and to survey the Martian landscape, over 3,000 photographs from NASA's rover Curiosity Gale Crater image depository were examined by a team of established experts in the fields of astrobiology, astrophysics, biophysics, geobiology, lichenology, phycology, botany, and mycology. As presented in this report, specimens resembling terrestrial algae, lichens, microbial mats, stromatolites, ooids, tubular-shaped formations, and mineralized fossils of metazoans and calcium-carbonate encrusted cyanobacteria were observed and tentatively identified. Forty-five photos of putative biological specimens are presented. The authors were unable to precisely determine if these specimens are biological or consist of Martian minerals and salt formations that mimic biology. Therefore, a review of Martian minerals and mineralization was conducted and the possibility these formations may be abiogenic is discussed. It is concluded that the overall pattern of evidence is mutually related and that specimens resembling algae-like and other organisms may have colonized the Gale Crater, beginning billions of years ago. That some or most of these specimens may be abiotic, cannot be ruled out. Additional investigation targeting features similar to these, should be a priority of future studies devoted to the search for current and past life on Mars.

1. Life on Mars and in the Gale Crater

The authors of this report provide visual evidence of specimens, photographed within the Gale Crater, which resemble terrestrial algae, lichens, microbial mats, stromatolites, ooids, tubular-shaped and calcium carbonate encrusted organisms, and fossil-like formations reminiscent of metazoans. The authors' interpretations are speculative and Martian minerals and salts may have contributed to these observations as detailed in this report. Our purpose in presenting these observations is to identify and target specimens for future research and additional examination by other investigators and for possible extraction and analyses by robotic missions to Mars.

There is substantial evidence for life on Mars (Dass, 2017; Joseph 2014; Joseph et al. 2019, 2020a; Krupa 2017; Levin and Straat, 1976, 2016; Rabb, 2018; Small, 2015; Thomas-Keprta et al. 2002, 2009). For example, vast colonies of mushroom-shaped, lichen-like organisms, possibly engaged in photosynthesis, have been observed, attached by thin stems, to hundreds of rocks photographed in Eagle Crater by the rover Opportunity. In addition, over a three-day period, 23 puffball-shaped organisms grew up out of the ground and expanded in size in surrounding areas of Meridiani Planum in the absence of wind or other abiotic contributing factors (Joseph et al. 2020a). No other living features have been reported or observed in Meridiani Planum, which includes Eagle Crater. By contrast, green specimens have been observed in Utopia Planitia and Chryse Planitia (Levin et al. 1978), and those with features similar to algae and stromatolites have been photographed in Gusav Crater and Gale Crater in particular (Joseph, 2014, 2016; Krupa 2017; Noffke 2015; Rabb, 2015, 2018, Rizzo and Cantasano 2016; Rizzo 2020; Ruffi and Farmer 2016; Small, 2015).

2. Water and Gale Crater

Gale Crater may have and may still provide a moist environment conducive to the proliferation of life with temperatures as high as 6°C (43°F), as measured by the rover Curiosity's onboard Remote Environment Monitoring Station. Gale Crater has all the characteristics of a dried lake that is periodically replenished with water (Williams et al. 2013; Grotzinger et al. 2014; Vaniman et al. 2014; Bridges et al. 2015; Steele et al. 2017) as indicated by its numerous fluvial valleys and water pathways (Grotzinger et al. 2015). An analysis by the rover Curiosity's suite of sampling instruments also indicates that minerals, clays and mudstones have been repeatedly hydrated (Williams et al. 2013; Grotzinger et al. 2014; Vaniman et al. 2014); and there is also evidence that moisture may be available, on a daily basis, at ground level, during the summer months (Castro et al. 2015; Martin-Torres et al. 2015; Steele et al. 2017) and that pure water ice has been trapped inside the Curiosity rover wheels (Joseph et al. 2020b).

Two layers of clouds have also been observed above Gale Crater (Moores et al. 2015). On Earth, clouds consist of water with saturation ranging from 81% to 100% (Pruppacher and Klett 2010; Hu et al. 2010). Above Earth, water content levels of at least 95% have also been measured in ice-clouds and supercooled clouds with temperatures between −40°C and 0°C (Hu et al. 2010). Therefore, the clouds above Gale Crater also likely have a significant water content, despite even subzero temperatures, thereby inducing precipitation and providing moisture for Martian organisms especially when temperatures rise above freezing.

Columns of water vapors also form every spring and summer as based on data collected from orbital observations (Smith, 2004; Read and Lewis, 2004), and these vapors are transported from the north toward the equator by southerly winds, and thus, toward Gale Crater (Harri et al. 2014). Moreover, these vapors have a precipitable water content of at least 10–15 pr μm (Smith, 2004). The vapors also appear to reach saturation, depending on humidity (Harri et al. 2014). Hence, given these vapors and layers of clouds, precipitation may moisten the surface on a daily basis, especially during the summer (Martin-Torres et al. 2015; Steele et al. 2017).

Martin-Torres and colleagues (2015), based on data from the Rover Curiosity's suite of sampling instruments, including measurements of ground temperature, air pressure and humidity, have reported the possible "formation of night-time transient liquid brines in the uppermost 5 cm of the subsurface that then evaporate after sunrise." Steele et al (2017) have come to similar conclusions and have argued that during "the evening and night, local downslope flows transport water vapour down the walls of Gale crater. Upslope winds during the day transport vapour desorbing and mixing out of the regolith up crater walls, where it can then be transported a few hundred metres into the atmosphere" (Steele et al. 2017).

Therefore, as reviewed here, there is evidence that the Martian water cycle, within the Gale Crater, follows the standard hydrologic cycle which characterizes water recycling on Earth (Fedorova et al. 2020), i.e. evaporation, vaporization, precipitation, condensation, freezing, melting, evaporation, thereby periodically forming frost, ice, and surface moisture, depending on temperature and relative humidity.

It is also believed that Martian water may be stored in underground aquifers (Malin and Edgett 2000), and is sequestered in rocks and hydrated minerals, or is locked within frozen ground (Biemann et al. 1977; Plaut et al. 2007; Mustard et al. 2012; Kieffer et al. 1976; Farmer et al. 1977), only to melt and pool upon the surface when temperatures rise above freezing. Geological features also indicate that flowing water, streams and run-off have carved fluvial pathways down the hillsides and across the surface of Gale Crater, only to eventually seep beneath the surface, evaporate, or freeze (Grotzinger et al. 2015; Steele et al. 2017).

Gale Crater may have first filled with water and became "Gale Lake" nearly 4 billion years ago, soon after the crater was formed. Based on morphological observations, Fairen et al. (2014) described what they believed to be "evidence for ancient glacial, periglacial and fluvial (including glacio-fluvial) activity within

Gale crater, and the former presence of ground ice and lakes." Likewise, based on morphology, Oehler (2013) argued that landforms indicate a "major history of water and ice in Gale crater, involving permafrost, freeze-thaw cycles, and perhaps ponded surface water." Masson et al. (2001) have come to similar conclusions.

The substantial evidence of hydrated minerals also indicates that Gale Crater has been repeatedly inundated with large bodies of water which could have sustained a variety of species, some of which may have become fossilized or mineralized (Grotzinger et al. 2012, 2015). These possibilities are supported by the findings of Eigenbrode and colleagues (2018) who reported organic matter preserved in three billion year old mudstones at Gale crater--and which may represent the residue of Martian organisms.

3. Water, Minerals, the Search for Life in Gale Crater

It appears that there is sufficient water available in the Gale Crater to support the growth and proliferation of a variety of species, beginning billions of years ago when the crater was first formed and filled with water. Moreover, early in its history, when Gale Crater was "Gale Lake," this watery environment may have facilitated the construction of stromatolites (Noffke, 2015; Joseph et al. 2019; Rizzo 2020), and could have induced the fossilization of various species (Grotzinger et al. 2012, 2015).

To determine if Gale Crater may be host to living organisms and to search for evidence of fossils, over three thousand images photographed by the rover Curiosity were examined. The authors of this report include experts in algae, cyanobacteria, lichens, microbial mats, stromatolites, microfossils, mineralogy and geology.

As detailed here, the authors have observed Martian specimens which resemble fossilized or trace fossils of various organisms including metazoans, and those with features similar to or which closely resemble green algae, cyanobacteria, lichens, fungi, and mat-forming organisms. The authors also identified three domical-shaped and mat-like structures which are morphologically similar to microbial mats and the living stromatolites of Lake Thetis, Australia, a possible analogue environment for Mars and the Gale Crater when (and if) flush with water (Baldridge et al. 2009; Bridges et al. 2015; Graham et al. 2016; Nguyen et al. 2014).

Various terrestrial minerals may have a green color similar to algae, and minerals and clumps of sand and salt may come to have unusual shapes secondary to wind, moisture, and weathering. Therefore, to assist in determining possible abiogenic contributions to these life-like observations, all reports related to the discovery of salts and relevant minerals in the Gale Crater are reviewed in Section IV of this article.

Methods and Results

Image Search: The rover Curiosity landed on Aeolis Palus inside Gale Crater, Mars, on August 6, 2012 (Sol 1) and has since transmitted over 20,000 images to Earth. Over 3000 images photographed between Sol 1 and Sol 1000, by the rover Curiosity's suite of cameras, and over 200 images photographed between Sol 1800 to 1900 were examined for features suggestive of living and fossilized organisms. Two hundred and fifty-three photographs, selected from these 3000+ images based on the target features were subject to additional visual analyses which included enlarging these photos by 300%. Of these 253 NASA-Curiosity photographs, 95 were judged to clearly depict the targeted features, 87 of which were in color and subsequently subject to color extraction protocols and additional visual inspection. It should be noted, that often, when obvious biological features were photographed, dozens to several hundred photos photographed immediately before or after these specimens were imaged were blanked out and could not be viewed on the NASA website.

Color Extraction: According to NASA, the rover Curiosity cameras have been calibrated to take "approximate true color" photos of the Martian landscape which are "imaged in 3 science color filters." They are "integrated" via an "RGB Bayer pattern filter" and the "RGB filter array" which makes possible "RGB imaging" (see https://msl-scicorner.jpl.nasa.gov/Instru-ments/Mastcam). However, NASA has admitted the emphasis given to the colors red, green and blue, are not equal, and certain (unidentified by NASA) color "pixels are poorer than in other pixels," and there is color leakage such that certain colors NASA has not identified were given less prominence in the resulting RGB images. Moreover, colors are adjusted based on "color calibration targets," such that the assumed colors of certain targeted abiotic features became the standard color which were given prominence over all other colors. As detailed by R. L. Levin of Lockheed Martin, NASA's Mars cameras have been calibrated such that there is an "excess of red" in the color photos transmitted back to Earth (Levin 2016) and that blues and greens become less evident or obvious. Hence, there is good reason to suspect that reds are emphasized in all NASA Curiosity color photos, and greens and blues which are more likely to depict living organisms or water on the surface, are deemphasized by the RGB pattern filter array.

For the purposes of this study, and to emphasize contrasting biogenic vs abiogenic surface features, the 87 NASA-Curiosity color photos judged to depict biological features, were subject to Red-Blue-Green color extraction, employing Fotor High Dynamic Range (HDR) image enhancement software which restored color balance and emphasized the natural colors within the original NASA images. No false colorization or tinting was applied to these NASA photos by the authors or their associates. Rather, greens and blues which were present in the original NASA photos were extracted and given near-equal balance compared to reds.

To ensure the accuracy of the HDR extraction process, six color photos of the Negev Desert, photographed by one of the authors (GJK) were subject to color extraction protocols. In five of the Negev Desert photographs, where no greens or blues were evident in the original photo or observed by the photographer when viewing the landscape, no greens or blues emerged following HDR extraction. In the sixth photo the HDR process was found to emphasize in equal measure, the green color of a Negev Desert bush and the red-brown color of the reddish-brown Negev surface.

Biological Features Analyses. These 87 HDR color photos and the eight black and white photos were subject to additional visual inspection by the authors. Based on the presumed biological features depicted (e.g. algae, lichen, fossil), relevant photos were distributed to outside experts with an expertise in those specific areas of biology and who were asked to provide feedback.

Of the 87 HDR photos and eight black and white photos, 45 were determined to depict the most obvious biological features. These 45 photos were then subject to additional visual analyses by the authors of this study, each of whom wrote summaries of their observations and which included identification of the specimens depicted.

Results: The authors of this study, each of whom has an expertise in the relevant subject matter, identified specimens that clearly resemble terrestrial algae, lichens, microbial mats, stromatolites, ooids, and tubular-shaped and calcium-carbonate encrusted fossil-like organisms. In addition, specimens resembling calcium-carbonate biosignatures and fossilized open-cone-like gas-bubble vents which are associated with photosynthesis-oxygen respiration (Bengtson et al. 2009; Sallstedt et al. 2018) were identified, as well as masses of tubular-shaped formations which resemble colonies of mat-forming filamentous organisms (Figures 1, 12, 16, 24-28). In total, these putative biological specimens were photographed by the rover Curiosity, in various locations of Gale Crater, on Sols 122, 173, 271, 298, 304, 305, 309, 319, 322, 528, 529, 809, 840, 853, 871, 880, 890, 1905.

The putative microbial mats presented in this report were also compared with terrestrial mats. All were found to have a macro-morphology nearly identical to terrestrial thrombolites and nodular and microbial mats photographed in Lake Thetis (Earth) in particular.

Domical Stromatolite Image Search: The ancient lakes of Gale Crater and Mars have been likened to the lakes of Western Australia, including Lake Thetis (Baldridge et al. 2009; Bridges et al. 2015; Graham et al. 2016; Nguyen et al. 2014) which host a variety of algae and living domical stromatolites. Therefore, a search was conducted for terrestrial domical stromatolites in Lake Thetis, and for specimens in Gale Crater which resemble these Australian stromatolites. Six concentric Martian formations were identified and their micro- and macro-structure, including laminae, fenestrae (gas bubbles) and concentric organizational geometry were visually examined, traced, and measured.

Results: Six concentric formations photo- graphed by the rover Curiosity's Mast Camera on Sol 122, 173, 309, 528, 529, were identified as resembling concentric and domical stromatolites (Figures 36, 38, 38-45). Three were found to be comparable, morphologically, to concentric stromatolites growing in Lake Thetis (Figures 34, 36, 38, 39, 42). Attached and/or adjacent to these specimens, photographed on Mars and in Lake Thetis, are numerous nodular microbial mat-like formations and thrombolites with the characteristic "peanut brittle" appearance of fossilized terrestrial mats (Figures 1, 8, 9, 16, 36, 37, 44). These putative Martian microbial mats were compared to their terrestrial counterparts and those in Lake Thetis and similarities noted.

Three of the domical specimens, photographed on Sol 122, 308, 528, 529, were found to meet most of the criteria for a biological formation (see Buick et al. 1981, 1995; Lowe, 1994), whereas a fourth (Sol 173) met all the criteria (Figures 40-43) which includes five layers of laminae, a lack of continuity of the stromatolite laminae, the presence of fine-scale crinkly and wavy laminae with several orders of curvature, an abundance of detrital material, the presence of a central (albeit collapsed) axial zone, the presence of what appears to be numerous fenestrae/gas bubbles, and numerous nodular mats attached to the walls and the inner collapsed surface and as part of adjacent structures.

Discussion:
4. Water, Algae, Lichens, Fossils, Microbial Mats, Stromatolites in Gale Crater

The Gale Crater (Aeolis Palus), located near the equator, was formed around 3.7 bya and presumably filled with water (Le Deit et al. 2013; Thomson et al. 2011). With a diameter of 154 km (96 mi), Gale Crater has the appearance of a dry lake, at the center of which rises a 5.5 km (18,000 ft) high mountain. There are foothills, a vast plain, and numerous gullies and channels that flow down from the hills and which appear to have been carved by flowing water. There are fluvial valleys rich in potassium (Grotzinger et al. 2015), mounds and peaks (Le Deit et al. 2013; Palucis et al. 2014; Schwenzer et al. 2012; Thomson et al. 2011) sedimentary rocks dated to 4.2 bya (Farley et al. 2014), a wide range and assortment of cobbles, small stones, and alkaline igneous rocks (Sautter et al. 2014; Schmidt et al., 2014; Stolper et al. 2013) including basalt tholeitic and alkaline basalts (Sautter et al. 2014), and rocks comparable to Earth rocks which formed around 2.7 bya or later (see Blichert-Toft et al. 1996; Taylor and McLennan, 2009), i.e. during the Archean-Proterozoic transition.

Throughout this time period leading into the present, the Gale Crater appears to have repeatedly filled with water (Bibring et al. 2006; Cabrol et al. 1999; Fairen et al. 2014; Murchie et al. 2009; Siebach and Grotzinger 2014; Buz et al. 2017). With daytime air temperatures as high as 6°C (43°F) and the likelihood of daily exposure to moisture (Martin-Torres et al. 2015; Steele et al.

2017) the Gale Crater may have provided a habitable environment for the last 3.7 billion years.

As detailed in this report, the authors have identified specimens with features similar or identical to green algae, cyanobacteria, lichens, fungi, stromatolites and mat-forming organisms, and those which resemble fossilized or trace fossils of various organisms including metazoans. Many of the algae-like specimens appear to be moist or covered by a thin layer of ice.

Masses of filamentous vegetative-like material (Figure 10) and tubular shaped formations and those which resemble colonies of mat-forming filamentous organisms (Figures 1, 12, 15, 16, 24-28) were also observed in Gale Crater. In addition, formations resembling calcium-carbonate biosignatures (Figure 19), nostoc balls (Figures 2-23) and open-cone-like gas-bubble-vents (Figures 3, 5, 7, 13, 17, 18) which are associated with photosynthesis-oxygen respiration (Sallstedt et al. 2018) were identified. In support of this interpretation: it's been reported that colonies of hundreds of skyward oriented mushroom-shaped specimens, observed in Eagle Crater, and attached by thin stems and jutting upward from rocks, may also be engaged in photosynthesis (Joseph et al. 2020a).

We are unable to provide scale-bars of the exact size of these specimens as that information was not provided by NASA or made available following inquiries. Therefore, only size-estimates are provided based on generalized camera pixel and focal length specifications and other parameters made available by NASA.

5. Algae

The authors of this study, ten of whom are established experts in algae and algae-fungi symbiotes, identified Martian specimens resembling algae, photographed at ground level, atop and alongside rocks, mudstone and sand (Figures 1-9, 12-16). The algae-like formations were often associated with or adjacent to specimens similar to ooids, lichens, fungi, microbial mats, and stromatolites.

These algae-like specimens, depending on substrate, appear as green clumps, spherules, cake-like layers, thin sheet-like layers and thick layered leafy vegetative masses of material which partially cover Martian rocks, sand, and fungi-like surface features. Many specimens appeared to be moist or covered by a thin layer of ice (Figures 2-7, 40-43). Although their exact identity is unknown, if biotic, they may include blue-green algae (cyanobacteria), green algae, diatoms, and mosses.

Figure 1. Sol 890: Greenish and yellow substances covering Martian sand and rock and specimens resembling microbial mats and a carpet of yellow-green substances within which are embedded patterns similar to mat-making filamentous cyanobacteria and which have also been found in association with stromatolites within Cambrian strata in Northern China (Latif, 2010). Although some features resemble abiotic exfoliation processes, it's been determined that exfoliation is not entirely abiotic, but due to the endolithic growth of unicellular cyanobacterium--thereby giving these surfaces a green color--as determined in Antarctic Beacon Sandstone (Friedmann 1980, Friedman and Weed 1987) and by Büdel and colleagues in South Africa and Antarctica (Weber et al. 1996, Büdel et al. 2004, 2008). Chroococcidiopsis and other species of algae including Saphophyte cyanobacteria also display changing color patterns (Ferrari et al. 2002), particularly in response to insufficient water (Pattanaiki et al. 2007). Cyanobacteria or blue-green algae contain chlorophyll and phycobiliproteins, the latter comprising pigments ranging from the blue phycocyanin to the red phycoerythrin. The early-diverging, rock-dwelling cyanobacterium genus Gloeobacter is often colored purple (Graham et al. 2016). In high irradiance environments, cyanobacteria are often colored brown or black due to their protective scytonemin sheaths. They also produce carotenoids (yellow to orange pigments) that provide protection against irradiance (Pattanaiki et al. 2007). Therefore, if these specimens are in fact algae/cyanobacteria, they would have adapted to this high radiation environment. It should be stressed that the chemical composition of these mosaics is unknown. However, mineral and salt concentrations so far detected in the Gale Crater are insufficient to account for these patterns, colors and morphologies.

Figure 2. Sol 871. Green sphericals upon Martian sand, soil, rocks and pinnicle-columnar structures resembling terrestrial stromatolites and thrombolites (see Graham et al. 2014) and algae growing in shallow water (see Ray and Thomas 2012). On Earth, the greenish-coloration of sand and rock is due to green cryptoendolithic cyanobacteria (Büdel et al. 2004, 2008; Jung et al. 2019; The darkening in soil coloration may indicate moisture. Thin sheets of sphericals covering Martian sand, rock and stromatolite-like structures/protrusions, similar to the layered structures of terrestrial stromatolites and thrombolites resemble fossilized as well as living stromatolites and thrombolites on Earth (see Graham et al. 2014). These sheets are also similar to algae growing in shallow water, but may be frozen – the arrangement of sandy substance in between rocks is the typical nature of shallow rocky pools (see Ray and Thomas 2012).

Figure 3. Sol 840: Algae-like spherules and open cone-like protrusions similar to fossilized gas bubbles (upper left) on broken rocks photographed in the Gale Crater. These specimens are similar to those subjected to stressed conditions (Gaysina et al. 2019) and appear to have been recently moistened. Terrestrial algae similar to these specimens may assume these features when subjected to highly stressful conditions (Gaysina et al. 2019)

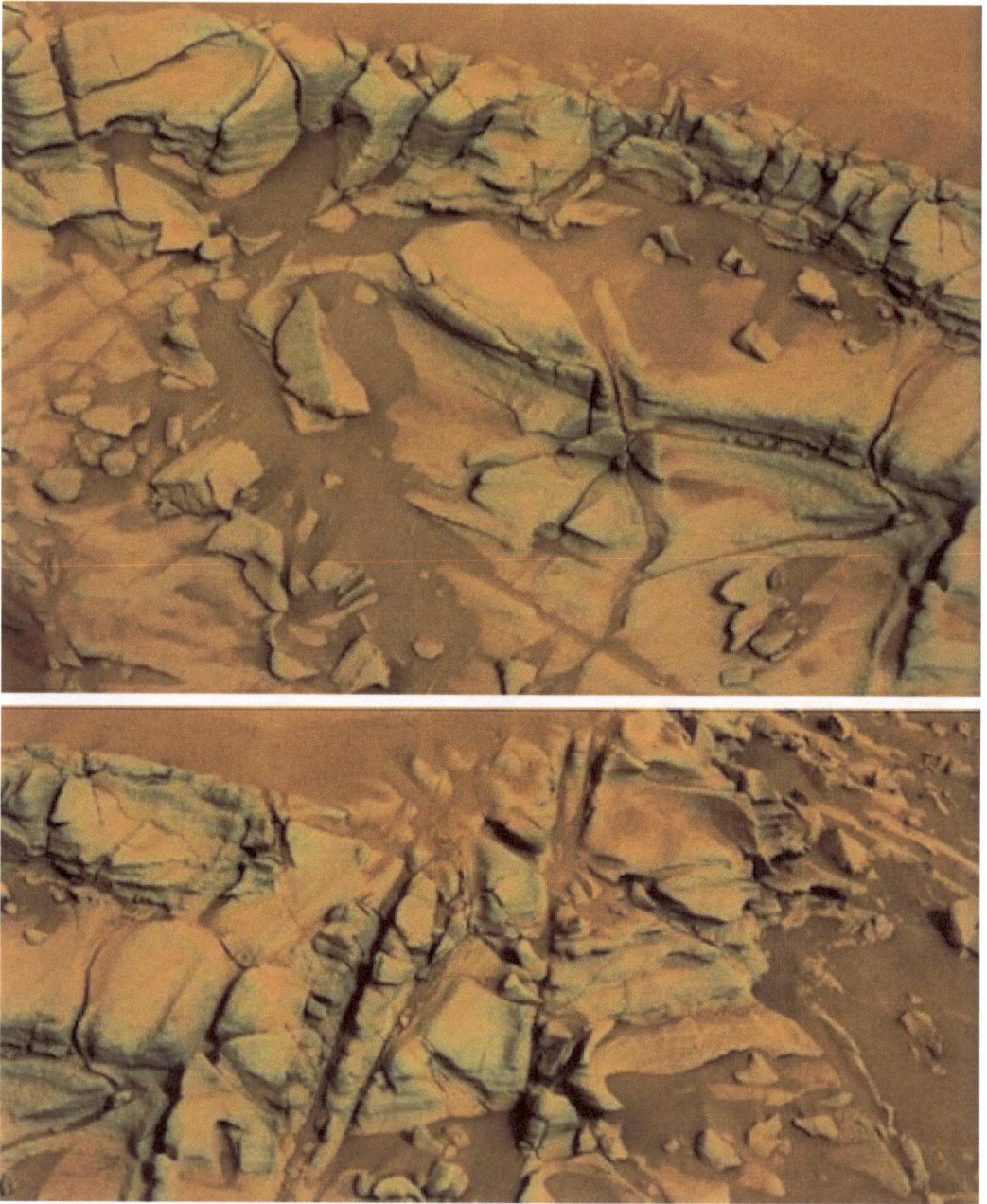

Figure 4. Sol 812 (top and bottom). The greenish-coloration may be due to green cryptoendolithic cyanobacteria (see (Büdel et al. 2004, 2008; Jung et al. 2019). The darkening in soil coloration may indicate moisture. Thin layer of green-colored substances on layered formations photographed in the Gale Crater. The overall appearance is similar to sand strata layers with fluvial water pathways. The contrasts between upper vs lower soils indicates the presence of moisture. Speculation: these strata may be harboring epipelic cyanobacteria (Hasler and Poulickova, 2010; Luesova, 2001) and algae recently exposed to water (see Ray et al. 2009). The possibility these are hydrated minerals is not likely due to the locations and insufficient quantities of minerals so far detected vs

the widespread and somewhat homogenous distribution of the specimens depicted here. Surface weathering of sandstone, such as occurs in some arid and desert environments, also support a biological interpretation, as colonies of green cryptoendolithic cyanobacteria are typically found in association (Büdel et al. 2004; Wierzchos et al. 2011). Enhanced with HDR processing.

Figure 5. Sol 305: Epilithic algae-like crust-like layers covering fungi-like specimens and microbial mat-like substance and mudstone marked by numerous apertures resembling oxygen-vents secondary to photosynthetic activity. The surface appears to be moist or covered by a thin layer of ice.

Figure 6. Sol 305. Epilithic algae-like material covering Martian sand, rock and fungi-like specimen (lower center).

Figure 7. Sol 305 (Top): Epilithic algae-like substances and fungi-like specimens. (Bottom) Sol 298: Specimens resembling green algae, lichens, and ooids.

Figure 8 Sols 840, 319, 271, 853. (Clockwise from the top) Thick-layered clumps of algae-like substance on the top of an overturned, multi-layered Martian rock. Bacterial mats with fluvial water pathways between them. Mushroom-shaped Algae-Symbiotes. Stromatolite-like protrusions. On Earth, cyanobacteria (microalgae) are the most abundant of all aerial and sub-aerial epilithic and hypo-lithic organisms that grow on rocks and which produce multiple mat formations (Eldrin, 2016), as well as stromatolites, and which form symbiotic relationships with fungi.

Figure 9. (Top): Algae-Microbial Mat. Image Credit, University of Waterloo. (Bottom) Sol 309: Formations resembling layers of elevated and possibly fossilized microbial mats dusted with algae.

Figure 10. Sol 322: Possible flowing tangles of filamentous cyanobacteria/ algae or Pseudoparenchymatous (plant roots pressed tightly to rock surfaces marked by porosity. The filamentous algae Klebsormidium and Oedogonium are comparable to the specimens seen here.

Within scientific literature, the term "algae" may be used broadly to include cyanobacteria (aka blue-green algae) plus oxygenic-photosynthetic protists such as green algae and non-photosynthetic relatives with a wide range of metabolic and reproductive processes, or more narrowly to exclude cyanobacteria (Graham et al. 2016). For our purposes, and as the exact identity of the specimens presented here cannot be determined with precision, these specimens (depicted or included in Figures 1-9, 12-16) will be referred to as "algae."

As will be explained (Section IV. The Geology of Martian Minerals...), it is not likely that these algae-like specimens are exfoliated weathered rocks containing masses of green minerals or salts, as no gross surface features resembling these abiotic substances have been previously identified by other scientists. The concentrations of the minerals and salts detected are insufficient, lack crystallization and are not visible in Gale Crater photos. By contrast, these algae-like and related specimens have been photographed in numerous locations and resemble their presumed terrestrial counterparts.

Nor can weathering or other geological processes account for these algae-like surface features. In fact, the greenish-coloration associated with surface weathering of terrestrial sandstone or other rocks, as occurs in some arid and desert environments on Earth, is evidence of biology, as colonies of green cryptoendolithic cyanobacteria are typically found in association (Büdel et al. 2004, 2008; Jung et al. 2019; Wierzchos et al. 2011). On Earth, minerals and desert salts and rocks are infested with cyanobacteria (Stivaletta, Barbieri and Billi, 2012; Finstad et al. 2017; Wierzchos, Ascaso and McKay, 2011), which help bind these substances together (reviewed by Joseph et al. 2019), and provide greenish coloration (Büdel et al. 2004, 2008; Jung et al. 2019). Therefore, although salt formations have not been visually observed on the surface of Gale Crater and salts detected are of insufficient quantity to account for these observations, even if some of these algae-like formations were masses of greenish-salts, the green coloration would be an indication of biology.

Other investigators have also observed what may be green algae growing on Mars. In 1978, and based on specimens photographed during the 1976 Mars Viking Missions, Levin, Straat and Benton published evidence of changing patterns on "greenish rock patches" which were "green relative to the surrounding area." Levin et al (1978) speculated that these greenish areas may represent "algae" growing on Mars. Thirty-five years later Joseph (2014), Rabb (2015), and Small (2015) published photos taken by the Mars rovers Spirit and Curiosity, depicting what they believed to be green algae. Soon thereafter, at the Lunar and Planetary Society, Krupa (2017) presented evidence of what appear to be green photosynthetic organisms growing in the Columbia Hills area of Gusev Crater, adjacent to what may be water pathways which intermittently fill with water. He noted that "the hillside...is covered by a very thin layer of green material" and "green spherules" which resembles algae in the soil. Thus, seven different investigators have observed and provided evidence that algae may be flourishing on the Red Planet.

6. The Lichens of Gale Crater vs Eagle Crater

Lichens are a symbiotic organism of at least one alga that can be green algae or cyanobacterium (photobiont) and at least one fungus (mycobiont), the latter of which is largely responsible for the lichens' mushroom shape, bulbous cap, thallus, and fruiting bodies (Armstrong 2017, 2019; Brodo et al. 2001; Tehler & Wedin, 2008). Molecular analyses, however, indicate that lichen consortia include a wide range of bacterial communities on the lichen-surface and within the photobiont zone, such as "Sphingomonas, Methylobacterium, and Nostoc," as well as "eukaryotic representatives of Rhizaria, Amoebozoa, Alveolata, Metazoa, and Viridiplantae" and the rhizarian protist Protaspa (Graham et al. 2018).

Eagle Crater Mushrooms: Colonies of thousands of skyward-oriented mushroom-shaped, lichen-like organisms, up to 8 mm in length, topped by bulbous mushroom caps, with 1 mm in diameter stems attached to rocks and have been identified in Eagle Crater (Joseph et al. 2020a). The upward orientation of these formations was interpreted as possible evidence that these organisms are engaged in photosynthesis.

Joseph and colleagues (2020a) did not find any evidence of algae in Eagle Crater. However, 23 specimens identical to "puffball" fungi--photographed by the rover Opportunity--were observed to increase in size with 12 emerging from beneath the soil during a three-day period in the absence of any contributing wind or weathering process. Other scientists have also identified spherical "puffballs" on the surface of Meridiani Planum (Dass, 2017; Joseph 2016; Rabb, 2018) whereas evidence that fungi may have contaminated the rover Opportunity has been reported (Joseph et al. 2019). Nevertheless, if the formations observed in Eagle Crater are fungi, lichens (algae-fungal symbiotes) or non-lichenized fungi, is unknown.

It's been falsely claimed that these Eagle Crater mushroom-shaped specimens consist of hematite. However, Burt, Knauth and Woletz (2005) in a presentation at the Lunar and Planetary Society and paper published in the journal Nature, dismissed the spherical hematite claims as "inappropriate." In addition, Joseph et al (2020a) have shown that the hematite hypothesis was based on inference and speculation, that no spheres and not a single mushroom-shaped specimen were examined for hematite, that the data was a "poor fit" for hematite, and the lichen-like formations of Eagle Crater are a different color from, and do not look anything like hematite, but instead resemble fungi or the lichen, Dibaeis baeomyces (Joseph et al. 2020a), and that the environment was never conducive to the formation of hematite.

Gale Crater Lichens: With few exceptions (Figure 8) the lichen-like formations observed in Gale Crater do not resemble the lichen Dibaeis baeomyces or any of the specimens photographed in Eagle Crater and Meridiani Planum.

Many of the putative Gale Crater lichens (Figures 11-18) are similar to those growing on the west coast Ireland cliffs of Moher whereas others are similar to agaric lichens (Lichenomphalia umbellifera).

The lichens of Meridiani Planum and Gale Crater form vast colonies in the absence of any other biological features. By contrast many of the Gale Crater lichens were found in association with specimens resembling fungi and spherical ooids (Figures 11-12). On Earth, ooids (ooliths), are formed in shallow seas and generally consist of calcite, bacteria and residues of degraded mucous produced by filamentous cyanobacteria (Devaud and Girardclos, 2001). Specifically, ooids are fashioned by the binding action of exopolysaccharides (Kumar et al. 2018) secreted by algae on sedimentary grains followed by calcification (Farbrius, 1977).

There are numerous photographs taken by the rover Curiosity of ooid-like specimens in association with lichen-like surface features. The authors have also observed what appears to be colonies of filamentous cyanobacteria which have been photographed in association with what appear to be spherical ooids and "dimpled" lichens (Figures 11, 12).

The putative Gale Crater lichens are also found in close proximity to veins and deposits of what appears to be gypsum which has been identified in the Gale Crater (Rapin et al. 2016). Gypsum is a favored substrate for extremophile algae/cyanobacteria (Bothe, 2019). Terrestrial veins of gypsum are derived, in part, in association with water via the dissolution of surface or near surface sulfate-rich veins (Anthony et al. 2003a; Bock, 1961; Garcia-Ruiz et al. 2007; Van Driessche et al. 2012). Water plus gypsum also promotes biological activity. The green alga Closterium (and its various relatives), for example, accumulates gypsum and stores it as crystals (Sr and/or Ba sulfates) in vacuoles located in the apices of their spindle-shaped cells (He et al. 2014).

Alternatively, the veins of white-colored substances and veins found in association with these lichen-like formations (Figures 11, 13-15) may consist of quartz and hydrated calcium sulfates (Nachon et al. 2014; McLennan et al. 2013), all of which have been identified in the Gale Crater where these specimens were photographed. Quartz is also associated with biological activity, and quartz, dominated by lichens and biocrusts, has been identified in the Atacama Desert. Furthermore, cyanobacteria, green algae and microfungi have colonized the inner structures of small quartz gravel (Jung, et al. 2019).

These putative Martian lichens have also been photographed in association with what appears to be ooids and calcium carbonate (Figure 11, 13-15), both of which are byproducts of cyanobacterial activity (Devaud and Girardclos, 2001). The Gale Crater lichens and the putative ooids are often densely packed together.

Many of the Gale Crater lichen-shaped formations have a central depression or nuclei giving them a "donut shape," as well as subsurface mycelium/hypa (Figures 11, 13-15, 17,18), similar to fungi and agaric lichens (Lichenomphalia umbellifera). Other specimens photographed in Gale Crater assume a diversity of forms similar to lichens and ooids discovered in Cambrian era deposits in northern China (Mei et al. 2019; Riaz et al. 2019), including "concentric and radial, rounded or elliptical, and with or without nuclei." By contrast, the lichen-like "Martian mushrooms" of Eagle Crater have one distinct form: a thin stem topped by a bulbous cap, colonies of which are oriented skyward as if engaged in photosynthesis, but without any evidence of ooids.

Therefore, hundreds of diverse lichen-like formations have been observed and photographed in various regions of Gale Crater which bear no resemblance to specimens photographed in Eagle Crater; whereas those of Eagle and Gale Crater have obvious counterparts on Earth. Eagle Crater is part of a vast plain (Meridiani Planum) just south of the Equator and is 5,200 miles (8,400 kilometers) from Gale Crater. There is no evidence that Eagle Crater was ever a lake or held large amounts of water. Therefore, differences in morphology may represent the on and off lake-like environment of Gale Crater vs the much dryer Eagle Crater. Or, like terrestrial lichens (Armstrong 2017, 2019), there may be hundreds of different varieties of Martian lichens.

Photosynthesis is a trait that the putative lichens of Gale Crater may share in common with those of Eagle Crater (see Joseph et al. 2020a) and with their counterparts on Earth. This interpretation is supported by the discovery of what may be gas bubble apertures produced secondary to photosynthesis and oxygen release. These apertures are often found in association with the putative lichens of Gale Crater (Figures 17, 18).

7. Gas Bubbles and Photosynthesis

Lichens are photosynthesizing organisms which respire oxygen--activity associated with its photobiont; respiration being dominated by fungi and bacteria in the consortium. Photosynthesis generates oxygen. Numerous studies have demonstrated that lichens remain viable and maintain photosynthetic activity when exposed to simulated Martian temperatures, atmosphere, humidity, and UV radiation (de Vera 2012; De la Torre Noetzel, et al. 2017; Sanchez et al. 2012). For example, De la Torre Noetzel and colleagues (2017) exposed the lichen Xanthoria elegans to simulated Mars-analogue conditions for 1.5 years including direct exposure to ultraviolet (UV) irradiation, cosmic radiation, temperatures and vacuum conditions. It was found that the lichen photobiont showed an average viability rate of 71%, whereas 84% of the lichen mycobiont survived--although they were adapted to and evolved on Earth. Moreover, 50-80% of alga and 60-90% of the fungus symbiote demonstrated normal functioning (Brandt et al. 2015). This included the ability to engage in photosynthetic activity post-exposure to these harsh environments with minimal impairment (Meesen et al. 2014).

Likewise, the lichens of Gale Crater may be engaged in photosynthesis. Bubble-like open-cone apertures have also been observed in association with the lichen-like specimens of Gale Crater (Figures 17, 18). It is well established that photosynthesizing organisms, such as cyanobacteria, respire oxygen and release gas bubbles via the surrounding matrix (most noticeable in water). These gas-bubble-formations may become mineralized and fossilized (Bengtson et al. 2009; Sallstedt et al. 2018). Therefore, we hypothesize that the open-cone apertures observed in Gale Crater serve to ventilate oxygen respired during photosynthesis and constitute additional indications of biology. Evidence suggestive of photosynthesis was also observed in Eagle Crater (Joseph et al. 2020a). Moreover, gas bubble release is a function of green algae (cyanobacteria) mat-producing behavior. Formations similar to microbial mats have been identified in this report.

8. Calcium Carbonate Encrusted Cyanobacteria and Nostoc Balls

Evidence of calcite and calcium carbonate has been detected on Mars (Boynton et al. 2009; Sutter et al. 2012; Wray et al. 2016; Leshin et al. 2013; Archer et al. 2014). Calcite may be formed via geological weathering coupled with fluid evaporation (Anthony et al. 2003b); and this watery environment would promote biological activity, whereas calcium carbonate is almost exclusively a biological byproduct. However, the detection of large amounts of calcium carbonate have been elusive due most likely to the destruction of biomolecules by weathering and UV and other forms of ionizing radiation (Ertem et al. 2017; Bibring et al. 2006). Therefore, if large amounts of carbonate are being produced by algae currently or in the ancient past on Mars, much would likely be destroyed by abiogenic processes.

Yet, despite destructive UV rays, a number of published reports indicate the presence of calcium and calcium carbonate on Mars (Archer et al. 2013; Boynton et al 2009; Cannon et al. 2012; Krall et al. 2014). Wray et al. (2016), for example, found evidence of calcium-rich Martian carbonates as detected by the Compact Reconnaissance Imaging Spectrometer aboard the Mars Reconnaissance Orbiter. Likewise, the Mars Global Surveyor Thermal Emission Spectrometer detected carbonate in global Martian dust (Bandfield et al. 2003). Evidence of carbonate was also detected by the Thermal and Evolved Gas Analyzer on the Phoenix polar lander, which measured $CO2$ release consistent with breakdown of Ca☐ rich carbonate in Martian soils (Boynton et al. 2009; Sutter et al. 2012), an interpretation supported by results from the Phoenix Chemistry Lab (Kounaves et al. 2010). Likewise, Curiosity's Mars Science Laboratory found evidence of $CO2$ release indicative of Fe/Mg-rich carbonates, albeit at minimal levels (Leshin et al. 2013; Archer et al. 2014; Ming et al. 2014).

Calcium carbonate is a biological byproduct of algae/cyanobacterial photosynthetic activity and their mucous secretions of polysaccharides which act as binding sites for Ca2+ thereby producing carbonate minerals and concentrations of calcium (Dittrich and Sibler, 2010; Kupriyanova et al. 2007; Samylina et al. 2016). Supporting the hypothesis that photosynthesizing, algae-like organisms have colonized Gale Crater, is evidence of what appear to be calcium biosignatures (Figure 19) and calcium encrusted cyanobacterial "Nostoc balls" (Figures 20-

22). Calcium carbonate is precipitated in the mucous of cyanobacteria via photosynthetic CO_2 or HCO_3- uptake and which cements together microbial mats and ooids (Barnes and Chalker, 1990; Dittrich and Sibler, 2010; Graham et al. 2016; Kupriyanova et al. 2007; Samylina et al. 2016). Diverse cyanobacterial taxa form carbonates intracellularly via Ca-Mg-Sr-Ba carbonate inclusions measuring several hundreds of nanometers in diameter and increasing the density of the cells by as much as 12% (Couradeau et al. 2012; Benzerara et al. 2004, 2014). The production of calcium carbonate also acts to bind colonies of cyanobacteria and sedimentary structures thereby producing macroscopic microbial mats, stromatolites, ooids, and nostoc balls (Barnes and Chalker, 1990; Graham et al. 2014; Mei et al. 2020); evidence of which is presented here.

Figure 11. Sol 298: Specimens similar to dimpled spherical ooids and algae (cyanobacteria) on the surface of a Martian rock. On Earth, ooids (ooliths), are formed in shallow seas and generally consist of calcite, bacteria, and residues of degraded mucous produced by filamentous cyanobacteria. The white deposits may consist of

calcium, gypsum or quartz. Cyanobacteria are found in association with gypsum. The green alga Closterium, for example, accumulates gypsum and stores it as crystals (Sr and/ or Ba sulfates) in vacuoles located in the apices of their spindle-shaped cells. Calcium carbonate is precipitated in the mucous of cyanobacteria via photosynthetic CO_2 or HCO_3- uptake and which cements together microbial mats and ooids. It is also possible that the "dimpled" spheres may include lichens.

Figure 12. Sol 298: Specimens resembling spherical ooids and green algae on the surface of a Martian rock or mudstone.

Figure 13. Sol 298: Specimens resembling dimpled lichens with what may be hyphae along the surface/subsurface. Note hollow apertures in the upper right corner and lower center of photo, and which resembles an oxygen-gas vents typically produced by photosynthesizing organisms.

Figure 14. Sol 298: Specimens similar to spherical ooids and/or dimpled lichens. The thin, tube-like formations may consist of calcium and the green coloration may represent green algae (cyanobacteria) which produce calcium via their mucous secretions.

Figure 15. Sol 298. Specimens similar to dimpled ooids and green algae (cyanobacteria) on the surface and interior of a Martian rock. The white deposit may consist of calcium or gypsum, which are associated with cyanobacteria.

Figure 16. Sol 298: Specimens similar to dimpled spherical ooids, dimpled lichens and green algae (cyanobacteria) on the surface and interior of a Martian rock and what appears to be a microbial mat (upper left) jutting out from the cavity. The micro-patterns embedded within and throughout the surface of this entire formation resemble colonies of mat-forming micro-organisms.

Figure 17. Sol 232: Specimens similar to dimpled lichens and gas-vent apertures for the release of oxygen secondary to photosynthesis. Photosynthesizing organisms, such as cyanobacteria, respire oxygen and release gas bubbles via the surrounding matrix (most noticeable in water). These gas bubble formations may become mineralized and fossilized. The white substance within the matric may consist of calcium which is secreted via the mucous of cyanobacteria which, along with algae, comprise the lichen organism.

Figure 18. Sol 232 (Top): Specimens similar to gas-vent apertures for the release of oxygen secondary to photosynthesis. Photosynthesizing organisms respire oxygen and release gas bubbles via the surrounding matrix (most noticeable in water). (bottom) Fossilized gas bubble-formations. (Bottom) Open globular structures, interpreted as formed by gas bubbles via cyanobacteria oxygen respiration within microbial mats (from Bengtson et al. 2009).

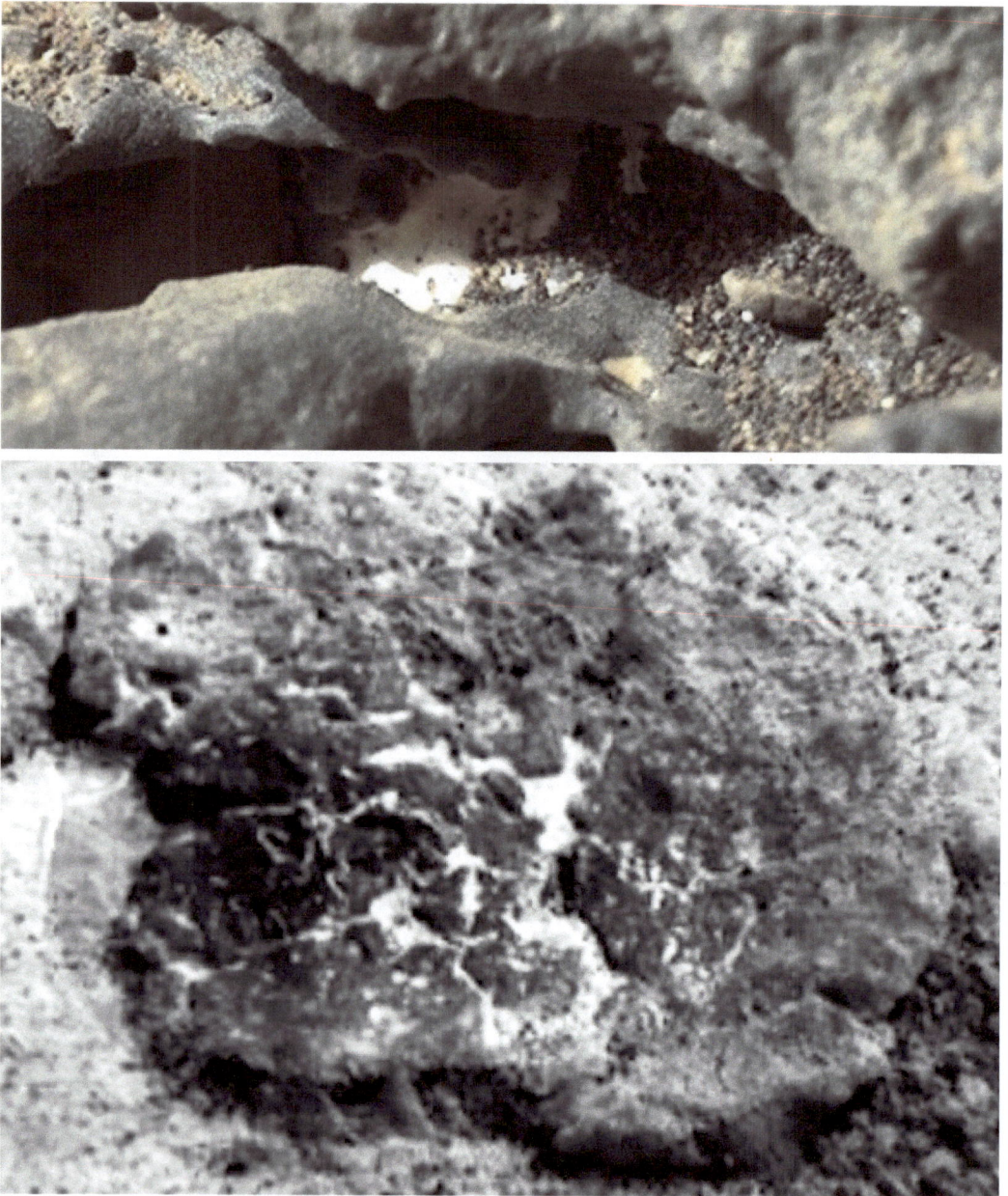

Figure 19. (Top) Sol 304: Possible calcium biosignature within a rock crevice. Alternate explanations include fungi. (Bottom) Sol 890: Possible calcium deposits within a specimen which resembles a microbial mat and thick bacterial crust.

Figure 20. CR0_473216607PRC_F0442414CCAM02853L1. Photographed by Curiosity Micro-Imager Camera. These specimens resemble calcium incrustations and calcium carbonate encrusted cyanobacteria--perhaps similar to Nostoc flagelliforme, N. parmelioides, N. verrucosum, N. pruniforme as well as spherical "Nostoc balls," and vesiculous thalli (see Aboal et al. 2016). Each of these interconnected "balls" are estimated to be less than 0.1 mm in diameter, based on the microscopic camera specifications.

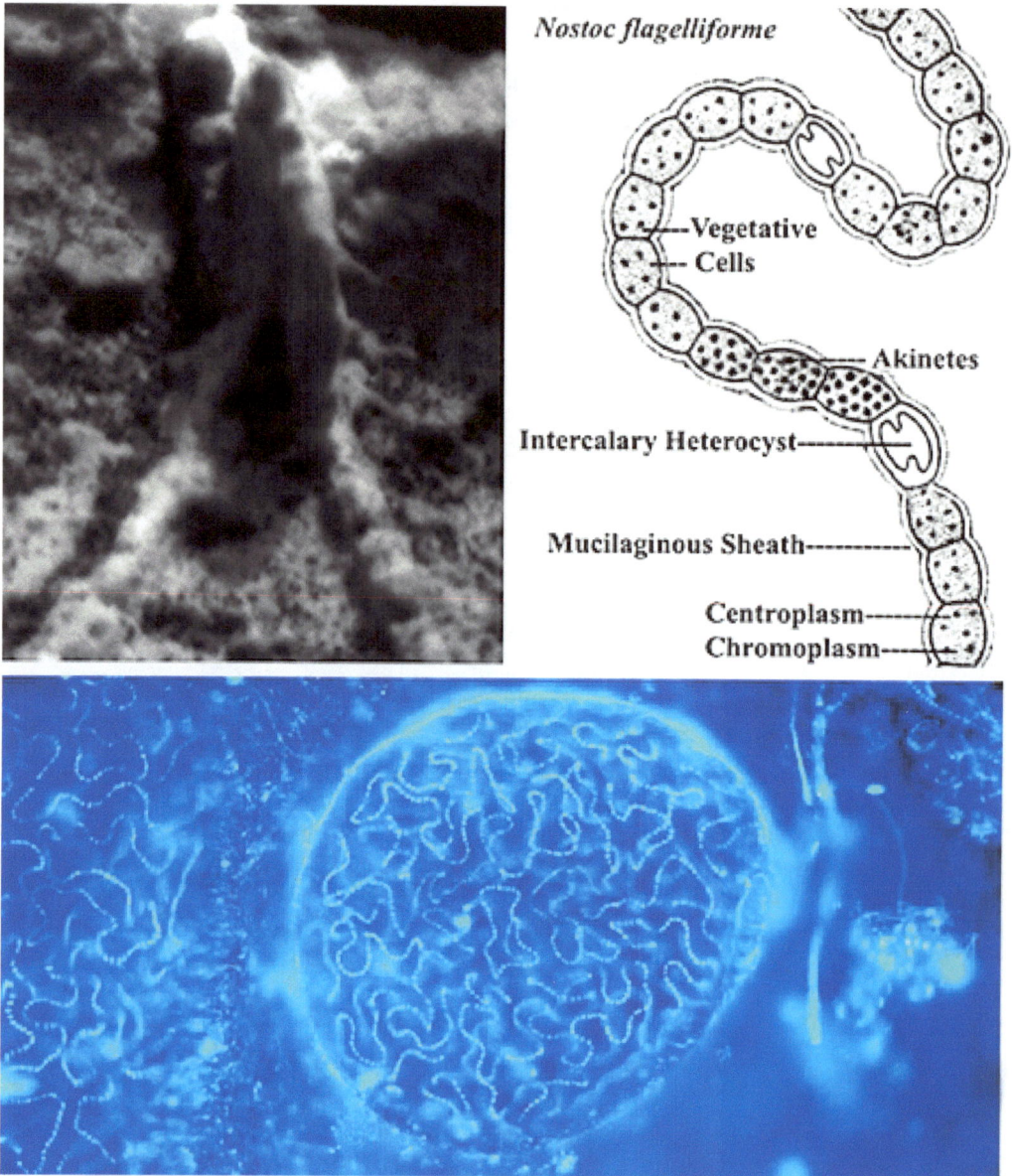

Figure 21. (Top): CR0_473216607PRC_F0442414CCAM02853L1. Calcium carbonate encrusted cyanobacteria, perhaps similar to Nostoc flagelliforme as well as spherical "Nostoc balls," and vesiculous thalli (see Aboal et al. 2016). Species such as Subtifloria (a fossil member of the Girvanella group) also create spherical thrombolites (large and small micritic clots) via calcification of extracellular polymeric substances making up multiple biofilms and within which calcified sheaths of cyanobacteria are typically embedded (Mei et al. 2020). These specimens also resemble the bio-mineralization pattern of Candidatus Gloeomargarita lithophora, which form intracellular Ca-carbonates bio-mineralization and which may be scattered and/or arranged in one or several

chains within its cells (Couradeau et al. 2012; Jinhua et al. 2016). Abiogenic calcium mineralization is an alternate explanation, except that mineral veins are planar and not bulged as seen here. (Bottom) photomicrograph of a modern "Nostoc ball," stained with the DNA-binding fluorochrome DAPI and viewed in UV epifluorescence (photo by L. Graham).

Figure 22. CR0_464779288PRC_F0421020CCAM02758L1. Photographed by Curiosity Micro-Imager Camera. These specimens resemble calcium incrustations and calcium carbonate encrusted cyanobacteria--perhaps similar to Nostoc flagelliforme, N. parmelioides, N. verrucosum, N. pruniforme, along with spherical "Nostoc balls," and vesiculous thalli.

"Nostoc balls" are formed by the calcification of extracellular polymeric substances making up multiple biofilms and reflect fundamental features of microbial growth and can form thrombolites by calcification (Graham et al. 2014; Mei et al. 2020). Fossils of calcified sheaths of filamentous cyanobacteria are typically found preserved within each clot (Mei et al. 2020). The specimens depicted in Figures 20-21, consist of what appear to be numerous calcium encrusted Nostoc balls, a calcium encrusted byproduct directly related to the photosynthetic activity of cyanobacteria.

The Geobiology Of Martian Minerals And Mineralized Fossils
9. The Geobiology of Weathering

In contrast to the putative lichens, ooids and gas vents which appear to be attached to mud stone, some of the algae-like specimens are associated with mafic and ultramafic rocks and soil which may have been subject to considerable weathering. Hence it could be argued that these specimens represent alteration surfaces for specific mineral assemblages that mimic algae. For example, Earthly chlorites, when weathered, may strain volcanic outcrops with a dark greenish to brownish color. The weathering and alteration of Fe and Ti-oxides may be responsible for the greenish appearance of rocks and soils as well. The bluish color of some specimens could be due to weathered surface of Fe-Mn rich rocks.

However, it's been determined that the greenish-coloration associated with surface weathering is usually secondary to biology and colonies of green algae are usually found in association (Büdel et al. 2004; Jung et al. 2019; Wierzchos et al. 2011). On Earth, the weathering of rock and minerals is often due to biological activity (Cecchi et al. 2019; Gadd et al. 2011; Wilson 2004).

10. Martian Minerals Do Not Mimic Algae

The Mars rover Curiosity, via its multiple cameras and Mars Science Laboratory, has been exploring and investigating the Gale Crater since August 2012. Among Curiosity's suite of sampling and analysis instruments is the mineral-detecting CheMin X-ray diffractometer (Bish et al. 2013; Vaniman et al. 2014). Trace amounts of chlorite, actinolite, epidote, olivine, and serpentine--which may appear green on Earth-- have also been detected on Mars via the analyses of spectral signals recorded from space (Schwenzer & Kring, 2009; Fairén et al. 2010; Ehlmann et al. 2009; Carter et al. 2010) and via the rover Curiosity in the Gale Crater (Michalsk and Niles, 2010; Bristow et al. 2015; Vaniman et al. 2014; Treiman et al. 2016). Those minerals detected in the Gale crater have been found in mixtures of clay and embedded in basaltic and phyllosilicate-bearing Martian rocks (Bristow et al. 2015; Vaniman et al. 2014). If large masses, sheets and clumps of these minerals crystalized and accumulated on the sides and tops of Martian rocks they could be misidentified as green algae. However, there is no evidence of crystallization or significant accumulation on the surface of rocks to support this possibility. And yet, the contribution of minerals should not be dismissed.

Terrestrial olivine is the primary mineral component in terrestrial basalt mantle rock and may be green in color when not weathered (Anthony et al. 2003c). Martian rocks in the Gale Crater have been subject to considerable weathering. Terrestrial pyroxene (a ferromagnesian mineral), is usually found in basalt and metamorphic rocks and although it can appear as a dark green, it is typically black. On Earth, serpentine may appear yellow or olive green in color but is usually black or brown. By contrast, actinolite may assume a fibrous, life-like appearance once it crystalizes (Anthony et al. 2003c). However, no crystalized actinolite has been detected in the Gale Crater.

Terrestrial chlorite, if crystalized, can also assume a micro-vegetative appearance, forming flaky or gray-green olive colored layers and clumps and is often found in a variety of rocks including slate, schist and phyllite. However, crystalized chlorite has not been detected. Moreover, in contrast to the wide surface areas that appear to be colonized by clumps and layers of green algae, with the exception of olivine, only trace amounts of these minerals have been detected.

Martian actinolite, for example, could only be discerned by complex algorithm analyses of spectral and hydration features (Lin et al. 2016); whereas chlorite has been found primarily deep within crater walls and central peaks or mixed with clay, but not on the surface of rocks or mudstone (Schwenzer and Kring, 2009; Michalski & Niles, 2010; Milliken et al. 2010; Vaniman et al. 2014) where the algae-like masses have been photographed. Basaltic rocks are abundant in the Gale Crater, but "linear modeling," or rock spectra, indicates an average olivine composition of only Fo45, which the authors of that study admit is an "overestimation" and which could be detected only indirectly because rocks containing olivine were covered with dust (Hamilton and Ruff, 2012). Moreover, there is no evidence of olivine crystallization on Mars, which, on Earth, gives this mineral a green coloration.

In fact, the relevant minerals in Gale Crater (i.e. chlorite, serpentine, olivine, actinolite, pyroxene) can only be detected via analysis of spectral absorption rates, or via chemical analyses of drill samples, and are typically embedded as a mixture in light- and dark bedrock and basalt (Michalsk and Niles, 2010; Bristow et al. 2015; Vaniman et al. 2014; Treiman et al. 2016) or volcanic and coarse grains of sand (Ehlmann, 2016). With the exception of olivine, the relevant mineral concentrations are often just above the detection thresholds and none have a crystalline shape (Bish et al. 2013). Even more importantly, these Martian minerals are not visible to the naked eye and were not detected on those rocks and sands where these algae-like specimens were photographed.

By contrast to the lack of minerals visible on the surface, these algae-like specimens are quite extensive and form thin sheets, layers and clumps, and tend to grow in large patchy masses on the sides and tops of rocks and sand. However, although minerals cannot account for these observations, these same (and other) minerals may provide an energy source which sustains and promotes the growth of the putative biological specimens depicted in this report.

For example, Bish et al. (2019) argued that XRD data from Mars revealed a rich mineralogy significantly similar to that found in the mineral-rich soils on the flanks of Mauna Kea volcano, Hawaii--an area in which algae, lichens, mosses and fungi proliferate (Jackson 1971). Terrestrial microorganisms, as well as fungi and algae-fungal symbiotes, feed on, metabolize, respire, excrete, or store these minerals internally or on the surface of their cell walls. Hence, when these organisms die, they often become mineralized fossils.

It is not likely that the algae-like specimens presented here reflect the accumulation of green-colored minerals (i.e. chlorite, actinolite, epidote, olivine, serpentine) which somehow massed together on the surface of Martian rocks and sand and crystalized together to take vegetative-like forms.

11. Martian Mineral Microbiology

Algae, fungi, lichens, and a variety of microorganisms have played a major role in the genesis of numerous minerals, including chlorite, serpentine, olivine, actinolite, and pyroxene (Bender et al. 2002; Maixner et al. 2008; Gadd et al. 2011; Gazze, et al. 2012; Jones et al. 1981; Templeton et al. 2011; Wilson & Jones 1984, Wilson 2004). Microorganisms have been active in the formation and decomposition of minerals in the earth's crust since soon after life appeared on this planet, accomplished via their well-developed ability to bind metallic ions. Ion binding allows their cells to serve as nucleation sites for initiating and mediating biomineralization, as well as respiration, expiration, decomposition and secretion, and to accumulate mineral and metal byproducts along the cell wall (Polgari et al. 2006; Gadd et al. 2011; Bender et al. 2002; Ayupova and Maslennikov 2012).

Chemolithotrophs are well adapted for life on Mars and function at optimal oxidizing and decomposing capacity in high CO_2 anoxic environments by autotrophically fixing CO_2 from the atmosphere (Rawlings et al. 2003; Todaro and Vogel, 2014). Thus, Mars, with its mineral deposits and atmosphere that consists of 95% CO_2, provides an ideal environment for these species.

Although biomineralizing species may also obtain energy via photosynthesis, chemolithotrophs generally employ ferrous iron or reduced inorganic sulfur compounds or chlorite, as an electron donor (Gadd et al. 2011; Bender et al. 2002)--substances widespread on Mars. Microbes that grow in mineral-rich environments may utilize a wide range of metal ions (Leduc and Ferroni 1994) and they may interact with algae, fungi, and lichens thereby forming mutualistic mineral-dependent relationships (Gadd et al. 2011).

Generally, it is within terrestrial (vs aquatic) environments where these organisms exert their most profound influence, including bioweathering, mineral transformations and mycogenic biomineral formation (Cecchi et al. 2019; Gadd et al. 2011). For example, algae, fungi and lichens can affect changes in metal speciation, toxicity and mobility, as well as induce mineral formation or mineral dissolution or deterioration (Cecchi et al. 2019; Rawlings et al. 2003; Wilson 2004).

These organisms are also responsible for the biochemical weathering and decomposition of rocks and minerals which they may transform into calcium or other minerals (Ascaso et al. 1990; Duchafour 1979; Jones and Wilson 1985, 1986). This includes the formation of calcium in the interface between basalt and other rocks (Jones and Wilson 1986), for which there is evidence on Mars as shown in Figures 11, 13-15, 17-22. Moreover, cyanobacteria also trap fine minerals which are incorporated into their crusts and networks (Rahmonov and Piatek 2007).

12. Perchlorates and the Geobiology of Chlorite

Chlorite is one of two potentially green-colored minerals so far detected on Mars which, on Earth, may take a clumpy, vegetative shape when crystalized or bio-weathered. Terrestrial perchlorate-reducing bacteria, dwelling in high CO_2 anaerobic environments (Chaudhuri et al. 2002; Maixner et al. 2008) feed on and metabolize chlorite, which is transformed into chloride and O_2, which provide energy to the cell and is respired. There is evidence, as presented in this report, of biologically produced apertures serving to ventilate oxygen respiration during photosynthesis (Figures 5, 13, 15).

The synthesis of oxygen plays a significant role in the reductive pathway of perchlorate--mediated by the enzyme chlorite dismutase and a chlorite dismutase gene cld (Bender et al. 2002; Maixner et al. 2008). Perchlorate also appears to be widespread in Martian soils with concentration ranging from 0.5 to 1% (Bishop et al. 2014; Kounaves, et al 2014; Wilson et al. 2016). Dismutase and the chlorite dismutase gene appear to have an ancient phylogeny, appearing very early in evolution, and are widely distributed among bacteria and archaea (Maixner et al. 2008). By analogy, Martian organisms may have acquired or evolved the same or similar genes, thereby providing a biochemical means of synthesizing oxygen billions of years ago, as evidenced by putative Martian stromatolites dated to 3.7 bya (Noffke 2015).

Many species reduce perchlorate and chlorate into chlorite, whereas yet others transform chlorite into chloride and O_2 (Coates et al. 1999; Kengen et al. 1999; Bender et al. 2002)--a transformational process which not only serves as an energy source but which can prevent self-toxification due to the accumulation of chlorite (Maixner et al. 2008). Thus, some microbes may be metabolizing perchlorates and secrete chlorates as a waste product. Whereas other species produce chlorite and reduce chlorite or chlorates--depending on the availability of mutualistic relations with other species and the presence of and concentrations of chlorate or perchlorate in the surrounding environment. In so doing they produce molecular oxygen (Hagedoorn et al. 2002) which is a metabolic energy source. The availability of these minerals determines the biomass of these organisms and the extent to which they are able to chemically and physically alter these and other minerals (Leake et al. 2008).

Thus, a variety of microorganisms, including algae, fungi and algae-fungal-lichen symbiotes, dwelling in an anoxic, high CO_2 environments, are dependent on chlorite and perchlorates which they may transform, reduce, and recycle. The same is true of serpentinite (Wilson et al. 1981), olivine (Wilson 2004) and actinolite (Gablina et al. 2016). The minerals detected on Mars would provide oxygen and the energy needs for a variety of organisms. Likewise, the detection of these Martian minerals may reflect biological activity.

By contrast, although perchlorates and related salts have been detected, their surface quantities are insufficient and they have not been observed to assume clumps or shapes, much less formations that might be mistaken for algae, lichens, and other organisms.

13. Microbiological Tunneling of Martian Rocks and Minerals

Various organisms burrow within olivine, chlorite and basalt rocks (Aspandiar and Eggleton 2002; Delvigne et al. 1979; Fisk et al. 1006; Haggart & Bustin 1999; Jones et al. 1981, Wilson 2004), creating etchings or tunnels and channels within which, for example, fungal hyphae (Gazze et al. 2012) or filamentous and tubular cellular organisms may be embedded (Sanz-Montero and Rodriguez-Aranda, 2009; Polgari et al. 2006; Ayupova et al. 2016; Ayupova and Maslennikov, 2012; Kilias et al. 2007; Sakakibara et al. 2014). Cyanobacteria and lichens burrow into rocks through the secretion of oxalic acid and cause weathering. They may derive their carbon source from the dissolution of limestone rocks. The tunneling depicted in Figure 23 (and others like it), photographed in the Gale Crater, has all the characteristics of a tunneling biosignature.

Biological vs abiogenic tunneling can be distinguished by asymmetrical vs symmetrical apertures (McLaughlin et al. 2010). For example, if due entirely to a chemical process, crystalline millstones, or pebbles driven against a relatively soft rock surface by wind or fluid pressure, produce a smooth and rounded cavity and with numerous adjacent cavities which are nearly identical in shape. These abiogenic tunnels are often linked together by longitudinal interconnected surface channels and overlapping holes (Tyler & Barghoorn, 1963; Knoll & Barghoorn, 1974; Wacey et al. 2008).

By contrast, biological tunneling is linked to watery environments, typified by jagged asymmetrical entry holes, with each tunnel, or hole, largely separate from one another. The tunneling and surface hole is due to microbial dissolution, demineralization, and removal of substrate, the fluids providing a transport system for the chemicals needed by these organisms for dissolution and as nutrients (Staudigel et al. 2008; McLaughlin et al. 2010). However, biological tunneling eventually ceases in the absence of water (McLaughlin et al. 2010). Thus, as water levels wax and wane--as appears to be the case for the Gale Crater-- biological tunneling would be affected.

Martian minerals and biological tunneling (Velbel, 2012; Fisk et al. 2006; White et al. 2014), as well as biological activity (McKay et al. 2009; Thomas-

Keprta et al. 2002, 2009), have been detected in several Martian meteorites. For example, channels and tunnels with similar biosignatures, and in association with olivine, have been discovered in Martian meteorite Nakhla (Fisk et al. 2006) and Yamato (White et al. 2014).

Likewise, numerous substrates in the Gale Crater also show evidence of biological tunneling. As exemplified by Figure 23 (which is just one example of many) almost all the perforations on the surface of this rock meet the criteria for biological tunneling. And this perforated substrate was photographed in the Gale Crater which was inundated at various times by water.

Mineralized Trace Fossils
14. Hypotheses and Speculation: Trace Fossils?

It can be argued that Martian minerals have been subject to considerable biological activity, bio-weathering and tunneling, and serve as an energy source for a variety of chemolithotrophic and biomineralizing organisms including those which engage in photosynthesis, i.e. algae, fungi, and lichens (Cecchi et al. 2019; Rawlings et al. 2003; Wilson, 2004).

A variety of organisms accumulate minerals internally and externally (Polgari et al. 2006; Gadd et al. 2011; Bender et al. 2002; Ayupova and Maslennikov, 2012), including chlorite (Ayupova et al. 2016; Haggart & Bustin, 1999). Many Martian minerals have been formed, hydrated, in recessive bodies of water (Xue & Jin, 2013; Schwenzer & Kring, 2009; Carter et al. 2010; Michalski & Niles, 2010; Lin et al. 2016), which, on Earth, are host to innumerable organisms.

Terrestrial microorganisms not only feed on minerals but bind metallic ions and act as nucleation sites for initiating and mediating biomineralization. This causes minerals to accumulate within and on the exterior of these organisms (Polgari et al. 2006; Gadd et al. 2011; Bender et al. 2002; Ayupova and Maslennikov, 2012). In consequence, when fossilized in regressive bodies of water, only the accumulated minerals may be detected (see Ayupova et al. 2016; Haggart & Bustin, 1999; Ran et al. 1999; Sanz-Montero and Rodriguez-Aranda, 2009; Sakakibara et al. 2014).

Gale Crater is believed to have been a water-rich environment that underwent early mineralization, conditions which would have preserved the denizens of this ancient habitat (Grotzinger et al. 2012, 2015). Therefore, the presence of various minerals on Mars may serve as a food and energy source and byproduct of biological activity and could represent the biomineralized, fossilized remains of these organisms as is true on Earth (Ayupova et al. 2016; Haggart and Bustin, 1999; Ran et al. 1999; Sanz-Montero and Rodriguez-Aranda, 2009; Sakakibara et al. 2014). In addition, Eigenbrode and colleagues (2018) reported organic matter preserved in 3-billion-year-old mudstones at Gale crater, which may represent the residue of Martian organisms. This hypothesis is supported by Figures 28-33 which depict fossil-like specimens, many similar to metazoans and which may have formed in regressive bodies of water.

15. Speculations: Evolution, Trace Fossils, Tube Worms, Metazoans?

The Gale Crater is marked by potassium-rich fluvial valleys and water pathways (Grotzinger et al. 2015b). The Gale Crater appears to have repeatedly filled with water (Bibring et al. 2006; Cabrol et al. 1999; Fairen et al. 2014; Murchie et al. 2009; Siebach and Grotzinger, 2014; Buz et al. 2017) and could have sustained a variety of species and promoted their evolutionary development due to major changes in their watery environment and availability of liquids. Wet followed by dry spells may also account for what could be fossilized impressions of water dwelling and subsurface Martian organisms including metazoans (Figures 27-33).

To speculate, some of the fossil-like specimens are similar to complex metazoans and burrowing worms, which evolved on Earth 500 million years ago. These include formations which resemble Cambrian fauna that first appeared during the Ordovician (e.g. Calymene callicephala, Flexicalymene meeki, Homotelus bromidensis, Isotelus sp., Pseudogygites canadensis, Streptelasma sp.). Our speculative interpretation of the organisms represented (pro and con) are detailed in the captions of Figures 33-37. For example, Foraminfera live under the effects of water currents and if they accumulate, and die, they are disposed isoriented. In fact, what appears to be "fossilized" Foraminfera-like specimens (Figures 33-34) could also be explained as an accumulation of plagioclase crystals.

Unfortunately, it is impossible to make precise determinations as to the identity or exact nature of these mineralized, fossil-like specimens, and this might be true if these same "fossils" were discovered on Earth (Graham 2019). Microfossil-like shapes are ubiquitous in Earth's geological record and despite careful extraction, preparation and microscopic examination there is great debate as to their authenticity (Knoll 2015, Brazier et al. 2003, De Gregorio et al. 2011; Marshall et al. 2011; Wacey et al. 2016). Samples have been repeatedly challenged as abiotic and a consequence of mineralization.

However, as to the controversy over the authenticity of the earliest terrestrial microfossils, it may also be that accumulated minerals are all that remains of these fossilized organisms. This would explain their fossil-like biogenic shape and why all that remains are the accumulated minerals.

Although the fossil-like impression presented in Figures 28-33 may represent the biosignatures of various organisms including metazoans, at best, we can only speculate and offer hypothesis and arguments both pro and con. We therefore offer up these fossil-like formations as targeted examples for additional investigation by the scientific community, as well as extraction and analyses by future robotic missions to Mars.

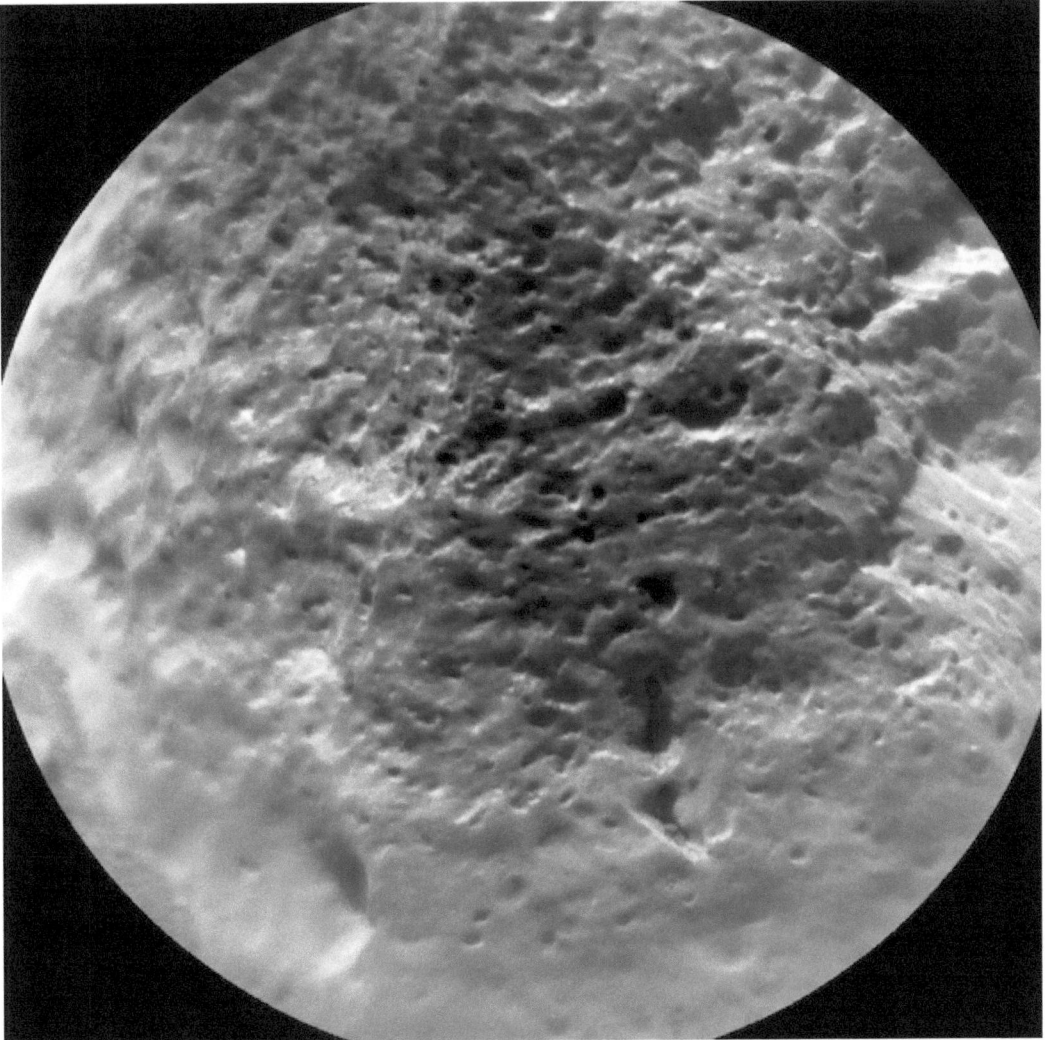

Figure 23. CR0_425453645PRC_F0060704CCAM01315L1: Microscopic detail of rock with tunnels/holes possibly secondary to biological activity. Photographed by Curiosity's Microscopic Chemistry Camera. Biological (vs abiogenic) tunnels and perforations are typically asymmetrical in shape, with each hole separate from one another, and are due to microbial dissolution, demineralization, and removal of substrate (McLaughlin et al. 2010). The overall pattern is consistent with a biological interpretation.

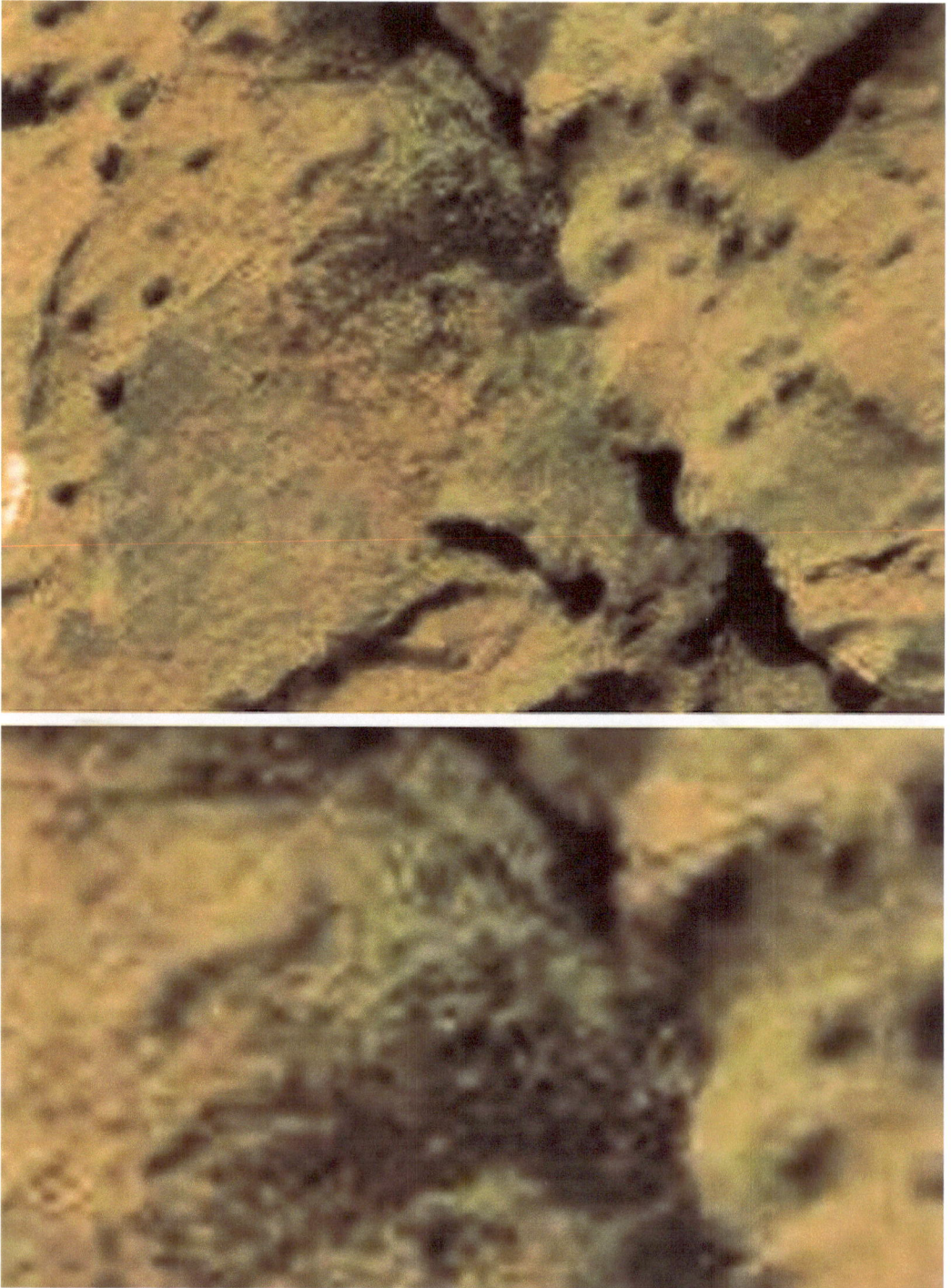

Figure 24. Sol 298: Algae and lichen-like-mushroom-shaped specimens, surrounded by light colored ovoid and tubular specimens embedded within the mudstone.

Figure 25. Sol 298: Dark colored specimens with crevice which resemble colonies of tubular organisms. See also Figures 25-27. These and the light-colored specimens embedded in the matric, adjacent to what may be ooids and dimpled lichens, are also similar to fossilized cyanobacteria which have been found in association with ooids within Cambrian strata in Northern China (Latif, 2019).

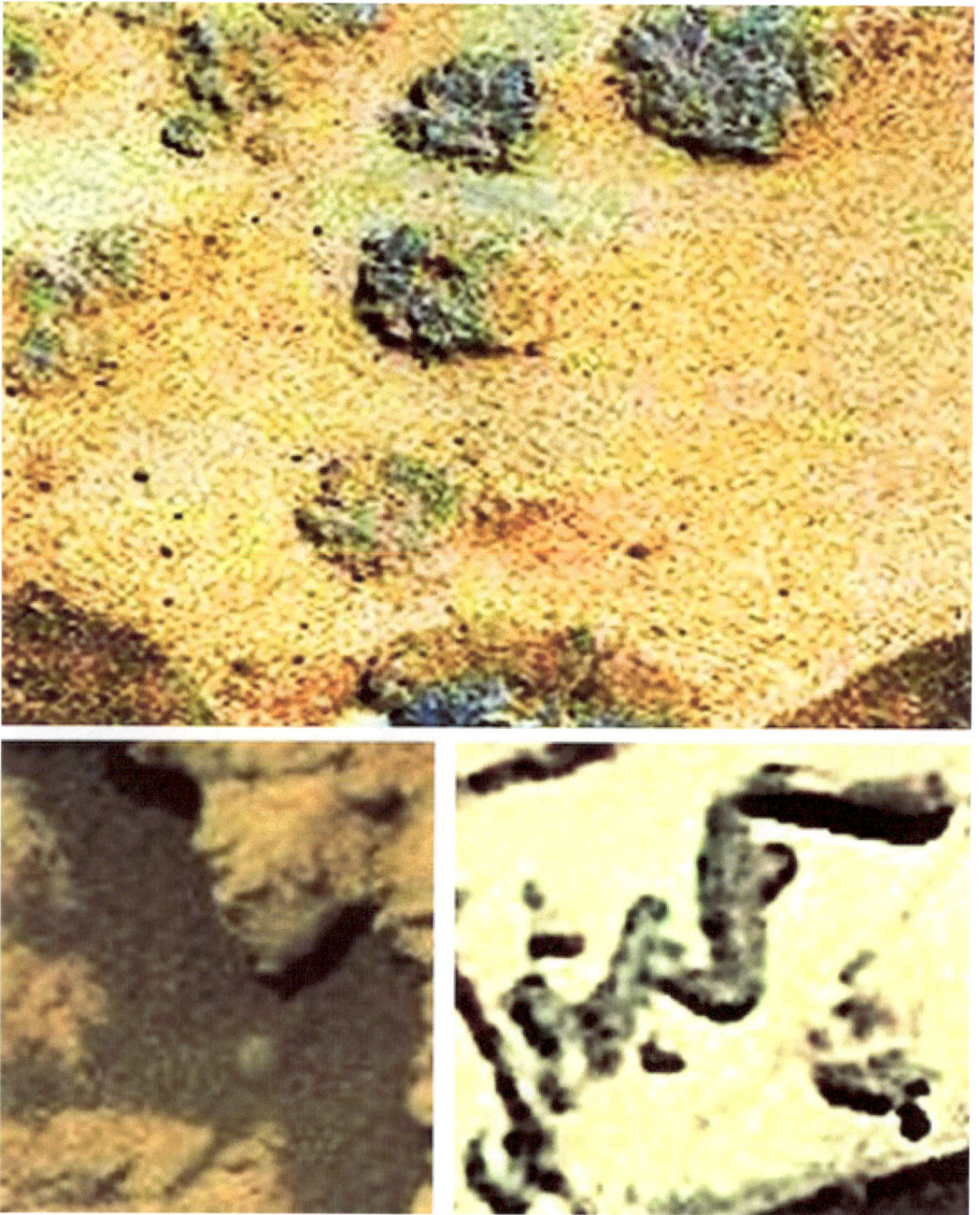

Figure 26. (Clockwise from top) Sol 890, 298, 1905: Specimens resembling colonies of micro-organisms and tube worms (bottom right and see Figure 27).

Figure 27. Sol 1905: Specimens resembling colonies of tube worms (see Figure 26) and mushroom-lichens (bottom right). Contrast with Figure 28 the colors of which have been desaturated by NASA, thereby giving these same specimens (top photo) a gray-rock-like fossilized appearance (see Figure 28).

Figure 28. Sol 1905: Specimens resembling mineralized trace fossils. These specimens are approximately 1 to 5 mm in length on average. In Figures 26, 27, and 28, one of these "worm-like" specimens has an open aperture at one end, suggesting either an orifice (Figures 26, 27) or that they are hollow (Figure 28). Whereas the specimens depicted in Figure 27 have colorization suggesting biology, the gray specimens depicted here are similar to trace fossils of burrowing tube worms and the tunneling and burrowing systems made by worms in Cambrian sedimentary rocks (see Vannier, 2010), with the tunnels filling with sediment then mineralizing and fossilizing after waters receded in the Gale Crater. As depicted in this NASA desaturated photo, these specimens also resemble the trace fossils of treptichnids, priapulid worms, and their burrowing curved branching beneath and along the mud of the seafloor on Earth, circa 545 mya and which filled with sediment and mineralized (Vannier ,2010). As the NASA image curator did not reply to inquiries, it is unknown if NASA/JPL desaturated the colors of these "worm-like" specimens and their surroundings, or if NASA applied false colors to these same specimens which are much darker in color in Figure 27 and have other biological-features as also depicted in other photos taken on Sol 1905.

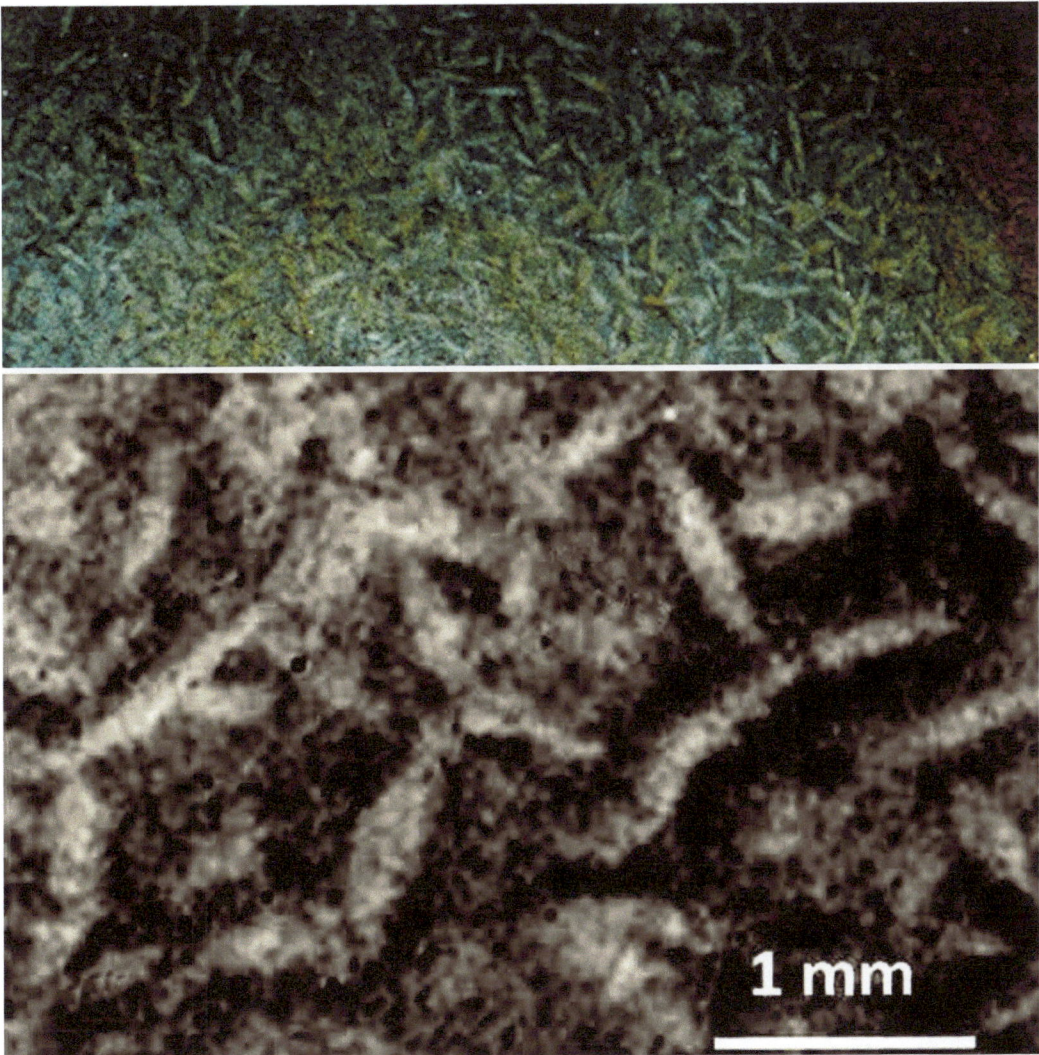

Figure 29. Sol 809: (Bottom enlarged, desaturated). These specimens resemble mineralized trace fossils formed, on Earth, during the Cambrian Ordovician transition, by microbes (Girvanella) and siliceous filamentous sponges, with lime, mud and biocasts between them (see Lee et al. 2014). Specimens similar to these have been found on the sediment surface of a relatively flat seashore in northern China during the late Cambrian extinction and the Great Odovician Biodiversification Event (Lee et al. 2014, 2016; Lou and Reitner 2015). Specimens similar to a variety of metazoans can also be viewed in Figures 30, 31. That these represent a pure mineralogical argumentation appears implausible as the orientation of the spindle-like, whitish structures are not organized like mineral structures (Graham, 2019; Graham et al 2016).

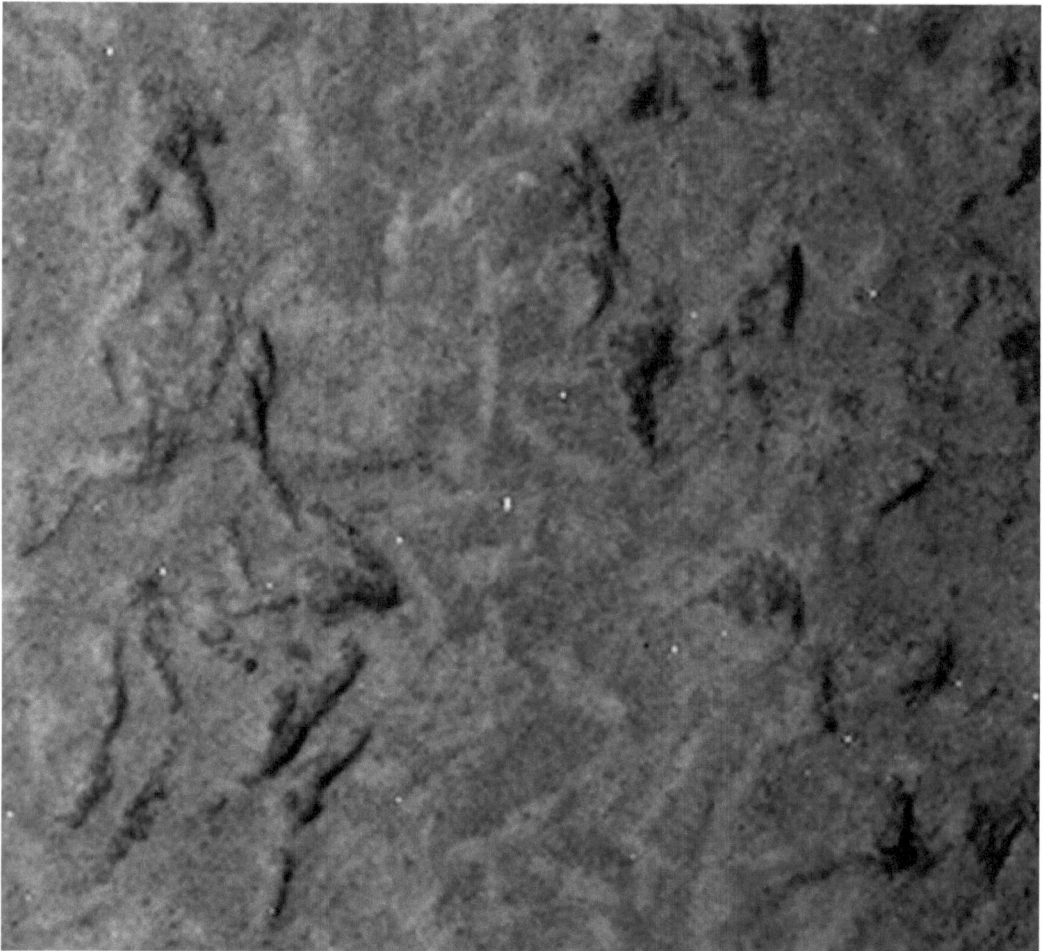

Figure 30. Sol 880: Specimens resembling trace fossils of metazoans (e.g. Calymene callicephala, Flexicalymene meeki, Homotelus bromidensis, Isotelus sp., Pseudogygites canadensis, Streptelasma sp.) including filamentous silica sponges upon and impressed into the surface of this Martian rock lying face-up on the floor of the Gale Crater. These formations are approximately 2 -3 mm in length on average. They are similar to metazoan fossils formed at the end of Cambrian (Lee et al. 2016). Although some resemble foraminifera in shape, their patterns of organizational preservation do not. On Earth, dead foraminifera rain down on the sea floor and their mineralized tests often become preserved as fossils in the accumulating sediment. Foraminifera are believed to have evolved around 1.15 bya, and have left a rich biomineralized, fossilized assembly of their multiple forms which include globular, spiral, spherical, and tubular species (Tappen and Loeblich, 1998; McIlroy et al. 2001) with all these diverse forms appearing together in a single species (Pawlowski et al. 2002). It is noteworthy that nearly identical complex metazoan-like forms have been photographed in other nearby locations within Gale Crater, including and especially the "ice-cream-cone" specimen at center right (see Figures 31, 32).

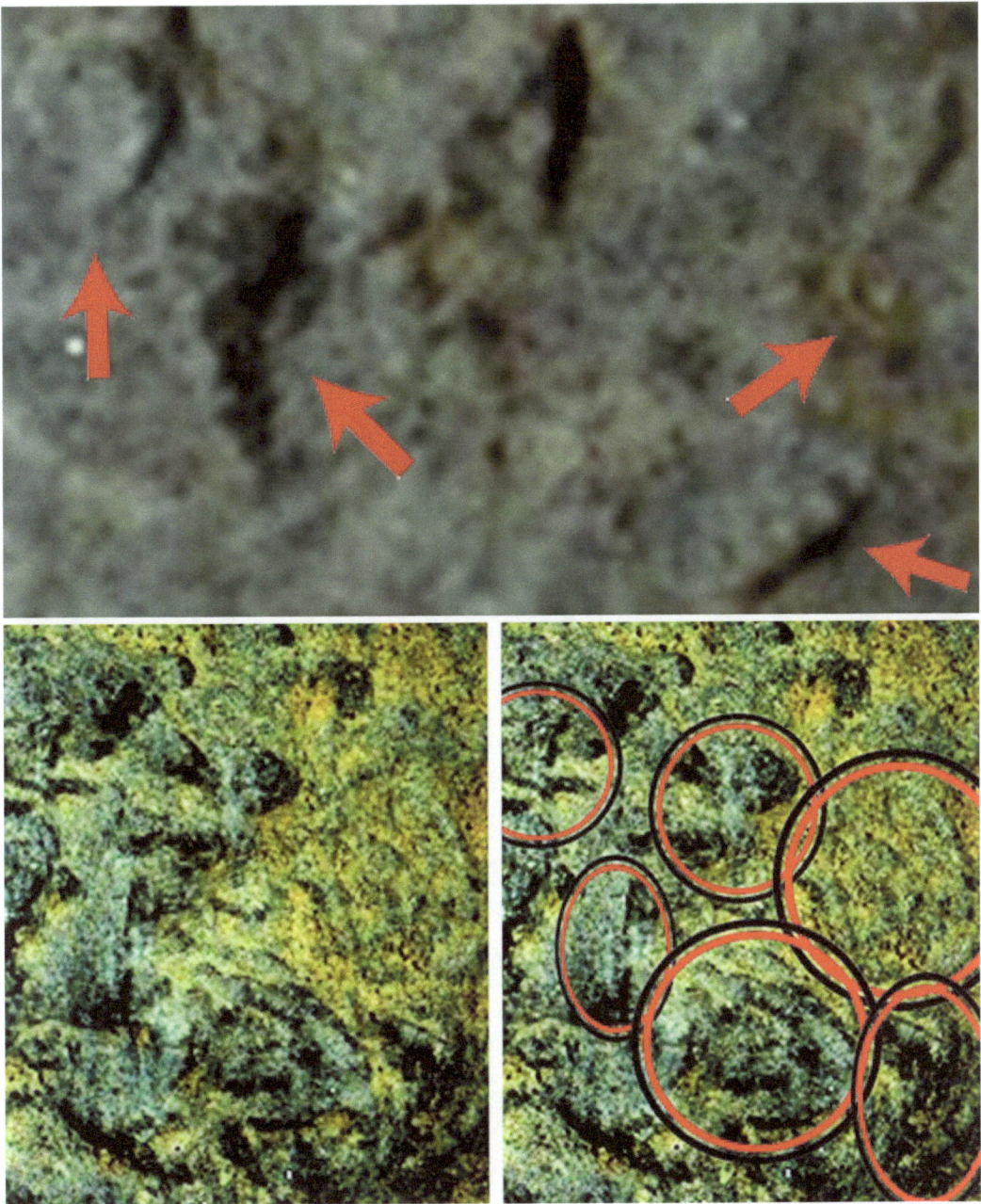

Figure 31. (Top) Sol 809. (Bottom) Sol 809. Similar specimens in two different locations, photographed alongside tubular, curved, and other fossil-like structures which resemble a variety of metazoans.

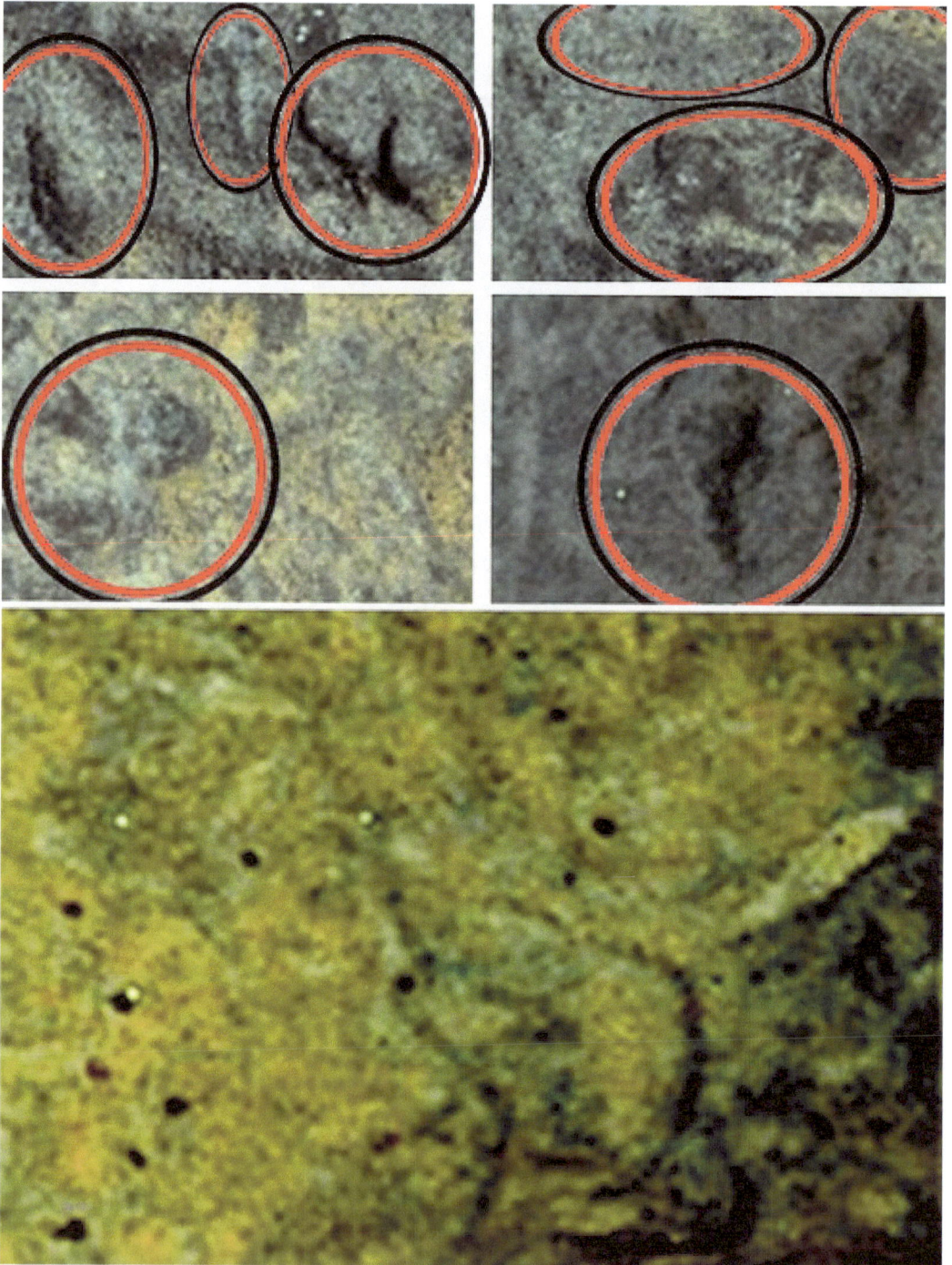

Figure 32. Sol 809, 880, 869: A variety of specimens, resembling the fossilized remains of metazoans (Calymene callicephala, Flexicalymene meeki, Homotelus bromidensis, Isotelus sp., Pseudogygites canadensis, Streptelasma sp.), photographed on rocks on the floor of the Gale Crater.

Figure 33. (Top) Sol 869: Specimens resembling mineralized fossils of worms and metazoans, approximately 1 to 2 mm in length. (Bottom): Sol 1905 compared with Sol 869.

Stromatolites, Microbial Mats
16. Fossilized Stromatolites and Microbial Mats

Sedimentary microstructures similar to the microstructure of stromatolites, microbial mats and microbialites, have been identified on Mars (Noffke, 2015; Joseph et al. 2019; Rizzo and Cantasano, 2016; Ruffi and Farmer, 2016; Small, 2015). If these Martian specimens are biological in origin is not known with certainty as not all terrestrial stromatolites are biological (Buick et al. 1995; Lowe, 1994; Riding, 2008; Wacy, 2010). The same could be said for what appear to be microbialites observed on Mars as the interpretation of their identity has been based on microstructure and not overall morphology. In terms of gross morphology, and until this report, there have been no reports of specimens which resemble terrestrial stromatolites in overall morphology.

Presented here are six Martian specimens, photographed in the dried lake bed of the Gale Crater, and three of which (Figures 36-45) closely resemble concentric-domical stromatolites growing in Lake Thetis, Australia. That the concentric Martian specimens presented here are similar to those in Lake Thetis is of great significance as the lakes of Western Australia have been identified as ideal analogs for Mars when it was presumably flush with rivers, lakes and streams (Baldridge et al. 2009; Bridges et al. 2015; Graham et al. 2016; Nguyen et al. 2014).

Figure 34. Stromatolites of Lake Thetis, winter (top) summer (bottom)

Figure 35. Sol 122, 654. 309: Concentric-domical specimens, Gale Crater.

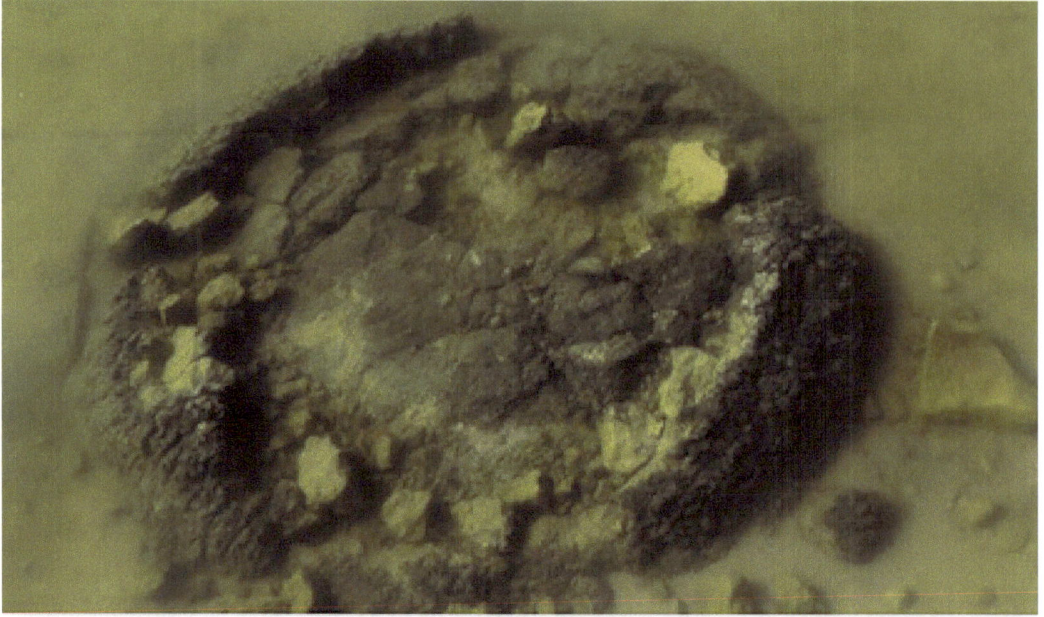

Figure 36. (Top) Submerged Lake Thetis stromatolite (permission to reproduce photo granted by TheTravelingLindfields.com). (Bottom) Sol 122: Martian specimen with collapsed dome and evidence of fossilized fenestrae within the upper portion of the walls. This specimen appears to be fossilized and displays the vertical and inward orientation typically caused by upward-migrating microbial colonies at the sediment-water interface. Several "peanut-brittle" specimens resembling thrombolite mats appear in the bottom portion of the photo (see Figure 37).

Figure 37. Sol 122 (A Top). "Peanut-brittle" specimens resembling microbial mats Sol 309 (Bottom): Domical-concentric-shaped specimen with concentric laminae (center right), which sits adjacent to three overlapping "peanut-brittle" specimens resembling thrombolite bacterial mats. (B): Terrestrial microbial Mat from Damer (2016) (permission to reproduce granted by Dr. Bruce Damer). (C) Algae-bacterial mat. Photo Credit, University of Waterloo.

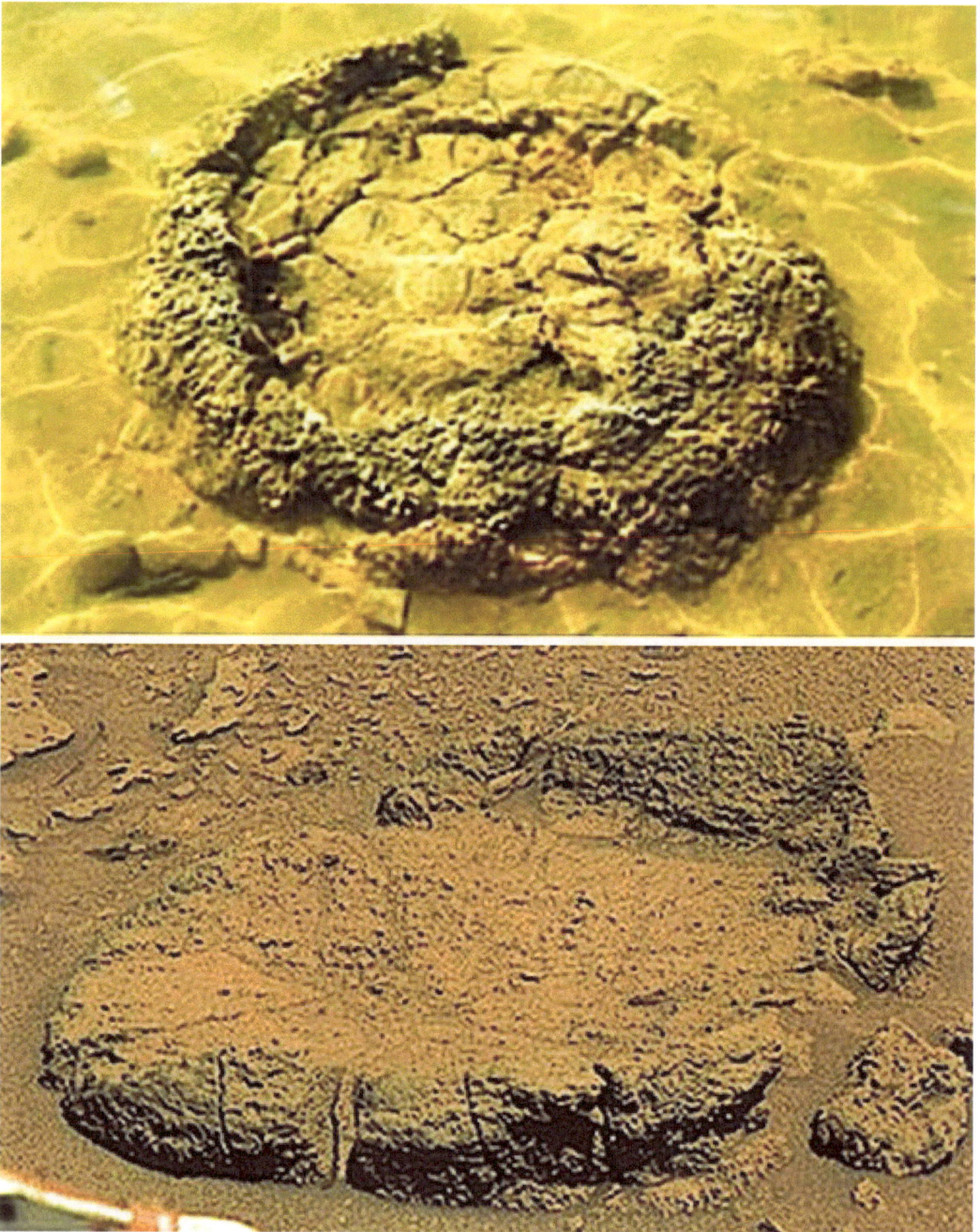

Figure 38. (Top): Lake Thetis under water stromatolite ((Photo Credit: Government of Western Australia Department of Mines and Petroleum, see also Grey and Planavsky 2009). (Bottom) Sol 529: Martian specimen with evidence of concentric lamination and fossilized fenestrae and "peanut-brittle" specimens resembling thrombolite bacterial mats.

Figure 39. Sol 308 (Top): Water pathways leading down and curving around a Martian Specimen resembling (Bottom): tthe remains of a Lake Thetis underwater stromatolite (Photo Credit: Government of Western Australia Department of Mines and Petroleum, see also Grey and Planavsky 2009).

A detailed examination indicates that two of the six domincal-concentric Martian specimens meet most and one meets all of the criteria for a stromatolite which is biological in origin as defined and detailed by Allwood et al. (2006), Buick et al. (1981, 1995), Lowe (1994), Riding (1999, 2008), and Wacy (2010).

Specimens which resemble microbial mats, and which are adjacent or attached to these concentric structures, were also compared to and found to have a morphology similar to terrestrial mats and those in Lake Thetis. The evidence and observations reported here, therefore, meet the criteria for biology.

Scale Bars, Size, and Magnification: Curiosity's Mast Camera is not a microscope and has only telescopic capabilities, thereby capturing images of numerous specimens, as depicted here, all of which are at varying and unknown distances from the camera and each other. Because of depth of field, it is impossible to apply accurate scale bars to any of these specimens and the same is true of those photographed at Lake Thetis, each of which are approximately 0.9 m (3 ft) in diameter and 0.4 m (1.2 ft) in height.

The estimated diameter (D) and height (H) of the putative Martian stromatolites are as follows: Sol 528 (Figure 44-45) D= 60 cm, H=25 cm; Sol 309 (Figure 37) D=5 cm, H=2 cm; Sol 122/529 (Figure 36,44-45), D=30 cm, H=20 cm; Sol 173 (Figures 40-43), D=30 cm, H=25 cm.

The domical specimens depicted in Sol 173 were subject to 300% to 400% magnification to emphasize their micro- and macro-structure, including laminae, fenestrae (gas bubbles) and concentric organizational geometry.

17. The Biology and Stromatolites of Lake Thetis

Lake Thetis is located 1.5 km inland from the Indian Ocean shoreline in Western Australia and has a total shore line of approximately 1000 m (Grey et al. 1990). The lake has a maximum water depth of 2.25 m and total alkalinity is slightly higher than 0.5% meq/L and slightly more basic than seawater with pH ranges from 8.28 – 8.6 (Grey and Planavsky, 2009). As such, the alkalinity may be similar to the waters which are presumed to have filled the Gale Crater (see Baldridge et al. 2009; Schwenzer et al. 2016; Grotzinger et al. 2014).

If "Gale Crater Lake" was slightly alkaline, similar to Lake Thetis, is unknown (see Balta and McSween, 2013; Filiberto et al. 2014). However, alkali igneous rocks and basalts are abundant in the Gale Crater (Payre et al. 2017; Schmidt et al. 2014; Trieman et al. 2016; Anderson et al. 2015; Sautter et al. 2014; Stolper et al. 2013). There is also evidence that these alkali Martian rocks were metasomatized by large bodies of water (Stolper et al. 2013), which may have caused alkali to leach into these waters (Schmidt et al. 2014).

Therefore, the waters of "Gale Crater Lake" may have been (and may still be) somewhat alkaline (Schwenzer et al. 2016; Grotzinger et al. 2014), not like sea-water, and may have been similar to the lakes of Western Australia, including the slightly alkaline Lake Thetis. Other investigators have come to similar conclusions (Baldridge et al. 2009; Benison and LaClair, 2003; Bridges et al.

2015; Cocks et al. 2014; Graham et al. 2012, 2016; Nguyen et al. 2014).

Lake Thetis is host to a variety of microorganisms, including green algae and cyanobacteria and several types of living microbial mats, thrombolites and concentric-domed stromatolites (Grey et al. 1990; Grey and Planavsky, 2009). Carbon isotopes indicate that photosynthesis is driving the formation of stromatolites in Lake Thetis (Grey et al. 1990). The microbial mats consist of a variety of bacteria and several types of cyanobacteria are the dominant primary producers (Grey and Planavsky, 2009; Latif et al. 2019). Microbialites with laminae (i.e. stromatolites), and nodular bacterial mats and thrombolites which have a non-laminted clotted "peanut brittle" structure, coexist often within the same matrix (Grey et al. 1990; Grey and Planavsky, 2009; Reitner et al. 1996). Because of the ebb and flow of summer vs winter water levels, the stromatolite domes of Lake Thetis are submerged in winter, whereas in summer most of these structures are above water and which can cause collapse of the domes (Figure 34).

18. Analysis of Concentric-Domical Specimens Sol 173

In the present study three concentric specimens which are morphologically similar to stromatolites growing in Lake Thetis were identified (Figures 36, 38-45). As only one of the three (Sol 173) was suitable for detailed comparative analyses, it was enlarged to various magnifications and compared with living and fossilized stromatolites from Lake Thetis.

As is evident upon examining Figures 40-43, this concentric Martian specimen consists of and includes five layers of crinkly and wavy nodular laminae with several orders of curvature, an abundance of detrital material, the presence of what appears to be numerous fenestrae/gas bubbles, a central (albeit collapsed) axial zone, and preferential vertical, upward and inward growth which is typically caused by upward-migrating microbial colonies at the sediment-water interface (Planavsky and Grey, 2008; Reid et al. 2000; Reitner et al. 1996). Moreover, extensive nodular biological mats and thrombolites were identified on and adjacent to this specimen. Moreover, the specimens depicted in Sol 173, do not have the appearance of fossils and appear to be moist.

The concentric specimen depicted in Sol 173 (Figures 40-43) meets the criteria for a biological vs an abiogenic formation (Allwood et al. 2006; Buick et al. 1981, 1995; Lowe, 1994) and should therefore be considered evidence of a Martian stromatolite. Based on geometry, it appears that this and the other putative Martian stromatolites underwent preferential vertical, upward and inward growth which is typically caused by upward-migrating microbial colonies at the sediment-water interface (Planavsky and Grey, 2008; Reid et al. 2000; Reitner et al. 1996). However, of the six formations presented in this report, only the specimen depicted in Sol 173 meets all the criteria for biology vs abiotic-mechanical activity (see Allwood et al. 2006). By contrast, Sol 308, 528, 529 appear to be fossilized and meet some but not all the criteria for a living stromatolite.

To speculate: Because the aquatic environment of "Gale Crater Lake" may have been similar to Lake Thetis, this may explain why the specimens resembling Martian stromatolites are morphologically similar to the stromatolites of Lake Thetis.

Figure 40. Sol 173: Concentric specimens with collapsed domes displaying the vertical and inward orientation which is typically caused by upward-migrating microbial colonies. Specimen at center (bottom) is adjacent to specimens covered with numerous nodular -mat like formations and fenestrae

Figure 41. Sol 173: This specimen has all the features of a terrestrial stromatolite from Lake Thetis, including microbialites with concentric lamina, nodular bacterial mats and thrombolites which have a non-laminated, clotted "peanut brittle" structure, as well as what appears to be numerous fenestrae/ gas bubbles. These specimens also appear to be moist and may represent living organisms.

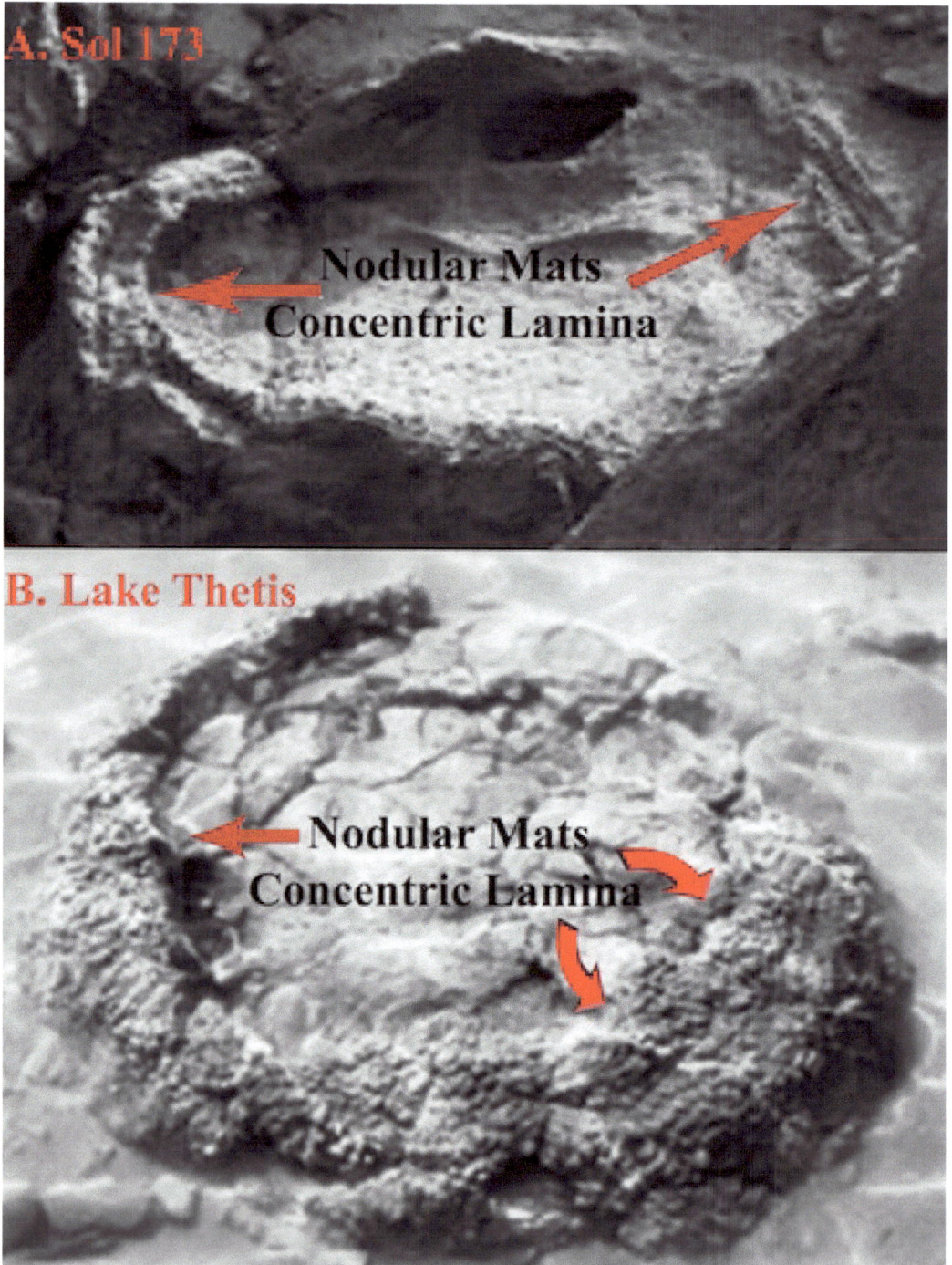

Figure 42. Sol 173 (A-top): This concentric specimen has all the features of a terrestrial stromatolite from Lake Thetis (B-bottom), including microbialites with concentric lamina, nodular bacterial mats and thrombolites which have a non-laminted clotted "peanut brittle" structure, as well as what appears to be numerous fenestrae/gas bubbles.

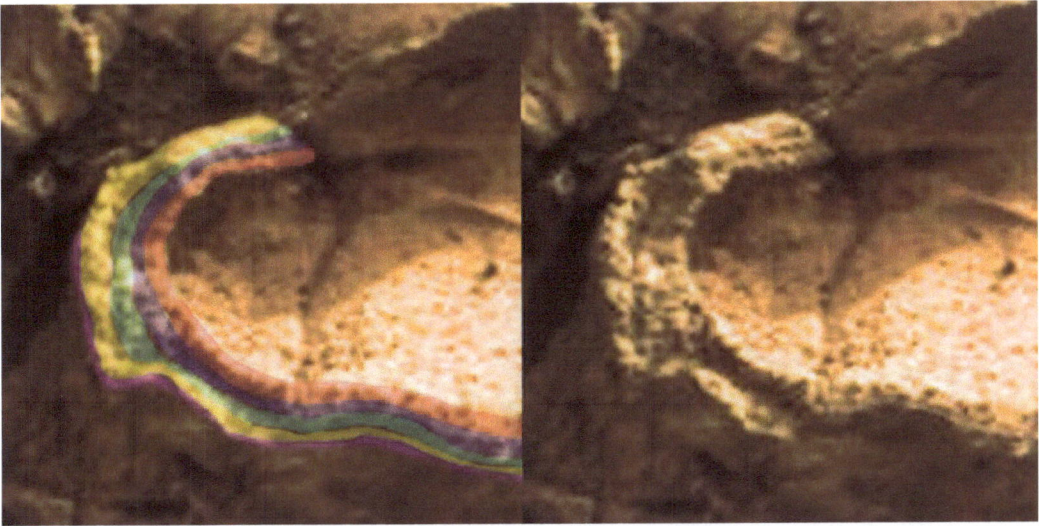

Figure 43. Sol 173: Five layers of concentric lamina and numerous fenestrae/ gas bubbles, thrombolites, and other features typical of stromatolites from Lake Thetis.

19. Visual Analysis of Concentric-Domical Specimens Sol 122, and 529

Martian specimen Sol 529, as depicted in Figure 37), is marked by concentric thrombolitic lamination, and fenestrae and surrounded by "peanut-brittle" specimens resembling fossilized bacterial mats and is nearly identical to the stromatolites of Lake Thetis.

This similarity to terrestrial stromatolites is even more evident upon examination of Sol 122 (see Figures 35, 44 and 45). The fenestrae within the upper portion of the walls appears to be fossilized, though this is unknown. When enlarged and portions are selectively examined, Sol 122 (Sol 528) displays most of the features of a biological formation which was constructed in a body of water, and thus, like Sol 173 and Sol 122/528, shares many of the characteristics of Lake Thetis stromatolites, including being surrounded by what appears to be fossilized bacterial mats and layers of yellowish masses which could be dried algae (see Figure 44).

Figure 44. Sol 528 (see Fig 45). Fossilized bacterial mats, thrombolitic fenestrae, and layers of dried algae?

Figure 45. Sol 528: This concentric specimen has all the features of a terrestrial stromatolite from Lake Thetis, including microbialites with concentric and wavy nodular laminae with several orders of curvature, an abundance of detrital material, and the presence of what appears to be numerous fenestrae/gas bubbles.

Conclusions

The authors, established experts in astrobiology, astrophysics, biophysics, geobiology, lichenology, phycology, botany, and mycology, have presented a sample of NASA photos depicting numerous specimens similar to terrestrial algae, microbial mats, stromatolites, filamentous tangles, algae-fungal-lichen symbiotes, colonies of what may be tubular-shaped organisms, calcium carbonate bio-secretions, ooids and metazoan-like fossils. It also appears that some specimens may be alive, moist, or covered by a thin sheet of ice. These impressions are speculative.

Arguments and data, both pro and con, have been presented and abiogenic explanations detailed, particularly in regard to mineralization. We cannot rule out the possibility that some of the specimens presented here are unusual formations of salts and minerals. However, terrestrial salts, sand, minerals and rocks are infested with cyanobacteria, and these and other organisms often act to bind these substrates together and give them a green coloration. The various shades of green on the Martian surface are indicative of biology.

Nevertheless, it is possible that some of the evidence presented may be due to unusual volcanic formations, geological weathering and the presence of green chlorite and other Martian minerals. What appears to be trace fossils of metazoans may represent mineralization. However, bio-weathering, the microbiology of mineralization and mineralized fossils may also be partly responsible for the existence of the minerals detected on Mars.

Our purpose in conducting this study and presenting this evidence was not to prove there is life on Mars. Our mission was to survey the landscape and identify specimens for future investigation and robotic examination, evaluation, extraction, analyses; and to identify specific features that may be targeted, collected and returned to Earth for study, or, preferably to the International Space Station, thus minimizing the dangers of exposing our planet to Martian organisms.

In conclusion: specimens photographed in Gale Crater were found to resemble algae, lichens, calcium-encrusted cyanobacteria, stromatolites, microbial mats, gas domes constructed by oxygen-respiring organisms, calcium biosignatures, fossilized metazoans, and colonies of micro-organisms. It is not likely that minerals, salts or other abiogenic features can account for all this evidence. The findings and observations detailed in this report, coupled with fact-based speculation and theorizing by the authors, support the hypotheses that, beginning billions of years ago, algae-like organisms and algae-fungal symbiotes may have colonized Gale Crater.

References

Aboal, M., Werner, O., García-Fernández, M. E., Palazón, J.A., Cristóbal, J.C., Williams, W. (2016). Should ecomorphs be conserved? The case of Nostoc flagelliforme, an endangered extremophile cyanobacteria. Journal for Nature Conservation 30, 52–64.

Adey. W. R. (1993). Biological Effects of Electromagnetic Fields. Journal of Cellular Biochemistry 51:410-416.

Allwood, A. et al. (2009). "Controls on development and diversity of Early Archean stromatolites". Proceedings of the National Academy of Sciences. 106, 9548–9555. PNAS. 106. 9548A.

Allwood, A.C., Walter, M.R., Kamber, B.S., Marshall, C.P., and Burch, I.W. (2006). Stromatolite reef from the Early Archaean era of Australia. Nature 441:714–718.

Anderson, R., Bridges, J. C., Williams, A. et al. (2015). ChemCam results from the Shaler outcrop in Gale crater, Mars, Icarus, 249, 2-21.

Anthony, J. W. et al. eds. (2003a). Gypsum" (PDF). Handbook of Mineralogy. V (Borates, Carbonates, Sulfates). Chantilly, VA, US: Mineralogical Society of America. ISBN 978-0962209703.

Anthony, J. W. et al. eds. (2003b). "Calcite" (PDF). Handbook of Mineralogy. V (Borates, Carbonates, Sulfates). Chantilly, VA, US: Mineralogical Society of America. ISBN 978-0962209741.

Anthony, J. W. et al. (2003c). Handbook of Mineralogy. V (Borates, Carbonates, Sulfates). Chantilly, VA, US: Mineralogical Society of America. ISBN 978-0962209741.

Archer Jr. P.D., et al. (2013). The effects of instrument parameters and sample properties on thermal decomposition: interpreting thermal analysis data from Mars, Planetary Science, 2, 2, doi:10.1186/2191-2521-2-2

Archer, P. D., Jr., et al. (2014). Abundances and implications of volatile-bearing species from evolved gas analysis of the Rocknest aeolian deposit, Gale Crater, Mars, J. Geophys. Res. Planets, 119, 237– 254.

Armstrong, R.A. (1976). The influence of the frequency of wetting and drying on the radial growth of three saxicolous lichens in the field. New Phytologist 77: 719-724.

Armstrong, R.A. (1981). Field experiments on the dispersal, establishment and colonization of lichens on a slate rock surface. Environmental and Experimental Botany 21: 116-120.

Armstrong R.A. (2017). Adaptation of Lichens to Extreme Conditions. In: Shukla V., Kumar S., Kumar N. (eds). Plant Adaptation Strategies in Changing Environment. Springer, Singapore.

Armstrong, R. A. (2019). The Lichen Symbiosis: Lichen "Extremophiles" and Survival on Mars Journal of Astrobiology and Space Science Reviews, 1, 378-397.

Ascaso, C., et al. (1990). The weathering action of saxicolous lichens in maritime Antarctica, Polar Biology 11, 33-39

Aspandiar, M.F., Eggleton, R. A. (2002). weathering of chlorite: ii. reactions and products in microsystems controlled by solution avenues, clays and clay minerals (2002). 50 (6).: 699-709.

Ayupova, N. R., Maslennikov, V. V. (2012). Biomineralization in Ferruginous–Siliceous Sediments of Massive Sulfide Deposits of the Urals. Doklady Earth Sciences; 442, 193-195.

Ayupova, N. R. et al. (2016). Evidence of Biogenic Activity in Quartz-Hematite

Rocks of the Urals VMS Deposits, Frank-Kamenetskaya et al. (eds.). Biogenic—Abiogenic Interactions in Natural and Anthropogenic Systems, Lecture Notes in Earth System Sciences, DOI 10.1007/978-3-319-24987-2_10

Azua-Bustos, A., Urrejola, C. and Vicuma, R. (2012). Life at the dry edge: Microorganisms of the Atacama Desert. FEBS Letters 586 SI:2939-2945.

Balta, J. B., and H. Y. McSween (2013). Water and the composition of Martian magmas, Geology, 41(10). 1115–1118.

Bandfield, J. L., Glotch, T. D.m Christensen, P. R. (2003). Spectroscopic identification of carbonate minerals in the Martian dust, Science, 301, 1084– 1087.

Baque, M., de Vera, J.P., Rettberg, P. and Billi, D. (2013). The BOSS and BIOMEX space experiments on the EXPOSE-R2 mission: Endurance of the desert cyanobacterium Chroococcidiopsis under stimulated space vacuum, Martian atmosphere, UVC radiation and temperature extremes. Acta Astronautica 91:180-186.

Baque, M., Verseux, C., Boettger, U. et al. (2016). Preservation of biomarkers from cyanobacteria mixed with Mars like regolith under stimulated Martian atmosphere and UV flux. Origin of Life and Evolution of Biospheres 46:289-310.

Baque, M. et al. (2017). Preservation of carotenoids in cyanobacteria and green algae after space exposure: a potential biosignature detectable by Raman instruments on Mars. EANA17, 14-18 August 2017, Aarhus, Denmark.

Barnes, D.J., Chalker, B, E. (1990). Calcification and photosynthesis in reef-building corals and algae. In: Dubinsky Z, editor. Coral Reefs. Ecosystems of the World. Vol. 25. Amsterdam: Elsevier. pp. 109–131.

Basset C.AL. (1993). Beneficial effects of electromagnetic fields. J Cell Biochem 31:387-393.

Becker RO. (1984). Electromagnetic controls over biological growth processes. Journal of Bioelectricity 3:105-118.

Becker RO, Sparado JA. (1972). Electrical stimulation of partial limb regeneration in mammals. Bull NY Acad Med 48:627- 641.

Beech, M., Comte, M., Coulson, I. (2018). Lithopanspermia – The Terrestrial Input During the Past 550 Million Years, American Journal of Astronomy and Astrophysics, 7(1).: 81-90.

Bekker A, Holland H, Wang PL, Rumble D, Stein H, Hannah J, et al. Dating the rise of atmospheric oxygen. Nature. 2004;427(6970).:117–120.

Bender, K.S. et al. (2002). Sequencing and Transcriptional Analysis of the Chlorite Dismutase Gene of Dechloromonas agitata and Its Use as a Metabolic Probe. Applied and Environmental Microbiology, p. 4820–4826 68,

Bengtson, S., Belivanova, V., Rasmussen, B., Whitehouse, M. (2009). The controversial "Cambrian" fossils of the Vindhyan are real but more than a billion years older, PNAS106 (19). 7729-7734.

Benison, K. C., D. A. L Clair (2003). Modern and ancient extremely acid saline deposits: Terrestrial analogs for Martian environments? Astrobiology, 3, 609–618.

Benzerara, K., Menguy, N., Guyot, F., Skouri, F., de Lucca, G., Heulin, T. (2004). Biologically Controlled precipitation of calcium phosphate by Ramlibacter tataouinensis. Earth Planet. Sci. Lett. 2004, 228, 439–449.

Benzerara, K. et al. (2014). Intracellular Ca-carbonate biomineralization is widespread in cyanobacteria. Proc. Natl. Acad. Sci. 111, 10933–10938.

Bertsch, A. (1966). CO2 Gaswechsel der Grünalge Apatococcus lobatus. Planta, 70, 46 –72.

Bibring, J.-P., Langevin, Y., Mustard, J.F., Poulet, F., Arvidson, R., Gendrin, A., Gondet, B., Mangold, N., Pinet, P., Forget, F., and the OMEGA team. (2006). Global mineralogical and aqueous Mars history derived from OMEGA=Mars Express data. Science 312:400–404.

Biemann, K., J. Oro, P. Toulmin III, L. E. Orgel, A. O. Nier, D. M. Anderson, D. Flory, A. V. Diaz, D. R. Rushneck, and P. G. Simonds (1977), The search for organic substances and inorganic volatile compounds in the surface of Mars, J. Geophys. Res., 82, 4641–4658, doi:10.1029/JS082i028p04641.

Billi, D. (2009). Subcellular integrities in Chroococcidiopsis sp. CCMEE 029 survivors after prolonged desiccation revealed by molecular probes and genome stability assays. Extremophiles 13:49–57.

Billi, D., Verseux, C., Fagliarone, C., Napoli, A., Baque´, M., and de Vera, J.-P. (2019). A desert cyanobacterium under simulated Mars-like conditions in low Earth orbit: implications for the habitability of Mars. Astrobiology 19:158–169;

Bish, D. L., Blake, D.F., Vaniman D.T., Chipera S.J., Morris R.V., Ming D.W. Treiman A.H., et al.et al. (2013). X-ray diffraction results from Mars Science Laboratory: Mineralogy of Rocknest at Gale Crater, Science, 341 (6153). doi:10.1126/science.1238932

Bishop, J. L., et al. (2014). Spectral 519 properties of Ca-sulfates; gypsum, bassanite, and anhydrite, Am. Mineral., 99(10). 2105– 520 2115, doi:10.2138/am-2014-4756.

Blank, C. E. (2013). Origin and early evolution of photosynthetic eukaryotes in freshwater environments: reinterpreting proterozoic paleobiology and biogeochemical processes in light of trait evolution, Journal of Phycology, 49, 1040-1055.

Blichert-Toft, J., et al. (1996). Precambrian alkaline magmatism, Lithos, 37, 97–111.

Bock, E. (1961). On the solubility of anhydrous calcium sulphate and of gypsum in concentrated solutions of sodium chloride at 25 °C, 30 °C, 40 °C, and 50 °C. Canadian Journal of Chemistry. 39 (9).: 1746–1751. doi:10.1139/v61-228.

Bothe, H. (2019). The Cyanobacterium Chroococcidiopsis and Its Potential for Life on Mars, Journal of Astrobiology and Space Science Reviews, 2, 398-412.

Boynton, W. V., et al. (2009). Evidence for Calcium Carbonate at the Mars Phoenix Landing Site, Science, 325, 61– 64, doi:10.1126/science.1172768.

Brandt, A., de Vera, J-P., Onofri, S., Ott, S., (2014/2015). Viability of the lichen Xanthoria elegans and its symbionts after 18 months of space exposure and simulated Mars conditions on the ISS, International Journal of Astrobiology, 14, 411-425.

Brasier, M.D., Green, O.R., Jephcoat, A.P., Kleppe, A.K., Van Kranendonk, M.J., Lindsay, J.F., Steele, A., Grassineau, N.V. (2002). Questioning the evidence for Earth's oldest fossils. Nature 416, 76-81.

Bridges, J.C., Schwenze, S. P., Leveille, R., et al. (2015a). Diagenesis and clay mineral formation at Gale Crater, Mars, JGR Planets, 120, 1-19.

Bridges, N., Núñez, J. I., Seelos, F. P., IV., Hook, S. J., Baldridge, A. M., Thomson, B. J. (2015b). Mineralogy of evaporite deposits on Mars: Constraints from laboratory, field, and remote measurements of analog terrestrial acid saline lakes, American Geophysical Union, Fall Meeting 2015, abstract id. P31A-2022 Bristow, T. F., et al.

(2015). The origin and implications of clay minerals from Yellowknife Bay, Gale Crater, Mars, American Mineral 100, 824– 836.

Broady, P. A. (1996). Diversity, distribution and dispersal of Antarctic terrestrial algae.

Brodo, I.M. et al. (2001). Lichens of North America. Yale University Press. pp. 50, 55, 173-4

Büdel, B., Weber, B., Kühl, M., Pfanz, H., Sültemeyer, D., and Wessels, D.C.J. (2004). Reshaping of sandstone surfaces by cryptoendolithic cyanobacteria: bioalkalisation causes chemical weathering in arid landscapes. Geobiology 2: 261-268.

Büdel, B., Bendix, J., Bicker, F., and Green, T.G.A. (2008). Dewfall as a water source frequently activates the endolithic cyanobacterial communities in the granites of Taylor Valley, Antarctica. Journal of Phycology 44: 1415-1424.

Buick, R., Dunlop, J.S.R., Groves, D.I. (1981). Stromatolite recognition in ancient rocks: an appraisal of irregularly laminated structures in an early Archean chert-barite unit from North Pole, Western Australia. Alcheringa 5:161–181.

Buick, R., Groves. D.I., Dunlop, J.S.R. (1995). Comment on: Abiological origin of described stromatolites older than 3.2 Ga. Geology 23:191.

Bundeleva, I.A et al. (2014). Experimental modeling of calcium carbonate precipitation by cyanobacterium Gloeocapsa sp. Chem. Geol. 374, 44–60.

Burt, D.M., Knauth, L.P., Woletz, K. H. (2005). Origin Of Layered Rocks, Salts, And Spherules At The Opportunity Landing Site On Mars: No Flowing Or Standing Water Evident Or Required. Lunar and Planetary Science XXXVI.

Buz, J., et al. (2017). Mineralogy and stratigraphy of the Gale crater rim, wall, and floor units, J. Geophys. Res. Planets, 122, 1090–1118, doi:10.1002/ 2016JE005163.

Cabrol, N. A., et al. (1999). Hydrogeologic Evolution of Gale Crater and Its Relevance to the Exobiological Exploration of Mars, Icarus 139, 235–245.

Cannon, K. M. et. al. (2012). Perchlorate induced low temperature carbonate decomposition in the Mars Phoenix Thermal and Evolved Gas Analyzer (TEGA). Geophysical Research Letters, 39, 13, https://doi.org/10.1029/2012GL051952

Cardon, Z. G., Gray, D. W. and Lewis, L. A. (2008). The green algal underground: evolutionary secrets of desert cells. Bioscience, 58, 114 –122.

Carter, J., Poulet, F., Bibring, J-P. & Murchie, S. Detection of hydrated silicates in crustal outcrops in the northern plains of mars. Science 328, 1682–1686 (2010).

Castro, J.F.B., et al. (2015). Liquid water at crater Gale, Mars Journal of Astrobiology and Outreach, ISSN 2332-2519, Vol. 3, no 3.

Cecchi, G., et al. (2019). Interactions among microfungi and pyrite-chalcopyrite mineralizations: tolerance, mineral bioleaching, and metal bioaccumulation, Mycological Progress, 18, 3, 415-423.

Chafetz, H. F., Buczynski, C. (1992). Bacterially induced lithification of microbial mats. Palaois, 7, 277-293.

Chang, H. K., et al. (1986). Comparisons between the diagenesis of dioctahedral and trioctahedral smectite, Brazilian offshore basins. Clays Clay Miner. 34, 407–423 doi: 10.1346/ CCMN. 1986.0340408

Chaudhuri, S. K., et al. (2002). Environmental factors that control microbial perchlorate reduction. Appl. Environ. Microbiol. 68:4425–4430.

Chen, P. C. and Lai, C. L. (1996). Physiological adaptation during cell dehydration and rewetting of a newlyisolated Chlorella species. Physiologia Plantarum, 96, 453 –457.

Chen, J., Blume, H-P., Beyer, L. (2000). Weathering of rocks induced by lichen colonization — a review. Catena 39, 121–146.

Claus, G., Nagy, B. (1961). A Microbiological Examination of Some Carbonaceous Chondrites. Nature 192, 594 - 596.

Coates, J. D., et al. (1999). Ubiquity and diversity of dissimilatory (per).-chlorate-reducing bacteria. Appl. Environ. Microbiol. 65:5234–5241.

Cockell, C.S., Schuerger, A.C., Billi, D., Friedmann, E.I., and Panitz, C. (2005). Effects of a simulated martian UV flux on the cyanobacterium, Chroococcidiopsis sp. 029. Astrobiology 5:127–140.

Cocks, C, et al. (2014). Analysis of the Paleoenvironment of Gale Crater on Mars: Using Ephemeral Lakes in Western Australia as Analogs to the Mineral Assemblages of Gale Crater, American Geophysical Union, Fall Meeting 2014, abstract id.P41A-3894

Collins, S., Sültemeyer, D., Bell, G. (2006a). Changes in C uptake in populations of Chlamydomonas reinhardtii selected at high CO2. Plant, Cell and Environment, 29, 1812 –1819.

Collins, S., Sültemeyer, D., Bell, G. (2006b). Rewinding the tape: selection of algae adapted to high CO2 at current and Pleistocene levels of CO2. Evolution, 60, 1392 –1401.

Couradeau, E. et al. (2012). An early-branching microbialite cyanobacterium forms intracellular carbonates. Science, 336, 459–462.

Cumbers, J. and Rothschild, L. J. (2014). Salt tolerance and polyphyly in the cyanobacterium Chroococcidiopsis. Journal Phycology 50:472-482.

Damer, B. (2016). A Field Trip to the Archaean in Search of Darwin's Warm Little Pond, Life, 6, 21; doi:10.3390/life6020021

Dass, R. S. (2017). The High Probability of Life on Mars: A Brief Review of the Evidence, Cosmology, Vol 27, April 15, 2017.

Davaud, E., and Girardclos, S., (2001). Recent freshwater ooids and oncoids from western Lake Geneva (Switzerland).: indications of a common organically mediated origin. Journal of Sedimentary Research, 71: 423–429.

Davies, P. C. W. (2007). The Transfer of Viable Microorganisms Between Planets, (Editors Gregoy R. Bock Jamie A.). Novartis Foundation Symposia.

De Gregorio, B.T., Sharp, T.G., Rushdi, A.I., Simoneit, B.R. (2011). Bugs or gunk? Nanoscale methods for assessing the biogenicity of ancient microfossils and organic matter. In: Golding S., Glikson M. (Eds.). Earliest Life on Earth: Habitats, Environments and Methods of Detection. Springer, Dordrecht, 239-289.

De la Torre, R., Sancho, L. G., Horneck, G., de los Ríos, A., Wierzchos, J., Olsson-Francis, K., et al. (2010). Survival of lichens and bacteria exposed to outer space conditions – results of the Lithopanspermia experiments. Icarus 208, 735–748.

De la Torre Noetzel, R. et al. (2017). Survival of lichens on the ISS-II: ultrastructural and morphological changes of Circinaria gyrosa after space and Mars-like conditions EANA2017: 17th European Astrobiology Conference, 14-17 August, 2017 in Aarhus, Denmark.

De los Ríos, A., Ascaso, C., Wierzchos, J., Sancho, L. G. (2009). "Space flight effects on lichen ultrastructure and physiology," in Symbioses and Stress. Cellular Origin, Life in Extreme Habitats and Astrobiology, Vol 17, eds J. Seckbach and M. Grube (Dordrecht: Springer). doi: 10.1007/978-90-481-9449-0_30

De Vera, J.-P. (2012). Lichens as survivors in space and on Mars. Fungal Ecology, 5, 472-479.

De Vera, J.-P. et al. (2014). Results on the survival of cryptobiotic cyanobacteria samples after exposure to Mars-like environmental conditions, International Journal of Astrobiology, 13, 35-44.

De Vera, J.-P, M., Backhaus, T., et al. (2019). Limits of Life and the Habitability of Mars: The ESA Space Experiment BIOMEX on the ISS, 19, Astrobiology.

Delvigne J., Bisdom E.B.A, Sleeman J., Stoops G. (1979). Olivines, their pseudomorphs and secondary products. Pedologie, 29, 247_309.

Dhanya, V., Ray, J. G. (2015a). Ecology and Diversity of Cyanobacteria in Kuttanadu paddy wetlands, Kerala, India. American Journal of Plant Sciences; http://dx.doi.org/10.4236/ajps.2015.618288

Dhanya, V., Ray, J. G. (2015b). Green algae of a unique tropical wetland, Kuttanadu, Kerala, India in relation to Soil regions, Seasons and Paddy-growth stages, International Journal of Science, Environment and Technology, 4(3). pp.770-803.

Dittrich, M., Sibler, S. (2010). Calcium carbonate precipitation by cyanobacterial polysaccharides, Geological Society, 336, 51-63.

Dong, H., Rech, J.A., Jiang, H. et al. (2007). Endolithic cyanobacteria in soil gypsum: Occurrences in Atacama (Chile). Mojave (United States). and Al-Jafr Basin (Jordan). deserts. Journal Geophysical Research. Biogeosciences 112.

Dongyan, W., et al. (1998). Biomineralization of mirabilite deposits of Barkol Lake, China, Carbonates and Evaporites, 13, 86-89.

Duchafour, P. (1979). Alteration des roches cristallines en milieu superficiel. Sci Sol 2–3:87–89.   Ehlmann, B. L., et al. (2008). Orbital identification of carbonate-bearing rocks on Mars. Science, 322, 829- 832.

Ehlmann, B. L. et al. (2009). Identification of hydrated silicate minerals on Mars using MRO-CRISM: Geologic context near Nili Fossae and implications for aqueous alteration. J. Geophys. Res. 114, 1–33.   Ehlmann, B. L., Edwards, C. S. (2015). Mineralogy of the Martian surface. Annu. Rev. Earth Planet. Sci. 42, 291–315.   Ehlmann, B. L., Edgett, K. S., Sutter, B., et al. (2016). Chemistry, mineralogy, and grain properties at Namib and High dunes, Bagnold dune field, Gale crater, Mars: A synthesis of Curiosity rover observations, JGR Planets, 9 2510-2543.

Eigenbrode J.L., et al. (2018). Organic matter preserved in 3-billion-year-old mudstones at Gale crater, Mars. Science 360:1096.   Eldrin, D.A. (2016). Morphotaxonomic Account of Epilithic Microalgae and Cyanobacteria in Los Baños, Laguna (Philippines). IAMURE International Journal of Ecology and Conservation, 17:22-39.   Eric H. et al. (2016). Perchlorate formation on Mars through surface radiolysis□initiated atmospheric chemistry: A potential mechanism, JGR Planets, 121, 1472-1487.

Ertem, G., et al. (2017). Shielding biomolecules from effects of radiation by Mars analogue minerals and soils. International Journal of Astrobiology, 16, 280-285.

Fabricius, F.H. (1977). Origin of Marine Ooids and Grapestone. Contributions to Sedimentology, Volume 7, Stuttgart: E. Schweizerbart'sche Verlagsbuchhandlung.

Fagliarone, C., Mosca, C., Ubaldi, I., Verseux, C., Baque', M., Wilmotte, A., and Billi, D. (2017). Avoidance of protein oxidation correlates with the desiccation and radiation resistance of hot and cold desert strains of the cyanobacterium Chroococcidiopsis. Extremophiles 21:981–991.

Fairén, A. G. et al. (2010). Noachian and more recent phyllosilicates in impact craters on Mars. Proc. Natl Acad. Sci. USA 107, 12095–12100.

Fairén, A.G., et al. (2014). A cold hydrological system in Gale crater, Mars, Planetary and Space Science, 93–94, 101-118.

Farley, K. C., Malespin, P., Mahaffy, J., Grot-zinger, P. Vasconcelos, R. Milliken, M. Malin, K. et al. (2014). In situ radiometric and exposure age dating of the Martian surface, Science, 343 (6169). doi:10.1126/science.1247166.

Farquhar J, Zerkle AL, Bekker A. (2011). Geological constraints on the origin of oxygenic photosynthesis. Photosynthesis research. 107 (1).: 11–36. pmid:20882345.

Filiberto, J., A. et al. (2014). High-temperature chlorine-rich fluid in the Martian crust: A precursor to habitability, Earth Planet. Sci. Lett., 401, 110–115.

Finkel, Z. V., Katz, M. E., Wright, J. D., Schofi eld , O. M. E., Falkowski , P. G. (2005). Climatically driven macroevolutionary patterns in the size of marine diatoms over the Cenozoic. Proceedings of the National Academy of Sciences of the USA, 102, 8927 –8932.

Finkel, Z. V., Sebbo, J., Feist-Burkhardt, S. et al. (2007). A universal driver of macroevolutionary change in the size of marine phytoplankton over the Cenozoic. Proceedings of the National Academy of Sciences of the USA, 104, 20416 –20420.

Finstad, K.M., Probst, A.J., Thomas, B. et al. (2017). Microbial community structure and persistence of cyanobacterial populations in salt crusts of the hyperarid Atacama Desert from genome resolved metagenomics. Frontiers Microbiology 8: article 1435.

Fisk, M. R., et al. (2006). Iron-Magnesium Silicate Bioweathering on Earth (and Mars?). Astrobiology, 1.

Flechtner, V. R. (2007). North American microbiotic soil crust communities: diversity despite challenge. In Algae and Cyanobacteria in Extreme Environments, ed. J. Seckbach Dordrecht: Springer, pp. 539–551.

Fleming, E.D., Castenholz, R.W. (2007). Effects of periodic desiccation on the synthesis of the UV-screening compound, scytonemin in cyanobacteria. Environmental Microbiology 9:1448-1455.

Folk, R. L., Lynch, F. L. (1997). Nanobacteria are alive on Earth as well as Mars, in Proceedings of SPIE The International Society for Optical Engineering. 3111, 407-419.

Fouche, T., et al. (2007). Martian water vapor: Mars express PFS/LW observations. Icarus 190, 32-49.

Friedmann, E.I. (1980). Endolithic microbial life in hot and cold deserts. Origins of Life 10, 223–235.

Friedmann, E.., Weed, R. (1987). Microbial trace-fossil formation, biogenous, and abiotic weathering in the Antarctic cold desert. Science 236, 703–705.

Friedmann, E.I. (1992). Endolithic microorganisms in the Antarctic cold desert.

Science 215:1045-1053.

Gablina I.F., Dobretsova I.G., Popova E.A. (2016). Biomineralization Processes During the Formation of Modern Oceanic Sulfide Ore and Ore-bearing Sediments. In: Frank-Kamenetskaya O., Panova E., Vlasov D. (eds). Biogenic—Abiogenic Interactions in Natural and Anthropogenic Systems. Lecture Notes in Earth System Sciences. Springer.

Gadd, G. M., (2007). Geomycology: biogeochemical transformations of rocks, minerals, metals and radionuclides by fungi, bioweathering and bioremediation, Mycological Research, 111, 3-49.

Gadd, G. M. (2012). Geomycology: Fungi as Agents of Biogeochemical Change, Biology & Environment: Proceedings of the Royal Irish Academy, 113, 1-15.

Gadd, G. M., Rhee, Y. J., Stephenson, K., Wei, Z. (2011). Geomycology: metals, actinides and biominerals, Environmental Microbiology Reports, 4, 270-296.

García-Ruiz, J.M., et al. (2007). Formation of natural gypsum megacrystals in Naica, Mexico, Geology, 35 (4).: 327-330.

Garwood, R. J. (2012). Patterns In Palaeontology: The first 3 billion years of evolution. Palaeontology Online. 2 (11).: 1–14.

Gaysina, L.A, Saraf, A.and Singh, P. (2019). Cyanobacteria in Diverse Habitats. Cyanobacteria From Basic Science to Applications, Wiley.

Gazzè, S. A., et al. (2012). Nanoscale channels on ectomycorrhizal-colonized chlorite: Evidence for plant□driven fungal dissolution, Biogeosciences, 117.

Golubic, S., Seong-Joo, L. (1999). Early cyanobacterial fossil record: preservation, palaeoenvironments and identification. Eur J Phycol 34:339–348

Gomez-Silvo, B. (2018). Lithobiontic life: Atacama rocks are well and alive. Antonie van Leeuwenhoek International Journal General Molecular Microbiology 111:1333-1343.

Graham, H.; Baldridge, A. M.; Stern, J. C. (2015). Australian Acid Playa Lake as a Mars Analog: Results from Sediment Lipid Analysis, American Geophysical Union, Fall Meeting 2015, abstract id.P23C-02.

Graham, H. V.; Stern, J. C.; Baldridge, A. M.; Thomsen, B. J. (2016). Australian Acid Brine Lake as a Mars Analog: An Analysis of Preserved Lipids in Shore and Lake Sediments (In Biosignature Preservation and Detection in Mars Analog Environments, Proceedings of a conference held May 16-18, 2016 in Lake Tahoe, Nevada. LPI Contribution No. 1912, id.2063

Graham, L.E., Graham, J.M., Wilcox, L.W., Cook, M.E. (2016). Algae. LJLM Press, Madison.

Graham, L.E. et al. (2014). Lacustrine Nostoc (Nostocales). And Associated Microbiome Generate A New Type of Modern Clotted Microbialite, Journal of Phycology, 50, 280-291.

Graham, L.E, et al. (2018). Microscopic and Metagenomic Analyses of Peltig Ponojensis (Peltigerales, Ascomycota). International Journal of Plant Science, 179, 241-255.

Graham, L.E. (2019). Digging Deeper: Why We Need More Proterozoic Algal Fossils and How To Get Them, Journal of Phycology, 55.

Gray, D. W., Lewis, L. A., Cardon, Z. G. (2007). Photosynthetic recovery following desiccation of desert green algae (Chlorophyta). and their aquatic relatives. Plant, Cell and Environment, 30, 1240–1255.

Life on Venus and Mars

Grey, K., et al. (2003). Neoproterozoic biotic diversification: Snowball Earth or aftermath of the Acraman impact? Geology, 31, 459-462.

Grey, K., Planavsky, N. J. (2009). Microbialities of Lake Thetis Cervantes, Western Australia, Government of Western Australia Department of Mines and Petroleum, Geological Survey of Western Australia.

Grey, K., Moore, L.S., Burne, R.V., Pierson, B.K. and Bauld, J. (1990). Lake Thetis, Western Australia: an example of saline sedimentation dominated by benthic microbial processes: Australian Journal of Marine and Freshwater Research, v. 41, p. 275–300.

Grotzinger, J., et al. (2012). Mars Science Laboratory mission and science investigation, Space Sci. Rev., 170 (1-4). 5– 56.

Grotzinger, J. P., et al. (2014). A habitable fluvio-lacustrine environment at Yellowknife Bay, Gale Crater, Mars, Science, 343, doi:10.1126/science.1242777.

Grotzinger, J. P., Crisp, J. A., Vasavada, A. R., & Science Team, M. S. L. (2015a). Curiosity's mission of exploration at Gale crater,

Grotzinger, J. P., et al. (2015b). Deposition, exhumation, and paleoclimate of an ancient lake deposit, Gale Crater, Mars, Science, 350, 6257, doi:10.1126/science.aac7575.

Hagedoorn, P. L. et al. (2002). Spectroscopic characterization and ligand-binding properties of chlorite dismutase from the chlorate respiring bacterial strain GR-1, European Journal of Biochemistry, 269, 4905-4911.

Haggart, J. W., Busti, R.M. (1999). Selective replacement of mollusk shell by chlorite, Lower Cretaceous Longarm Formation, Queen Charlotte Islands, British Columbia, Canadian Journal of Earth Sciences, 1999, 36(3).: 333-338,

Hamilton, V. E., Ruff, S. W. (2012). Distribution and characteristics of Adirondack-class basalt as observed by Mini-TES in Gusev crater, Mars and its possible volcanic source. Icarus, 218, 917-949.

Hansen, H. (1979). Test structure and evolution in the Foraminifera, Lethaia 12, 173–181.

Hardie, L. A. (1967). The Gypsum-Anhydrite Equilibrium at One Atmosphere Pressure, Am. 540 Mineral., 52, 171–200.

Hasler, P., Poulickova, A. (2010). Diversity, taxonomy and autecology of autochtonous epipelic cyanobacteria of the genus Komvophoron(Borziaceae, Oscillatoriales).: a studyon populations from the Czech Republic and British Isles. Biologia, 65, 1: 7—16.

Häubner, N. et al. (2006). Aeroterrestrial algae growing in biofilms on facades: response to temperature and water stress. Microbial Ecology, 51, 285 –293.

He, H., et al. (2014). Physiological and ecological significance of biomineralization in plants. Trends in Plant Science 19:166-174.

Herburger, K., Holzinger, A. (2015). Localization and Quantification of Callose in the Streptophyte Green Algae Zygnema and Klebsormidium: Correlation with Desiccation Tolerance, Plant and Cell Physiology, 56, 2259–2270.

Hoham R.W., Ling H.U. (2000). Snow Algae: The Effects of Chemical and Physical Factors on Their Life Cycles and Populations. In: Seckbach J. (eds). Journey to Diverse Microbial Worlds. Cellular Origin and Life in Extreme Habitats, vol 2. Springer, Dordrecht.

Holmes, N.T.H., Whitton, B.A. (1975). Notes on some macroscopic algae new\ or seldom recorded for Britain: Nostoc parmelioides, Heribaudiella fluviatilis, Cladophora

159

aegagropila, Monostroma bullosum, Rhodoplax schinzii. Vasculum 60:47-55.

Holzinger, A., Pichrtová, M., (2016). Abiotic Stress Tolerance of Charophyte Green Algae: New Challenges for Omics Technique, Front. Plant Sci., 20.

Huang, Y., Liu, X., Laws, E.A., Chen, B., Li, Y., Xie, Y., Wu, Y., Gao, K., Huang, B. (2018). Effects of increasing atmospheric CO2 on the marine phytoplankton and bacterial metabolism during a bloom: A coastal mesocosm study. Science of the Total Environment 633, 618–629.

Hughes, K. A. (2006). Solar UV-B radiation, associated with ozone depletion, inhibits the Antarctic terrestrial microalga Stichococcus bacillaris . Polar Biology, 29, 327 –336.

Jackson, T. A. (1971). A Study of the Ecology of Pioneer Lichens, Mosses, and Algae on Recent Hawaiian Lava Flows, Pacific Science, 25, 22-32.

Jansson, C.; Northen, T. Calcifying cyanobacteria-the potential of biomineralization for carbon capture and storage. Curr. Opin. Biotechnol. 2010, 21, 365–371.

Jinhua,L. et al. (2016). Biomineralization Patterns of Intracellular Carbonatogenesis in Cyanobacteria: Molecular Hypotheses. Minerals, 6, 10: 1-21.

Jones, D., Wilson M.J. & Tait J.M. (1980). Weathering of a basalt by Pertusaria, corallina. The Lichenologist, 12, 277-289.

Jones, D., et al. (1981). Lichen weathering of rock-forming minerals: application of scanning electron microscopy and microprobe analysis, Journal of Microscopy, 124, 95-104.

Jones, D, Wilson MJ (1985). Chemical activity of lichens on mineral surfaces-A review. Int Biodeterior Bull 21:99–104.

Jones, D, Wilson MJ (1986). Biomineralization in crustose lichens. In: Leadbeater BSC, Riding R (eds). Biomineralization in lower plants and animals. Clarenton Press, Oxford pp 91–101.

Joseph, R. (2014). Life on Mars: Lichens, Fungi, Algae, Cosmology, 22, 40-62.

Joseph, R. (2016). A High Probability of Life on Mars, The Consensus of 70 Experts, Cosmology, 25, 1-25.

Joseph, R. (2019). Life on Venus and the Interplanetary Transfer of Biota From Earth, Astrophysics and Space Sciences, 364, 11. DOI: 10.1007/s10509-019-3678-x

Joseph, R. G, Dass, R. S., Rizzo, V., Cantasano, N., Bianciardi, G. (2019). Evidence of Life on Mars? Journal of Astrobiology and Space Science Reviews, 1, 40–81.

Joseph R, Armstrong, R., Kidron, G., Gibson, C. H., Schild, R. (2020a). Life on Mars? Colonies of Mushroom-shaped specimens in Eagle Crater. Astrophysics and Space Science (revision under review).

Joseph, R., Gibson, C. H., Schild, R. (2020b). Water, Mud and Ice in Gale Crater, Mars (under review).

Jung, P., Baumann, K., Lehnert, L. W., Samolov, E., Achilles, S., Schermer, M., Karsten, U. (2019). Desert breath—How fog promotes a novel type of soil biocenosis, forming the coastal Atacama Desert's living skin. Geobiology, 18, 113-124.

Kaplan-Levy, R.N., et al. (2010). Akinetes: dormant cells of cyanobacteria. In: Dormancy and resistance in harsh environments. Springer; 5–27.

Karsten, U., et al. (2016). A. Living in biological soil crust communities of African deserts—physiological traits of green algal Klebsormidium species (Streptophyta). to cope with desiccation, light and temperature gradients. J Plant Physiol. 194, 2–12.

Karsten, U., Schumann, R., Mostaert, A. S. (2007a). Th e eff ects of ultraviolet radiation on photosynthetic performance, growth, and sunscreen compounds in aeroterrestrial biofi lm algae isolated from building facades. Planta, 225, 991 –1000.

Karsten, U., Schumann, R., Mostaert, A. S. (2007b). Aeroterrestrial algae growing on man-made surfaces: what are the secrets of their ecological success? In Algae and Cyanobacteria in Extreme Environments, ed. J. Seckbach . Dordrecht, Springer, 583–597.

Kengen, S. W., et al. (1999). Purification and characterization of (per).chlorate reductase from the chlorate-respiring strain GR-1. J. Bacteriol. 181, 6706–6711.

Kennett, J.P., Srinivasan, M.S. (1983). Neogene planktonic foraminifera: a phylogenetic atlas. Hutchinson Ross. ISBN 978-0-87933-070-5.

Kidron, G.J. (1999). Altitude dependent dew and fog in the Negev desert, Israel. Agric Forest Meteorol 96: 1-8.

Kidron, G.J. (2019). Cyanobacteria and Lichens May Not Survive on Mars. The Negev Desert Analogue Journal of Astrobiology and Space Science Reviews, 1, 369-377, 2019.

Kidron, G.J., Starinsky, A., Yaalon, D.H. (2014). Dewless habitat within a dew desert: Implications for weathering and terrestrial evolution. J Hydrology 519, 3606-3614.

Kilias, S.P., et al. (2007). Evidence of Mn-oxide biomineralization, Vani Mn deposit, Milos, Greece. In: Andrew, C.J., (ed.). Digging deeper: proceedings of the ninth biennial Meeting of the Society for Geology Applied to Mineral Deposits, Dublin, Ireland 20th-23rd August 2007. Dublin, Ireland, Irish Association of Economic Geologists, 1069-1072.

Klein, H.P., Horowitz, N.H., Levin, G.V., Oyama, V.I., Lederberg, J., Rich, A., Hubbard, J.S., Hobby, G.L., Straat, P.A., Berdahl, B.J., Carle, G.C., Brown, F.S., and Johnson, R.D. (1976). The Viking Biology Investigation: Preliminary Results. Science. 194, 4260, p. 92-105.

Knack, J. J., Wilcox, L. W. Delaux, P.-M. Ané, J.-M. Piotrowski, M. J. Cook, M. E. Graham, J. M. Graham, L.E. (2015). Microbiomes of streptophyte algae and bryophytes suggest that a functional suite of microbiota fostered plant colonization of land. International Journal of Plant Sciences 176, 405-420.

Knoll, A.H., Barghoorn, E.S. (1974). Ambient pyrite in Precambrian chert: new evidence and a theory. Proceedings of the National Academy of Sciences of the United States of America 71, 2329–2331

Kogarko, L. (2006). Alkaline magmatism and enriched mantle reservoirs: Mechanisms, time, and depth of formation, Geochem. Int., 44, 3– 10.

Kounaves, S. P., et al. (2010). Soluble sulfate in the Martian soil at the Phoenix landing site, Geophys. Res. Lett., 37, L09201.

Kral, T. A., et al. (2014). Potential use of highly insoluble carbonates as carbon sources by methanogens in the subsurface of Mars, Planet. Space Sci., 101, 181– 185.

Kremer, B., Kazmierzcak, J., Lukomska-Kowalczyk, M., Kemp, S. (2012). Astrobiology, 12(6).: 535-548.

Krupa, T. A. (2017). Flowing water with a photosynthetic life form in Gusav Crater on Mars, Lunar and Planetary Society, XLVIII.

Kumar, D., Kaštánek, P., Adhikary, S.P. (2018). Exopolysaccharides from cyanobacteria and microalgae and their commercial application. Current Science, 115,

234-241.

Kupriyanova, E., et al. (2007). Extracellular carbonic anhydrases of the stromatolite-forming cyanobacterium Microcoleus chthonoplastes , Microbiology, 153.

Latif, K., Xiao, E., Riaz, M., et al. (2019). Calcified cyanobacteria fossils from leiolitic bioherm in the Furongian Changshan Formation, Datong (North China Platform). Carbonates and Evaporites, 34(3). 825–843. DOI:10.1007/

Latif, K. (2019). Cambrian Stratigraphy and Sedimentology of the Cyanobacteria-Dominated Microbial Carbonates in the North China Platform. China University of Geosciences Beijing, 137.

Lange OL, Kidron GJ, Büdel B, Meyer A, Kilian E, Abeliovitch A. (1992). Taxonomic composition and photosynthetic characteristics of the biological soil crusts covering sand dunes in the Western Negev Desert. Func. Ecol. 6, 519-527.

Latif, K. (2019). High-energy oncoids within the ooid-grained bank built by photosynthetic biofilms: A case study of the Cambrian Zhangxia Formation at the Sandaogou section of Huludao City in the western part of Liaoning province, Acta Geologica Sinica 93:1-18.

Le Deit, L., et al. (2013). Sequence of infilling events in Gale Crater, Mars: Results from morphology, stratigraphy, and mineralogy, J. Geophys. Res. Planets, 118, 2439–2473.

Le Deit, L., E., (2013). Sequence of infilling events in Gale Crater, Mars: Results from morphology, stratigraphy, and mineralogy, J. Geophys. Res. Planets, 118, 2439– 2473, doi:10.1002/2012JE004322.

Leake, J. R., et al. (2008). Biological weathering in soil: The role of symbiotic root☐associated fungi biosensing minerals and directing phiotosynthate☐energy into grain☐scale mineral weathering, Mineral. Mag., 72,85– 89.

Lee, J.H, et al. (2016). The earliest reef-building anthaspidellid sponge Rankenella zhangxianensis n. sp. from the Zhangxia Formation (Cambrian series 3). Shandong Province, China. Journal of Paleontology, 90. 1-9.

Lee J H, Chen J, Choh S J, et al. (2014). Furongian (late Cambrian). sponge-microbial maze-like reefs in the North China Platform. Palaios, 29, 27-37.

Lee, J. J. (1992a). Symbiosis in Foraminifera. In: W. Reisser, W. (ed.). Algae and Symbioses, BioPress, Bristol, UK, 63–78.

Lee, J. J. (1992b). Taxonomy of algae symbiotic in Foraminifera. In: W. Reisser, W. (ed.). Algae and Symbioses, BioPress, Bristol, UK, pp. 79–92.

Lepot, K., Benzerara, K., Brown Jr, G. and Philippot, P. (2008). Microbially influenced formation of 2,724-million-year-old stromatolites, Nature Geosciences, online publication 28 January 2008.

Leshin, L. A., et al. (2013). Volatile, isotope, and organic analysis of Martian fines with the Curiosity rover, Science, 341, 1238-1937,

Levin, G.V., Straat, P.A., and Benton, W.D. (1978). Color and Feature Changes at Mars Viking Lander Site. J. Theor. Biol., 75: 381-390.

Levin, G., Straat, P. A. (1976). Viking Labeled Release Biology Experiment: Interim Results, Science, 194, 1322-1329.

Levin, G.V., Straat, P. A. (1977). Life on Mars? The Viking labeled release experiment, Biosystems 9 :2-3, pp. 165-174.

Levin, G.V., Straat , P. A. (1979). Completion of the Viking Labeled Release Experiment on Mars, J. Mol. Evol., 14, 167-183.

Levin, G.V., Straat, P.A., and Benton, W.D. (1978). Color and Feature Changes at Mars Viking Lander Site. J. Theor. Biol., 75: 381-390.

Levin, M. (2003). Review: Bioelectromagnetics in Morphogenesis. Bioelectromagnetics 24:295-315.

Lewis, L. A., Flechtner, V. R. (2002). Green algae (Chlorophyta). of desert microbiotic crusts: diversity of North American taxa. Taxon, 51, 443 –451.

Lewis, L.A. (2007). Chlorophyta on land: independent lineages of green eukaryotes from arid lands. In Algae and Cyanobacteria in Extreme Environments, ed. J. Seckbach, Dordrecht, Springer, pp. 571–582. L

Lewis, L.A., Lewis, P. O. (2005). Unearthing the molecular phylodiversity of desert soil green algae (Chlorophyta). Systematic Biology, 54, 936 –947. Lewis, L.A., McCourt, R. M. (2004). Green algae and the origin of land plants. American Journal of Botany, 91, 1535 –1556.

Leya, T. (2013). Snow Algae: Adaptation Strategies to Survive on Snow and Ice. In: Seckbach J., Oren A., Stan-Lotter H. (eds). Polyextremophiles. Cellular Origin, Life in Extreme Habitats and Astrobiology, vol 27. Springer, Dordrecht.

Lin, H.L. Zhang, X., Shuaia, T., Zhang, L., Suna, Y., (2016). Abundance retrieval of hydrous minerals around the Mars Science Laboratory landing site in Gale crater, Mars, Planetary and Space Science, 121, 76-82

Loeblich, A. R., Helen, T. (1964). Protista 2: Sarcodina Chiefly "Thecamoebians" and Foraminiferida. Moore, R.C. (ed.). Treatise on Invertebrate Paleontology. C (5th ed.). Geological Society of America. ISBN 978-0-8137-3003-5.

López-Bautista, J. M., Rindi, F., Casamatta, D. (2007). Th e systematics of subaerial algae. In Algae and Cyanobacteria in Extreme Environments, ed. J. Seckbach . Dordrecht, Springer, 601–617.

Lowe, D.R. (1994). Abiological origin of described stromatolites older than 3.2 Ga. Geology 22, 387–390.

Lueksova, A. (2001). Soil algae in brown coal and lignite post-mining areas in Central Europe (Czech Republic and Germany). Restor. Ecol., 9, 341–350.

Luo, C., Reitner, J. (2015). 'Stromatolites' built by sponges and microbes-a new type of Phanerozoic bioconstruction. Lethaia, 49, 1-16.

Luo, W., et al. (2006). Genotype versus phenotype variability in Chlorella and Micractinium (Chlorophyta, Trebouxiophyceae). Protist, 157, 315 –333.

Lutman, B.F. (1910). The Cell Structure of Closterium Ehrenbergii and Closterium moniliferum. Botanical Gazette, 49, 241-255, University of Chicago Press.

Maffei, M.E. (2014). Magnetic field effects on plant growth, development, and evolution, Front. Plant Sci., 04.

Maixner, F. et. al. (2008). Environmental genomics reveals a functional chlorite dismutase in the nitrite-oxidizing bacterium 'Candidatus Nitrospira defluvii' Environmental Microbiology, 10, 3043-3056.

Malin, M.C. K.S. Edgett, L.V. Posiolova, S.M. McColley, E.Z.N. Dobrea, Present-day impact cratering rate and contemporary gully activity on Mars. Science 314(5805), 1573 (2006)

Malin, M.C. K.S. Edgett, (2000). Evidence for recent groundwater seepage and surface runoff on Mars. Science 288(5475), 2330

Manning, C.E., Mojzsis, S. J., Harrison, T. M. (2006). Geology. age and origin of supracrustal rocks at Akilia, West Greenland. American Journal of Science, 306, 303-366.

Marshall, C.P., Emry, J.R., Marshall, A.O. (2011.). Haematite pseudomicrofossils present in the 3.5-billion-year-old Apex Chert. Nature Geoscience 4, 240.

Mansour, H.A., Shaaban, A. S. (2010. Algae of soil surface layer of Wadi Al-Hitan protective area (World Heritage Site). El-Fayum Depression, Egypt, Journal of American Science, 6, 243-255.

Martín-Torres, F. J. et al. (2015). Transient liquid water and water activity at Gale crater on Mars. Nature 8, 357–361.

Masson, P., Carr, M.H., Costard, F. et al. (20010. Geomorphologic Evidence for Liquid Water. Space Science Reviews 96, 333–364.

McCubbin, F. M., Boyce, J. W., Srinivasan, P., Santos, A. R., Elardo, S. M., Filiberto, J., et al. (2016). Heterogeneous distribution of H2O in the Martian interior: Implications for the abundance of H2O in depleted and enriched mantle sources. Meteoritics & Planetary Science, 51, 2036–2060.

McEwen, A. S. et al. (2013). Recurring slope lineae in equatorial regions of Mars. Nature Geosci. 7, 53–58.

McIlroy, D., Green, O. & Brasier, M. (2001). Palaeobiology and evolution of the earliest agglutinated Foraminifera: Platysolenites, Spirosolenites and related forms Lethaia 34, 13–2

McKay, D.S., et al. (1996). Search for past life on Mars: possible relic biogenic activity in Martian meteorite ALH84001. Science 273: 924-930.

McKay, D.S., Thomas-Keprta, K.L., Clemett, S.J., Gibson Jr, E.K., Spencer, L. and Wentworth, S.J. (2009). Life on Mars: new evidence from martian meteorites. In, Instruments and Methods for Astrobiology and Planetary Missions, 7441, 744102.

McLennan, S. M. et al. (2014). Elemental Geochemistry of Sedimentary Rocks at Yellowknife Bay, 558 Gale Crater, Mars, Science, 343(6169). 1244734, doi:10.1126/science.1244734

Mcloughlin, N., H. Staudigel H. Furnes B. Eickmann M. Ivarsson (2010). Mechanisms of Microtunneling In Rock Substrates: Distinguishing Endolithic Biosignatures From Abiotic Microtunnels, Geobiology, 8, 245- 255.

Meessen, J., Backhaus, T., Sadowsky, A., Mrkalj, M., Sanchez, F.J., de la Torre, R., Ott, S. (2014). Effects of UVC254 nm on the photosynthetic activity of photobionts from the astrobiologically relevant lichens Buellia frigida and Circinaria gyrosa. Int J Astrobiol 13: 340-352.

Mei, M., Latif, K., Mei, C., et al. (2020). Thrombolitic clots dominated by filamentous cyanobacteria and crusts of radio-fibrous calcite in the Furongian Changshan Formation, North China, Sedimentary Geology, 395.

Melosh, H. J. (2003). Exchange of Meteorites (and Life?). Between Stellar Systems. Astrobiology, 3, 207-215.

Michalski, J. R., Niles, P. B. (2010). Deep crustal carbonate rocks exposed by meteor impact on Mars, Nature Geoscience, 3, 751-755.

Miller, K. G., Kominz , M. A. , Browning , J. V. et al. (2005). The Phanerozoic record of global sea-level change. Science, 310, 1293 –1298.

Milliken, R.E., (2010). The case for mixed-layered clays on Mars. Lunar Planet. Sci. Conf. 41, 2030.

Ming, D. W., et al. (2014). Volatile and organic compositions of sedimentary rocks in Yellowknife Bay, Gale Crater, Mars, Science, 343, 1245-1267, doi:10.1126/science.1245267.

Miyachi, S. et al. (2003). Historical perspective on microalgal and cyanobacterial acclimation to low- and extremely high-CO2 conditions. Photosynthesis Research, 77, 139 –153.

Mojzsis, S.J., Arrhenius, G., McKeegan, K.D., Harrison, T.M., Nutman, A.P., Friend, C.R.L., (1996). Evidence for life on Earth before 3,800 million years ago. Nature 384, 55-59.

Molina, E., Arenillas, I., Arz, J.A. (1996). The Cretaceous/Tertiary boundary mass extinction in planktic foraminifera at Agost, Spain. Rev. Micropaléont., 39(3).:225-243.

Mollenhauer, D., et al. (1999). Macroscopic cyanobacteria of the genus Nostoc: a neglected and endangered constituent of European inland aquatic biodiversity. European Journal of Phycology 34, 349-360.

Moment, G.B. (1949). On the relation between growth in length, the formation of new segments, and electric potential in an earthworm. J Exp Zool 112:1-12.

Murchie, S. L., et al. (2009). A synthesis of Martian aqueous mineralogy after 1 Mars year of observations from the Mars Reconnaissance Orbiter, J. Geophys. Res., 114, E00D06,

Mustard, J.F., et al. (2012). Sequestration of volatiles in the Martian crust through hydrated minerals: A significant planetary reservoir of water, in 43rd Lunar and Planetary Sci. Conf., Abstract No. 1539, Lunar and Planetary Institute (LPI). Houston, Tex.

Nachon, M., et al. (2014). Calcium sulfate veins characterized by ChemCam/Curiosity at Gale crater, 575 Mars, J. Geophys. Res. Planets, 119(9). 2013JE004588, doi:10.1002/2013JE004588.

Nemchin, A.A., et al. (2008). A light carbon reservoir recorded in zircon-hosted diamond from the Jack Hills. Nature 454, 92-95.

Nguyen, A.V., et al. (2014). Microbial Analysis of Australian Dry Lake Cores; Analogs For Biogeochemical Processes, American Geophysical Union, Fall Meeting 2014, abstract id.P33C-4039.

Nicholson, W.L., et al. (2012). Growth of Carnobacterium spp. from permafrost under low pressure, temperature, and anoxic atmosphere has implications for Earth microbes on Mars. PNAS.

Nisbet, E.G. (1987). Young Earth: An Introduction to Archean Geology, Springer Science and Business Media, Netherlands.

Noffke, N. (2015). Ancient Sedimentary Structures in the < 3.7b Ga Gillespie Lake Member, Mars, That Compare in macroscopic Morphology, Spatial associations, and Temporal Succession with Terrestrial Microbialites. Astrobiology 15(2).: 1-24.

Nuding, D. L. et al. (2014). Deliquescence and efflorescence of calcium perchlorate: An investigation of stable aqueous solutions relevant to Mars. Icarus 243, 420–428.

Occhipinti, A., De Santis, A., and Maffei, M. E. (2014). Magnetoreception: an unavoidable step for plant evolution? Trends Plant Sci. 19, 1-4.

Oehler, D. Z. (2013). A Periglacial Analog for Landforms in Gale Crater, Mars. Technical Report, Lunar and Planetary Science Conference; March 18, 2013 - March 22, 2013; The Woodlands, TX; United States

Olsson-Francis, K. and Cockell, C.S. (2010). Use of cyanobacteria for in-situ resource use in space applications. Planetary and Space Science 58:1279-1285.

Olsson-Francis, K., de la Torre, R., Towner, M.C., Cockell, C.S. (2009). Survival of akinetes (resting-state cells of cyanobacteria). in low Earth orbit and simulated extraterrestrial conditions. Orig Life Evol Biosph 39:565–579.

O'Neil, J., Carlson, R. W., Francis, E., Stevenson, R. K. (2008). Neodymium-142 Evidence for Hadean Mafic Crust Science 321, 1828 - 1831.

Ong, B. L., Lim, M., Wee, Y. C. (1992). Effects of desiccation and illumination on photosynthesis and pigmentation of an edaphic population of Trentepohlia odorata (Chlorophyta). Journal of Phycolog , 28, 768 –772.

Onofri, S., et. al. (2012). Survival of rock-colonizing organisms after 1.5 years in outer space." Astrobiology. 12, 508-516.

Onofri, S., et. a.1 (2018). Survival, DNA, and Ultrastructural Integrity of a Cryptoendolithic Antarctic Fungus in Mars and Lunar Rock Analogues Exposed Outside the International Space. Astrobiology, 19, 2.

Onofri, S., Selbman, L., Pacelli, C. et al. (2019). Survival, DNA and ultrastructural integrity of a cryptoendolithic Antarctic fungus on Mars and lunar rock analogues exposed outside the International Space Station. Astrobiology, 19, 2, DOI:10.1089/ast.2017.1728.

Osman, S., Peeters, Z., La Duc, M.T., Mancinelli, R., Ehrenfreund, P., Venkateswaran, K., (2008). Effect of shadowing on survival of bacteria under conditions simulating the Martian atmosphere and UV radiation. Applied and Environmental Microbiology 74, 959-970.

Palucis, M. C., W. E. Dietrich, A. G. Hayes, R. M. Williams, S. Gupta, N. Mangold, H. Newsom, C. Hardgrove, F. Calef, and D. Y. Sumner (2014). The origin and evolution of the Peace Vallis fan system that drains to the Curiosity landing area, Gale Crater, Mars, J. Geophys. Res. Planets, 119, 705– 728, doi:10.1002/2013JE004583.

Pattanaik, B., Schumann,R., Karsten, U. (2007). Effects of Ultraviolet Radiation on Cyanobacteria and Their Protective Mechanisms. In: J. Seckbach (ed.). Algae and Cyanobacteria in Extreme Environments, 47–58. Springer.

Pawlowski, J., Holzmann, M., Berney, C., Fahrni, J., Cedhagen, T. & Bowser, S. S. (2002). Phylogeny of allogromiid Foraminifera inferred from SSU rRNA gene sequences J. Foraminiferal Res. 32, 334–343.

Pawlowski, J., Holzmann, M., Berney, C., Fahrni, J., Gooday, A.J., Cedhagen, T., Habura, A., and Bowser, S.S. (2003). PNAS 100 (20). 11494-11498.

Pelkey, S. M., et al. (2004). Surficial properties in Gale Crater, Mars, from Mars Odyssey THEMIS data, Icarus, 167, 244–270.

Payre, V., Fabre, C., Cousin, A., et al. (2017). Alkali trace elements in Gale crater, Mars, with ChemCam: Calibration update and geological implications, JGR Planets, 122, 650-679.

Pierangelini,M., et al. (2017). Terrestrial adaptation of green algae Klebsormidium and Zygnema (Charophyta) involves diversity in photosynthetic traits but not in CO2acquisition; Planta. 2017; 246(5).: 971–986.

Planavsky, N., Grey, K. (2008). Stromatolite branching in the Neoproterozoic of the Centralian Superbasin, Australia: an example of shifting sedimentary and microbial control of stromatolite morphology: Geobiology, v. 6, 33–45.

Plaut, J. J., et al. (2007). Subsurface radar sounding of the south polar layered deposits of Mars, Science, 316, 92–95.

Polgári, M., (2006). Geochemical Aspect of Chemolithoautotrophic Bacterial Activity in the Role of Black Shale Hosted Mn Mineralization, Jurassic Age, Hungary, Europe, Clay Science, 12, 233-239.

Rabb, H. (2015). Life on Mars - Visual Investigation. https://www.scribd.com/doc/288486718/Life-on-Mars-Visual-Investigation. Scrib D. publishers.

Rabb, H. (2018). Life on Mars, Astrobiology Society, SoCIA, University of Nevada, Reno, USA. April 14, 2018.

Rabbow, E., Rettberg, P., Parpart, A., Panitz, C., Schulte, W., Molter, F., Jaramillo, R., Demets, R., Wei, P., and Wilnecker, R. (2017). EXPOSE-R2: the astrobiological ESA mission on board of the International Space Station. Front Microbiol 8.

Raggio, J., et aal. (2011). Whole lichen thalli survive exposure to space conditions: results of Lithopanspermia experiment with Aspicilia fruticulosa. Astrobiology. 2011 May;11(4).:281-92.

Rahmonov, O., Piatek, J. (2007). Sand colonization and initiation of soil development byCyanobacteria and algae. Ekológia (Bratislava) Vol. 26, No. 1, p. 52–63.

Ramos GJP, Branco LHZ, Moura CWN (2019). Cyanobacteria from bromeliad phytotelmata: new records,morphological diversity, and ecological aspects from northeastern Brazil. Nova Hedwigia 118(1–2).: 51–72.

Ran, E., et al. (1999). Mineral deposits in cells of Hookeria lucens, Journal of Bryology, 21, 281-288

Rapina, W. P et al. (2016). Hydration state of calcium sulfates in Gale crater, Mars: Identification of bassanite veins, Earth and Planetary Science Letters, 452, 197-205.

Rawlings, R. E., Dew, D., Plessis, C., (2003). Biomineralization of metal-containing ores and concentrates, Trends in Biotechnology, 21, 38-44.

Ray, J.G., Krishnan, J., Unni K.S., Shobha, V. (2009). Physico-chemical environmental complex of a commercially exploited tropical freshwater system within a wildlife sanctuary, Kerala, India. Ecology and Noospherology, 20, (3-4). 124-144.

Ray, J. G., Thomas, B. (2012). Ecology and Diversity of Green-algae of Tropical Oxic Dystrustepts Soils in Relation to Different Soil Parameters and Vegetation, Research Journal of Soil Biology, 4 (3). 42-68

Reid, RP, Visscher, PT, Decho, AW, Stolz, JF, Bebout, BM, Dupraz, C, MacIntyre, IG, Paerl, HW, Pinckney, JL, Prufert-Bebout, L, Steppe, TF and Desmarais, DJ 2000, The role of microbes in accretion, lamination and early lithifi cation of modern marine stromatolites, Nature, 406, 989–992.

Reitner, J., Pual, J., Arp, G., Hause-Reitner, D. (1996). Lake Thetis Dome Microbialities - A Complex Framework of Calicdified Biofilms and Organomicrites. In Reitner J., et al. (Eds). Global and Regional Controls on Biogenic Sedimentation. Gottingen, 85-89.

Renno, N. O., et al. (2009). Physical and Thermodynamical Evidence for Liquid Water on Mars, Lunar and Planetary Science Conference, Houston, March 23-27.

Riaz, M., Xiao, E., Latif, K. et al. (2019). Sequence-Stratigraphic Position of Oolitic Bank of Cambrian in North China Platform: Example from the Kelan Section of Shanxi Province, Arab J Sci Eng (2019). 44: 391.

Riding, R.E. (1999). The term stromatolite: towards an essential definition. Lethaia 32:321–330.

Rindi, F. (2010). Terrestrial green algae: systematics, biogeography and expected responses to climate change, Climate Change, Ecology and Systematics, ed. Trevor, R. et al. Cambridge University Press.

Rindi, F. (2007). Diversity, distribution and ecology of green algae and cyanobacteria in urban habitats. In Algae and Cyanobacteria in Extreme Environments, ed. J. Seckbach, Dordrecht, Springer, 571–582.

Rindi, F. and Guiry, M. D. (2002). Diversity, life history and ecology of Trentepohlia and Printzina (Trentepohliales, Chlorophyta). in urban habitats in western Ireland. Journal of Phycology, 38, 39 –54.

Rindi, F., Guiry, M. D. (2004). Composition and spatial variability of terrestrial algal assemblages occurring at the bases of urban walls in Europe. Phycologia, 43, 225 –235.

Rizzo, V. (2020). Why should geological criteria used on Earthnot be valid also for Mars? Evidence of possiblemicrobialites and algae in extinct Martian lakeslakes. International Journal of Astrobiology, 1–12. https://doi.org/10.1017/S1473550420000026

Rizzo, V., Cantasano, N. (2009). Possible organosedimentary structures on Mars. International Journal of Astrobiology 8 (4).: 267-280.

Rizzo, V., Cantasano, N. (2016). Structural parallels between terrestrial microbialites and Martian sediments. International Journal of Astrobiology, doi:10.1017/S1473550416000355.

Robbins, S., Hynek, B. M. (2012). A new global database of Mars impact craters ≥1 km: 2. Global crater properties and regional variations of the simple-to-complex transition diameter, Journal of Geophysical Research Atmospheres 117, 6001-6011.

Rosing, M. T., (1999). C-13-depleted carbon microparticles in > 3700-Ma sea-floor sedimentary rocks from west Greenland. Science 283, 674-676.

Rosing, M. T., Frei, R., (2004). U-rich Archaean sea-floor sediments from Greenland - indications of > 3700 Ma oxygenic photosynthesis. Earth and Planetary Science Letters 217, 237-244.

Rossi F., De Philippis R. (2016). Exocellular Polysaccharides in Microalgae and Cyanobacteria: Chemical Features, Role and Enzymes and Genes Involved in Their Biosynthesis. In: Borowitzka M., Beardall J., Raven J. (eds). The Physiology of Microalgae. Developments in Applied Phycology, vol 6. Springer,

Ruffi, W., Farmer, J.D., (2016). Silica deposits on Mars with features resemblinghot spring biosignatures at El Tatio in Chile. Nature Communications, 7: 13554, DOI: 10.1038/Ncomms13554.

Sankaran, A. V., (2002). The controversy over early-Archaean microfossils, Current Science, 83, 15-17.

Sakakibara, M., et al. (2014). Filamentous microbial fossil from low-grade metamorphosed basalt in northern Chichibu belt, central Shikoku, Japan, Planetary and Space Science, 95, 84-93.

Sallstedt T., et al. (2018). Evidence of oxygenic phototrophy in ancient phosphatic stromatolites from the Paleoproterozoic Vindhyan and Aravalli Supergroups, India. Geobiology 16 (2).: 139-159; doi: 10.1111/gbi.12274

Samuel, P. et al. (2014). Identification of the perchlorate parent salts at the Phoenix Mars landing site and possible implications, Icarus, 232, 226-231

Samylina, O.S., Zaytseva, L.V. & Sinetova, M.A (2016). Participation of algal–bacterial community in the formation of modern stromatolites in Cock Soda Lake, Altai Region, Paleontol. J. (2016). 50: 635.

Sanchez, F. J., et al. (2012). The resistance of the lichen Circinaria gyrosa (nom. provis.). towards simulated Mars conditions-a model test for the survival capacity of an eukaryotic extremophile." Planetary and Space Science, 2012, 72(1). 102-110.

Sancho L. G., et al. (2007). Lichens Survive in Space: Results from the 2005 LICHENS Experiment Astrobiology. 7, 443-454.

Sanz-Montero, M. E., Rodríguez-Aranda, J.P. (2009). Silicate Bioweathering And Biomineralization In Lacustrine Microbialites: Ancient Analogues from The Miocene Duero Basin, Spain, Geological Magazine, 146, 527-539.

Sautter, V., et. al., (2014). Igneous mineralogy at Bradbury Rise: The first ChemCam campaign at Gale crater, J. Geophys. Res. Planets, 119, 30–46, doi:10.1002/2013JE004472.

Scalzi, G., et al. (2012). LIFE Experiment: Isolation of Cryptoendolithic Organisms from Antarctic Colonized Sandstone Exposed to Space and Simulated Mars Conditions on the International Space Station.

Schirrmeister, B.E, Gugger M., Donoghue, P.C.J. (2015). Cyanobacteria and the great oxidation event. Palaeontology 1-17.

Schmidt, M., J. Campbell, R. Gellert, G. Perrett, A. Treiman, D. Blaney, A. Olilla, F. Calef, L. et al. (2014). Geochemical diversity in first rocks examined by the Curiosity Rover in Gale Crater: Evidence for and significance of an alkali and volatile□rich igneous source, J. Geophys. Res. Planets, 119, 64–81.

Schmidt, M., J., et al. (2014). Geochemical diversity in first rocks examined by the Curiosity Rover in Gale Crater: Evidence for and significance of an alkali and volatile□ rich igneous source, J. Geophys. Res. Planets, 119, 64–81.

Schoenberg, R., Kamber, B.S., Collerson, K.D., Moorbath, S. (2002). Tungsten isotope evidence from approximately 3.8-Gyr metamorphosed sediments for early meteorite bombardment of the Earth. Nature 418, 403-405.

Schulze□Makuch, D., et al. (2005.). Scenarios for the evolution of life on Mars, Journal of Geophysical Research: Planets, 110, E12.

Schwenzer, S. P., Kring, D. A. (2009). Impact-generated hydrothermal alteration on Mars: clay minerals, oxides, zeolites, and more. In: 40th Lunar and Planetary Science Conference, 23-27 Mar 2009, Houston, TX, USA.

Schwenzer, S. P., et al. (2012). Gale Crater: Formation and post□impact hydrous environments, Planet. Space Sci., 70, 84–95.

Sen Gupta, B.K. (1999). Systematics of moder Foraminifera. In: Modern Foraminifera. Springer, Sen Gupta, B (Ed).. (Springer, Dordrecht, The Netherlands). pp. 7–36.

Sharma, M., Shukla, B.(2019). Akinetes from late paleoproterozoic Salkhan limestone (> 1600 Ma). of India: A proxy for understanding life in extreme conditions. Frontiers Microbiology 10: 397, DOI:10.3389/fmicb.2019.00397.

Sheath, R.G., Cole, K.M. (1992). Biogeography of stream macroalgae in North America. Journal of Phycology 28:448-60.

Siebach, K.L., Grotzinger, J. P. (2014). Volumetric estimates of ancient water on Mount Sharp based on boxwork deposits, Gale Crater, Mars, JGR Planets, 119, 189-198.

Small, L. W, (2015). On Debris Flows and Mineral Veins - Where surface life resides on Mars. https://www.scribd.com/doc/284247475/On-Debris-Flows-eBook

Smith, H.D., Baque, M., Duncan, A.G., et al. (2014). Comparative analysis of cyanobacteria inhabiting rocks with different light transmittance in the Mojave Desert: A Mars terrestrial analogue. International Journal of Astrobiology 13:27-277.

Staudigel H., et al. (2008). 3.5 Billion years of glass bioalteration: volcanic rock as a basis for microbial life? Earth-Science Research, 12, 4-9.

Steele, L. J. et al. (2017). The water cycle and regolith-atmosphere interaction at Gale crater, Mars. Icarus, 289 pp. 56–79.

Stivaletta, N., Barbieri, R. (2009). Endolithic microorganisms from spring mound evaporate deposits (southern Tunisia). Journal Arid Environment 73:31-19.

Stivaletta, N., Barbieri, R., Billi, D. (2012). Microbial colonization of the salt deposits in the driest place of the Atacama Desert (Chile). Origins of Life and Evolution of Biospheres 42:187-200.

Stolper, E.M., et al. (2013). The petrochemistry of Jake_M: A Martian mugearite, Science, 341 (6153). doi:10.1126/science.1239463.

Sutter, B., et al. (2012). The detection of carbonate in the Martian soil at the Phoenix landing site: A laboratory investigation and comparison with Thermal and Evolved Gas Analyzer (TEGA). data, Icarus, 218, 290–296.

Tan, Y.H., Lim P.E., Beardall J., Poong S.W., Phang S.M. (2019). A metabolomic approach to investigate the effects of ocean acidification on a polar microalga Chlorella sp. Aquatic Toxicology DOI: 10.1016/j.aquatox.2019.105349

Tang, C.M. and Roopnarine, P.D. (2003). Evaporites, water, and life, Part I: Complex morphological variability in complex evaporitic systems - Thermal spring snails from the Chihuahuan Desert, Mexico. Astrobiology 3 (3).: 597-607.

Tappan, H. & Loeblich, A. R., Jr. (1988). Foraminiferal evolution, diversification, and extinction J. Paleontol. 62, 695–697

Taylor, S. R., McLennan, S. (2009). Planetary Crusts: Their Composition, Origin and Evolution, Cambridge Univ. Press, Cambridge, U. K.

Templeton, A. S.; Mayhew, L.; McCollom, T.; Trainor, T. (2011). Microbial Fe biomineralization in mafic and ultramafic rocks, American Geophysical Union, Fall Meeting 2011.

Thomas-Keprta K.L., et al. (2002). Magnetofossils from Ancient Mars: A Robust Biosignature in the Martian Meteorite ALH84001. Applied and Environmental Microbiology 68, 3663-3672.

Thomas-Keprta, K. L., et al. (2009). Origins of magnetite nanocrystals in Martian meteorite ALH84001. Geochimica et Cosmochimica Acta, 73, 6631-6677.

Thomson, B. J., et al. (2011). Constraints on the origin and evolution of the layered mound in Gale Crater, Mars using Mars Reconnaissance Orbiter data, Icarus, 214, 413–432.

Titus, T.N., Hugh H. Kieffer1, Phillip R. Christensen (2003). Exposed Water Ice Discovered near the South Pole of Mars, Science. 299, 1048-1051, DOI: 10.1126/science.1080497

Trainor, F. R., Gladych, R. (1995). Survival of algae in desiccated soil: a 35-year study. Phycologia , 34 , 191 –192.

Treiman A.H. (2003). The Nakhla martian meteorite is a cumulate igneous rock: Comment on Varela et al. (2001). Mineralogy and Petrology 77 , 271-277.

Trieman, A. H., Bish, D. L., Vaniman, D. T., et al. (2016). Mineralogy, provenance, and diagenesis of a potassic basaltic sandstone on Mars: CheMin X☐ray diffraction of the Windjana sample (Kimberley area, Gale Crater). JGR Planets, 121, 75-106.

Treiman, A. H., & Essen, E. J. (2011). Chemical composition of magnetite in Martian meteorite ALH 84001: Revised appraisal from thermochemistry of phases in Fe-Mg-C-O. Geochimica et Cosmochimica Acta, 75, 5324-5335.

Treiman, A.H., et al. (2016). Mineralogy, provenance, and diagenesis of a potassic basaltic sandstone on Mars: CheMin X-ray diffraction of the Windjana sample (Kimberley area, Gale Crater). JGR Planets, 121, 75-106.

Tugay, T. Zhdanova, N.N., Zheltonozhsky, V., Sadovnikov, L., Dighton, J. (2006). The influence of ionizing radiation on spore germination and emergent hyphal growth response reactions of microfungi, Mycologia, 98(4). 521-527.

Tuller, S.E., (1968). World distribution of mean monthly and annual precipitable water. Mon Weather Rev 90, 785-797.

Twitchett, R.J. (2006). The palaeoclimatology, palaeoecology and palaeoenvironmental analysis of mass extinction events. Palaeogeogr., Palaeoclimat., Palaeoecol. 232, 190-213.

Tyler, S.A., Barghoorn, E.S. (1963). Ambient pyrite grains in Precambrian cherts. American Journal of Science 261, 424–432.

Uyeda, J. C., Harmon, L. J., Blank, C. E. (2016). A Comprehensive Study of Cyanobacterial Morphological and Ecological Evolutionary Dynamics through Deep Geologic Time, Plos One, 11.

Van Driessche, A. E. S., et al. (2012). The Role and Implications of Bassanite as a Stable Precursor Phase to 606 Gypsum Precipitation, Science, 336(6077). 69–72, doi:10.1126/science.1215648.

Vaniman, D. T., Bish, D. L., Ming, D. W., et al. (2014). Mineralogy of a mudstone at Yellowknife Bay, Gale Crater, Mars, Science, 343(6169). doi:10.1126/science.1243480.

Vannier, J., Gaillard, I., Christian, G., Żylińska A. (2010). Priapulid worms: Pioneer horizontal burrowers at the Precambrian-Cambrian boundary Geology 38(8).:711-714. DOI: 10.1130/G30829.1

Velbel, M. A., (2012). Aqueous Alteration In Martian Meteorites: Comparing Mineral Relations In Igneous-Rock Weathering Of Martian Meteorites Sedimentary Geology of Mars SEPM Special Publication No. 102, Copyright 2012 SEPM (Society for Sedimentary Geology). Print ISBN 978-1-56576-312-8, CD/DVD ISBN 978-1-56576-313-5, p. 97–117

Verseux, C., Baque, M., Cifariello, R. et al. (2017). Evaluation of the resistance of Chroococcidiopsis spp. to sparsely and densely ionizing radiation. Astrobiology 17:118-125.

Vitek, P., Jehlicka, J., Ascaso, C. et al. (2014). Distribution of scytonemin in endolithic microbial communities from halite crusts in the hyperarid zone of the Atacama Desert, Chile. FEMS Microbial Ecology 90:351-366.

Vítek, P., Ascaso, C., Artieda, O. et al. (2017). Discovery of carotenoid red-shift in endolithic cyanobacteria from the Atacama Desert. Sci Rep 7, 11116.

Wacy, D. (2010). Stromatolites in the ☐3400 Ma Strelley Pool Formation, Western Australia: Examining Biogenicity from the Macro- to the Nano-Scale, Astrobiology, 10, https://doi.org/10.1089/ast.2009.0423

Wacey, D., Saunders, M., Kong, C., Brasier, A., Brasier, M. (2016). 3.46 Ga Apex chert 'microfossils' reinterpreted as mineral artefacts produced during phyllosilicate exfoliation. Gondwana Research 36, 296-313.

Wacey, D., Kilburn, M.R., McLoughlin, N., Parnell, J., Brasier, M.D. (2008). Using NanoSIMS in the search for early life on Earth: ambient inclusion trails in a c. 3400 Ma sandstone. Journal of the Geological Society of London 165, 43–53.

Weber, B., Wessels, D.C.J., Büdel, B. 1996. Biology and ecology of cryptoendolithic cyanobacteria of a sandstone outcrop in the Northern Province, South Africa. Arch. Hydrobiol., Suppl. 117, Algological Studies, 83,565-579.

White, L. M., et al. Putative Indigenous Carbon-Bearing Alteration Features in Martian Meteorite Yamato 000593, Astrobiology, 14, No. 2

Wierzchos, J., Ascaso, C., McKay, C.P. (2006). Endolithic cyanobacteria in halite rocks from the hyperarid core of the Atacama Desert. Astrobiology 6:415-422.

Wierzchos, J., De los Ríos, A., Ascaso, C. (2011). Microorganisms in desert rocks: the edge of life on Earth, International Microbiology 15:171 DOI: 10.2436/20.1501.01

Williams, R. M. E., et al. (2013). Martian fluvial conglomerates at Gale Crater, Science, 340, 1068–1072.

Williams, W., Chilton, A., Schneemilch, M., Williams, S., Neilan, B., Colin Driscoll, C. (2019). Microbial biobanking – cyanobacteria-rich top soil facilitates mine rehabilitation. Biogeosciences, 16: 2189–2204.

Wilson, M. A., Taylor, P. D., (2017). Exceptional Pyritized Cyanobacterial Mats Encrusting Brachiopod Shells from The Upper Ordovician (Katian). of The Cincinnati, Ohio, Region, Palaios, 32m 673-677.

Wilson, M. J. (2004). Weathering of the primary rock-forming minerals: processes, products and rates, Clay Minerals, journal of Fine Particle Science, 3. 303-306.

Wilson MJ, Jones D, McHardy WJ (1981). The weathering of serpentinite by Lecanora atra. Lichenologist 13, 167–176.

Wilson M.J., Jones D. (1984). The occurrence and significance of manganese oxalate in Pertusaria corallina (Lichenes). Pedobiologia, 26, 373-379.

Wong, C-Y., Teoh, M-L., Phang, S-M., Lim, P-E., Beardall, J. (2015). Interactive Effects of Temperature and UV Radiation on Photosynthesis of Chlorella Strains from Polar, Temperate and Tropical Environments: Differential Impacts on Damage and Repair. PLoS ONE 10(10).

Wray, J.J., et al. (I2016). Orbital evidence for more widespread carbonate☐bearing rocks on Mars, JGR Planets, 121, 652-677.

Xue, Y., Jin, S. (2013). Martian minerals components at Gale crater detected by MRO CRISM hyperspectral images, 2013 2nd International Symposium on Instrumentation and Measurement, Sensor Network and Automation, DOI: 10.1109/IMSNA.2013.6743465.,

Zakharova, K., et al. (2014). Protein patterns of black fungi under simulated Mars-like conditions. Scientific Reports, 4, 5114.

Zhdanova, N.N., et al. (2004). Ionizing radiation attracts soil fungi. Mycol Res. 2004, 108: 1089-1096.

Zorzano, M-P., Mateo-Martí, E., Prieto-Ballesteros, O., Osuna, S., Renno, N. (2009). Stability of liquid saline water on present day Mars. Geophys. Res. Lett. 36, L20201.

IV. LIFE ON MARS: COLONIES OF PHOTOSYNTHESIZING MUSHROOMS IN THE EAGLE CRATER. THE HEMATITE HYPOTHESIS REFUTED

Rhawn Joseph1*, R. A. Armstrong2, G. J. Kidron3, Rudolf Schild4

1Astrobiology Associates of California

2Aston University, Birmingham, UK.

3Institute of Earth Sciences, The Hebrew University of Jerusalem, Israel

4Dept. of Astrophysics, Harvard-Smithsonian, Cambridge

Abstract

Throughout its mission at Eagle Crater, Meridiani Planum, the rover Opportunity photographed thousands of mushroom-lichen-like formations with thin stalks and spherical caps, clustered together in colonies attached to and jutting outward from the tops and sides of rocks. Those on top-sides were often collectively oriented, via their caps and stalks, in a similar upward-angled direction as is typical of photosynthesizing organisms. The detection of seasonal increases and replenishment of Martian atmospheric oxygen supports this latter interpretation and parallels seasonal photosynthetic activity and biologically-induced oxygen fluctuations on Earth. Twelve "puffball" fungal-shaped Meridiani Planum spherical specimens were also photographed emerging from beneath the soil and an additional eleven increased in size over a three-day period in the absence of winds which may have contributed to these observations. Growth and the collective skyward orientation of these mushroom and fungus-like specimens are indications of behavioral biology. Reports claiming these Eagle Crater spheres consist of hematite are reviewed and found to be based on inference as the instruments employed were not hematite specific. The hematite-research group targeted oblong rocks which were mischaracterized as spheres, and selectively eliminated spectra from panoramic images until what remained was interpreted to resemble spectral signatures of terrestrial hematite photographed in a laboratory, when it was a "poor fit." The Eagle Crater environment was never conducive to creating hematite and the spherical hematite hypothesis is refuted. By contrast, lichens and fungi survive in Mars-like analog environments. There are no abiogenic processes that can explain the mushroom-morphology, size, colors and orientation and growth of, and there are no terrestrial geological formations which resemble these mushroom-lichen-shaped specimens. The evidence strongly supports the hypothesis that mushrooms, algae, lichens, fungi, and related organisms have colonized the Red Planet and may be engaged in photosynthetic activity and oxygen production on Mars.

Key Words: Lichens; Fungi; Algae; Mushrooms; Eagle Crater; Life on Mars; Astrobiology; Extremophiles; Mars Simulated Environments; Water on Mars; Martian Mushrooms; Hematite

Mushroom-Lichen-Like Specimens, Oxygen on Mars: The Hematite Hypothesis Refuted

During the first 100 days of its mission in various locations in Eagle Crater (Meridiani Planum), Mars, the rover Opportunity photographed thousands of mushroom-shaped, lichen-like specimens, with features that include stems and bulbous caps, a sample of which are presented here (Figures 1-9). These specimens are attached by thin stalks to the sides and tops of rocks, and those top-side are often collectively oriented in a similar upward-angled direction, jutting above these rocks, as might be expected of colonies of organisms engaged in photosynthesis. Moreover, in subsequent photographs, some specimens on the top-sides appear to change their orientation and bend and arch downward (Figure 9). Those on the sides and some on the tops of these rocks or upon the soil were often oriented horizontally or were bent downward as if due to the pull of gravity on their top-heavy bulbous caps.

There are no terrestrial analogs or abiogenic or weathering processes which can sculpt high density masses of mushroom-shapes with thin stalks and bulbous caps out of rock, salt, or sand, and which orient skyward, above their substrates, in the same or similar upward angled direction—as documented by an extensive abiotic-image search, using relevant key words (see Methods). In addition, weathering and winds would be expected to destroy not sculp these specimens if they were abiogenic.

Mars is often buffeted by powerful winds, and is seismically active (Banerdt et al. 2020), whereas these thin stems are an estimated 1-2 mm in diameter and up to 6 mm in length with top-heavy spherical caps. If consisting of sand, minerals or salt, then powerful winds, Mars-quakes, meteor strikes, or the turbulence created by Opportunity's wheels or drill (See Figure 3) would cause these thin stalks to fracture and break and these bulbous caps to tumble to the surface. Instead, they have remained standing, and are oriented upward, which suggests they recently developed and are in a state of continual renewal and engaged in photosynthesis. In favor of this hypothesis, the authors provide photographic evidence of 23 spherical specimens, photographed in Meridiani Planum, 12 of which emerged from beneath the soil and 11 which increased significantly in size over a three-day period (Figure 10).

Furthermore, a team of 14 established experts in astrobiology, astrophysics, biophysics, geobiology, microbiology, lichenology, phycology, botany, and mycology have identified specimens resembling terrestrial algae, lichens, fungi, and mushrooms in the Gale Crater (Joseph et al. 2020a), also located near the equator. Also observed were what appear to be open-cone-like gas-bubble vents--associated with photosynthesis-oxygen respiration (Bengtson et al. 2009; Sallstedt et al. 2018)--and which were photographed adjacent to mushroom and lichen-like surface features (Joseph et al. 2020a).

Oxygen has also been detected in the atmosphere and within soil samples on

Mars (Leshin et al. 2013; Ming et al. 2014; Rahmati et al. 2015; Sutter et al. 2017; Valeille et al. 2010). Although a variety of hypothetical abiogenic scenarios have been proposed which "could have contributed.... could have contributed... could contribute... could be a candidate..." (Hogancamp et al. 2018) for the generation of Martian oxygen "such as abiotic photosynthesis" (Franz et al. 2020) it is well established that the primary source of oxygen, on Earth, is via the photosynthetic activity of cyanobacteria (blue-green algae) and water living and land-based plants (Canfield 2014; Hall and Rao, 1986) including lichens (Vinyard et al. 2018; ted Veldhuis et al. 2020) which are fungal-algae composite organisms. Hence, there is substantial evidence of oxygen in the atmosphere and soil of Mars whereas surface features which resemble oxygen-gas vents adjacent to lichen-like formations have been observed in Gale Crater (Joseph et al. 2020a), and as detailed in this report, vast colonies of lichen-like specimens possibly engaged in photosynthesis have been observed in Eagle Crater and which may be respiring oxygen.

It's been inferred that the spheres of Eagle Crater, and by extension, the colonies of lichen-mushroom-sphere-shaped specimens, consist of hematite (Squires et al. 2004). However, a number of investigators have rejected the spherical hematite hypothesis (Burt, et al. 2005; Dass 2017; Joseph 2014; Knauth et al. 2005; Rabb 2018; Small 2015). In a presentation at the Lunar and Planetary Society and paper published in the journal Nature, Burt, Knauth and Woletz (2005) referred to the spherical hematite claims as "inappropriate." According to these scientists the hematite "interpretation for features observed at the Opportunity landing site on Mars contains so many contradictions and problems that an alternative explanation seems necessary.... unlike all known terrestrial concretions... they are uniformly spherical... uniform in their size distribution (concretions have no implicit restrictions as to maximum or minimum size), and uniform in their distribution in the rocks... The frequent analogy to hematitic spheroids is inappropriate" (see also Knauth et al. 2005).

Terrestrial spherical hematite does not have a mushroom-lichen-like shape or a bulbous cap atop an elongated stalk jutting upward from rocks as if engaged in photosynthesis; and which are the defining features of the specimens presented here. Moreover, there is no evidence that the stalked-mushroom-shaped specimens or a single isolated sphere lying loosely atop the soil within Eagle Crater, were individually or selectively examined and analyzed by Opportunity's suite of sampling instruments for the presence of hematite. Instead, individual samples inferred to contain hematite consisted of oblong rocks (see Figure 6 in Belle et al. 2004). Claims about hematite were also based on the spectral signatures of false colors (Soderblom et al. 2004), panoramic images, and claims about the averaging of high and low "temperatures" (Klingelhöfer et al. 2004) when the temperature sensors had failed (Glotch and Bandfield 2006); and with spectra selectively eliminated until what remained was interpreted as similar to the spectral signature of hematite photographed in a laboratory (Christensen et al. 2004), when the results were a "poor fit" for hematite and there were significant problems with calibration (Glotch and Banfield, 2006).

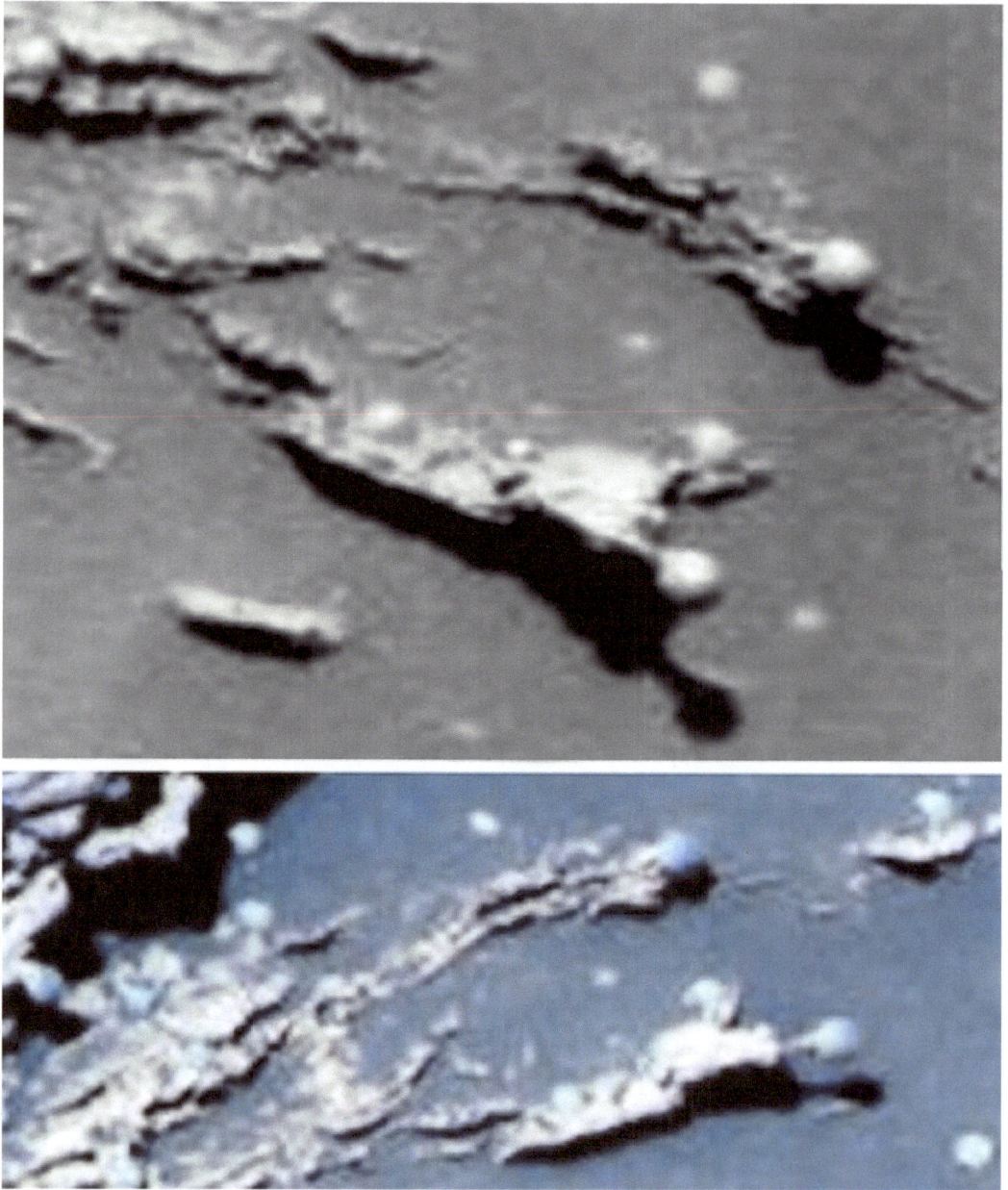

Figure 1. Opportunity - Sol 40 (top) Sol 37 (bottom). Note similar elevated angled orientation of mushroom-like specimens photographed growing on an unknown (fungi-like) substrate above the Martian surface in Eagle Crater. These "mushrooms" are up to 8 mm in length, with stems approximately 1 mm (or less) in width.

Figure 2. Opportunity - Sol 88. These "mushrooms" are up to 8 mm in length, with stems and apothecia approximately 1 mm to 3 mm in width, with what may be bulging hyphae along the rock surface. The bulbous cap may be a spore producing fruiting body. Note "bore hole" (see Figure 3).

As admitted by Glotch and Banfield (2006): "The gradual change of the instrument response function over the course of the mission combined with the failure of temperature sensors on the on-board calibration targets ...necessitated a change in... the instrument calibration... Figure 3b shows the Mini-TES hematite spectrum recovered using a magnetite-derived hematite target spectrum. There is a poor fit to the 450 cm 1band width and position of the emissivity minimum. Additionally, there is a poor fit to the 390 cm 1feature that is present in the test spectrum."

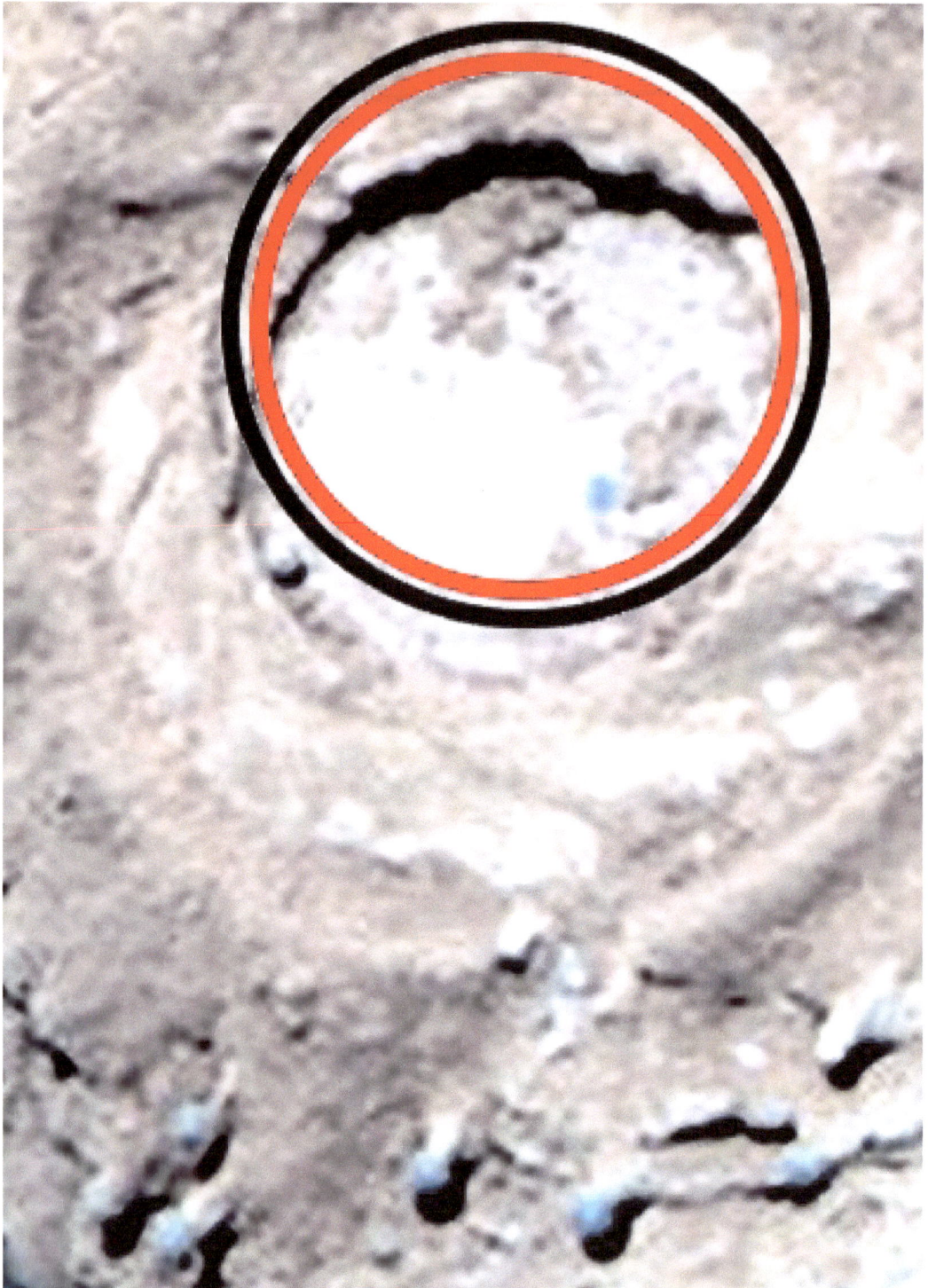

Figure 3. Opportunity - Sol 88. Bore hole drilled by the Opportunity's rotary blade (RAT) into the overlying rock. All but one of the "mushrooms" (lower left beneath the red circle) were destroyed by the RAT, except for their hollow stems/stalks 2-3 mm beneath the surface of the rock (Note center of red circle). The "mushroom" at the lower left of the circle protrudes from the surface (note shadow) indicating it was flexible and was pushed aside by the drill or it grew after the bore hole was fashioned.

178

The hematite hypothesis also rests upon the high concentration of iron detected within the soil (Bell et al 2004; Klingelhöfer et al. 2004, Squires et al. 2004). Lichens have high concentrations of iron (Bajpai et al. 2009; Hauck et al. 2007), and many species feed on iron (Bosea et al. 2009; Fredrickson et al. 2008; Gralnick & Hau 2007). The presence of iron does not prove the hematite hypothesis, but instead may provide a substrate for biological proliferation.

Furthermore, the Martian mushroom-shaped spheres atop rocks and upon the soil are a different color and smaller than terrestrial hematite (Bell et al. 2004; Soderblom et al. 2004), averaging 0.6 to 6 mm in size and diameter (Herkenhoff et al., 2004) which is also the characteristic size of a variety of terrestrial lichens (Armstrong 1981, 2017) including the specimens presented here. Nor does terrestrial hematite have a mushroom shape and stem and grow upward and outward from the tops of rocks.

Given the colors, size, favored locations, mushroom shapes, thin flexible stalks, large bulbous caps, evidence of growth, flexibility, and movement, and the collectively similar skyward angled-orientation of these colonies as if engaged in photosynthesis (Figures 2, 4-9), coupled with evidence of oxygen most likely produced secondary to photosynthesis, it is reasonable to conclude that the specimens presented in this report represent evidence of life on Mars.

Methods And Results

The Eagle impact crater is 22 meters in diameter, is likely several billion years in age, and located in a large plain known at Meridiani Planum. The rover Opportunity landed on Mars at 1.95°S 354.47°E, in Eagle Crater on January 25, 2004, 10 meters below the crater's rim. At near equatorial latitude there are about 12.4814 terrestrial hours of sunlight on the first day of summer and 12.2299 hours on the first day of Martian winter. Temperatures are estimated to reach highs of 20°C (68°F) during the summer to lows of -73°C (-99.4°F) at night (see http://quest.nasa.gov/aero/planetary/mars.html). However, the exact temperatures of Eagle Crater, are unknown due to failure of the Opportunity's temperature sensors.

Search of the Eagle Crater Image Data Base For Mushroom-Shaped Spheres

Methods: Throughout the first 100 Sols (Martian days) of its surface mission at Eagle Crater, the rover Opportunity transmitted to Earth several thousand images captured via its Microscopic Imager and Navigation and Panoramic Camera (see https://mars.nasa.gov/mer/gallery/all/opportunity.html). These included photographs of soil, crevices, rocks and thousands of mushroom-shaped and other spherical specimens.

Based on morphology and location, three different types of spherical specimens were observed; A) Thin stemmed specimens, topped with spherical caps (AKA Martian mushrooms) which (based on parameters provided by Herkenhoff et al. (2004) and Joseph et al, (2019) appear to be up to 6 mm in diameter, with stems up to 6 mm in length and less than 1 mm in diameter, attached to the tops of rocks jutting skyward and those on the sides of rocks, top-heavy with bulbous caps, oriented downward; B) Round and "lemon-shaped" spheres upon the soil surface, some with long stems, other with short stalks, and those with no discernible stalk (AKA "blue berries") up to 6 mm in diameter; C) Gray spheres embedded within thick wavy layers of what appears to be a calcium-cement-like matrix.

Unfortunately, neither NASA or the Opportunity team in their published reports provided any detailed metrics about these images or the specimens depicted, other than inferences and estimates as to the size of the surface spheres (Herkenhoff et al., 2004), and the estimated size of a few rocks and outcrops. Therefore, it was impossible to precisely determine the exact height, size, orientation, or density of the mushroom-like specimens which are the focus of this report.

Results: Based on surface features, 185 photos, photographed on 36 separate days in different locations, and depicting, collectively, several thousand stemmed-mushroom-shaped and other spherical specimens, were selected for detailed inspection. These 185 photos were enlarged by 300% and visually inspected to identify the presence of clearly discernible mushroom-lichen-like features which included a visible stalk topped with a spherical cap.

Several thousand specimens which resembled mushrooms and that were clustered together and attached via their stems to the tops of rocks, were photographed in various locations on 24 of the 100 days in Eagle Crater; i.e. Sol 28, 32, 35, 36, 37, 38, 39, 40, 41, 46, 50, 63, 69, 71, 73, 74, 80, 81, 84, 85, 86, 87, 88, 97. It was determined, based on a visual inspection of these photos, that the abundance of specimens with features similar to mushrooms appeared to be greatest in the west and crater rim facing the rising sun, and lowest in the crater floor. Thirty photos, photographed on Sol 37, 40, 81, 84, 85 and 88, were determined to depict the most obvious visual evidence of mushroom-lichen-like features. These were subject to additional visual inspection by all the authors of this report.

Illustrative examples photographed on Sol 37, 40, 85 and 88 via the rover Opportunity's Navigation and Panoramic Cameras are presented. Mushroom-shaped, lichen-like specimens, attached by stalks to the surface and upon rocks, photographed on Sols 37, 40 were observed to be oriented (pointed) in a similar upward-angled direction (Figure 1). Clusters of several dozen specimens, photographed on Sol 84 and attached by stalks atop a number or rocks, were also found to be directionally oriented at the same or a similar upward angle above these rocks. The same is true of thirty-six specimens photographed on Sol 88 on the topside of a single rock; and collectively, several hundred specimens photographed at various locations on Sol 85 and jutting upward above these rocks, are oriented, depending on location, at a similar skyward angle. These photos have been enlarged by 300% (Sol 37, 40) and 150% (Sol 85, 88). Based on published parameters (Herkenhoff et al., 2004; Joseph et al. 2019), these mushroom-lichen shaped specimens are estimated to range up to 8 mm in height and length.

Abiotic Image Search

Methods: To determine if there are any terrestrial abiotic structures which resemble these specimens, a Google and Bing image search was conducted by three of the authors, using A) key words "rocks" or "minerals" or "hematite" or "salt" or "sand" or "weathering" plus "mushroom" or "mushroom shape" or "domed" or "diapir" and B) by inserting Figures 1-2 into the Google "Search by image" function. Lastly, C) a "scholar.google" search was conducted, using the same key words, and the photos/figures from relevant articles examined.

Results: **Thousands of pictures of abiotic specimens were visually examine**d, including photos of salt diapir, hematite, serpentine, shale, and granitoid rock. Not one of these abiotic specimens resembled, in size, shape and form, the mushroom-lichen-like specimens photographed in Eagle Crater. The only terrestrial analogs for the specimens presented in this report are the fruiting bodies of mushrooms and lichens; i.e. living organisms.

Search for Life, Wind, Dust Storms, Dust Devils: Sol 1145 to 1148

Methods: It has been previously reported that 15 specimens similar to "puffballs" ("blue berries") have been photographed by the rover Opportunity in Meridiani Planum, increasing in size and emerging from beneath the coarse grained surface as based on comparisons of Sol 1145 and Sol 1148 (Joseph et al. 2019). Those authors interpreted this as biological growth but could not completely rule out wind. It's been estimated that the movement of coarse-grained Martian soil requires wind velocities of 70 m/s at least one m above the surface, but that velocities of 40 m/s may "occasionally" displace coarse-grained sand and soil (Jerolmack et al. 2006). Therefore, it's possible that a powerful wind may have uncovered these specimens and contributed to what appears to be growth.

To verify and replicate the observations of Joseph and colleagues (2019) and to rule out wind or other abiotic contributions to these observations, three of the authors searched the Opportunity Raw images data base. All photographs from the Panoramic Camera (Sol 1145, 1146, 1147, 1148), the Front Hazcam (Sols 1145, 1146, 1148), Navigation Camera (Sol 1146) and Microscopic Imager (Sols 1145, 1148) were visually examined for evidence of wind-blown dust in the air, dust devils, dust storms, or wind-driven soil displacement or buildup. NASA's data base was reviewed and a search was conducted for reports of any wind in Meridiani Planum on these dates.

Results: Comparing Microscopic Imager photographs on Sols 1145, 1148, reveals that 12 specimens emerged from beneath the coarse grained soil as they were not visible on Sol 1145; and that an additional 11 specimens increased in size. Therefore, in comparison to the 15 identified by Joseph et al (2019) an additional 8 specimens were observed to either emerge from beneath the soil or increase in size, for a total of 23. All surrounding soil in Sol 1145 and 1148 appears to be coarse (vs fine) grained with no evidence of displacement or buildup.

No winds or dust storms in Meridiani Planum were reported by NASA on Sols 1145, 1146, 1147, or 1148. Likewise, as based on a visual examination of all photos between Sol 1145 and 1148 there is no evidence or comparative evidence of wind, dust storms, or dust devils or the accumulation of dirt, sand, or dust. Nor was there any evidence of soil displacement, soil buildup or "filling in" or that soil is higher or lower on one side of any of the specimens as might be expected if subject to powerful directional winds.

Discussion
Martian Mushrooms, and Eagle Crater

Over forty experts have previously identified, by name, "puffballs," "mushrooms" and "lichens" that had been photographed in the Eagle Crater (Joseph 2016). In this report the authors have identified and presented over 200 specimens, a sample of thousands photographed within the Eagle Crater, which closely resemble mushroom-like organisms and lichens. These specimens range from 3 to 8 mm in length and diameter, have thin hollow stalks and bulbous caps; and colonies, including those on adjacent rocks, are angled upward, above these rocks via their stems, in a similar direction which is typical of photosynthesizing organisms. It was also noted that the density of mushroom-shaped spheres appears to be greatest in the northwest crater rim facing the rising sun, and lowest in the southeast crater floor as based on Sol-photograph dates as related to parameters provided by NASA. Moreover, typical of numerous stemmed/stalked plants and lichens these mushroom-like-stalks are hollow (Figure 3) and tubular; a finding incompatible with any abiotic explanation (e.g. hematite, salt, minerals), but which in terrestrial plants serves to draw up, distribute and store water and nutrients obtained from the soil.

In 1978, Levin, Straat and Benton reported "green patches" photographed during the 1976 Mars Viking Missions, which they believed might be lichens. The Viking Labeled Release experiments also detected activity consistent with biology at two locations, Utopia Planitia and Chryse Planitia, over 4,000 miles apart (Levin & Straat 1976, 1977); possibly that of lichens and algae (Levin et al. 1978). Condensation and sublimation of ground frost (Wall, 1981) and water within regolith was also detected via the Viking's suite of instruments (Biemann et al., 1977).

Joseph and colleagues (2020a) have also identified numerous specimens resembling green and blue-green algae, lichens, and open-cone-gas vents, photographed by the rover Curiosity in Gale Crater. This crater also appears to be subject to varying degrees of moisture and displays evidence of water pathways and a history of being filled with water.

The lichen-like species presented here were photographed by the rover Opportunity in Eagle Crater, located in Meridiani Planum which is 2 degrees south of the Martian equator in an area known as Terra Meridiani. Gale crater is also located near the equator. The equatorial region has a warmer climate than Utopia and Chryse Planitia perhaps reaching highs of 20°C (68°F) during the day to lows of -73°C (-99.4°F) at night (see http://quest.nasa.gov/aero/planetary/mars.html).

It's been hypothesized that Eagle Crater has been repeatedly exposed to flowing surface water and precipitation (Bell et al. 2004; Herkenhoff et al. 2004; Squyres et al. 2004). As theorized by Squyres and colleagues (2004): liquid water may have been abundant at Meridiani Planum which "suggests that conditions were suitable for biological activity for a period of time in Martian history." Thus, we see evidence of what may be Meridiani Planum stromatolites fashioned by cyanobacteria perhaps billions of years ago (Rizzo and Cantasano, 2009). In addition to bacteria, Squyres et al (2004) suggests that Eagle Crater could have been colonized by eukaryotic "filamentous microorganisms." The mushroom-shaped lichen-like formations presented in this report also appear to be "filamentous" as some have what may be hyphae extending along and bulging beneath the subsurface and which emerge as thin stalks topped by bulbous caps (Figures 2, 3).

The specimens presented in this report have been previously referred to as "Martian mushrooms" (Joseph 2014) and clearly resemble lichens (Dass, 2017, Joseph 2016, Joseph et al. 2019). Lichens are composite life forms consisting of a symbiotic relationship involving fungi (mycobiont) and algae/cyanobacteria (photobiont), the former of which is largely responsible for the lichens' mushroom shape, thallus, and fruiting bodies (Armstrong 1981, 2017, 2019; Armstrong and Bradwell 2010; Brodo et al. 2001). To speculate: the bulbous mushroom-like cap of the specimens presented here, may represent the fruiting body of the lichen whereas the remainder of the lichen inhabits the subsurface for which there is considerable evidence in the form of what may be bulging hyphae which snakes just beneath the surface (Figures 2-3).

Many of the mushroom-shaped specimens appear to lack a crustose thallus which is a lichen characteristic (Armstrong, 2017, 2019; Armstrong and Bradwell 2010; Kidron 2019). Martian organisms, however, would be expected to adapt and evolve in response to the unique Martian environment. The crustose thallus could be endolithic or buried in the surface layers of rock and soil (see Figures 1-3).

On the other hand, in contrast to the green-algae-like specimens of Gale Crater (Joseph et al. 2020a) and the observation of Levin et al (1978) who reported "green patches" in Viking photographs, algae-like specimens have not yet been observed in Eagle Crater or identified in Meridiani Planum. However, Figure 1(A) in Soderblom et al. (2004) depicts pools of "blue" completely surrounded by masses of compacted "green" sphericles on the floor of the Eagle Crater. If these were true colors, the obvious interpretation is the "blue" represents pools of water and the surrounding layers of "green" are green algae.

Figure 4. Opportunity - Sol 85. Top photo: 20-cm-wide rock (center, top) with specimens on sides of rocks oriented downward and those on tops of rocks oriented skyward; differential orientations possibly due to access to gravity, and direct sunlight and exposure to vs protection from wind. Bottom photo: sphere-shaped and mushroom-shaped specimens upon the Martian surface and on center-slab of rock, all directed oriented right-ward and perhaps top-heavy with fruiting bodies.

Figure 5. Opportunity - Sol 85. Colonies of lichen-mushroom-like specimens approximately 2 to 6 mm in length, photographed in Eagle Crater.

Figure 6. Opportunity - Sol: 85. Lichen-like specimens approximately 2 to 8 mm in length. Note similar orientation of specimens on tops of rocks, whereas those closer to the surface are also elevated and angled at a similar direction.

In contrast to Gale Crater (Joseph et al. 2020a) and the observation of Levin et al (1978) who reported "green patches" which could be lichens and green algae in Viking photographs, algae-like specimens have not been observed in Eagle Crater; and algae comprise the photobiont portion of the lichen (Armstrong, 2017). Therefore, these "Martian mushrooms" may be non-lichenised fungi. However, Figure 1(A) in Soderblom et al. (2004) depicts pools of "blue" completely surrounded by masses of compacted "green" sphericles on the floor of the Eagle Crater. If these were true colors, the obvious interpretation is the "blue" represents pools of water and the surrounding layers of "green" are green algae.

Sources of Water

There is every reason to suspect that Eagle Crater, where these mushrooms features were photographed, may be periodically exposed to ground water and water-mist precipitation. There are indications that water on Mars may be stored in underground aquifers (Malin and Edgett 2000), and sequestered in Martian rocks, hydrated minerals, or locked within frozen ground (Plaut et al., 2007; Mustard et al., 2012; Kieffer et al., 1976; Farmer et al., 1977). Martian rocks and regolith, which are porous with crevices, cracks, and voids, also appear to contain water ice (Biemann et al., 1977; Mellon and Phillips 2001).

Depending on seasonal-orbital and temperature variations, water frozen within top soils, rocks, and regolith, may melt. Humidity and rising temperatures may also increase subsurface/surface pressures thereby forcing ice to liquify and water to pool upon the surface (Mellon and Phillips 2001) as depicted in figure 1(A) of Soderblom et al. (2004) which shows pools of blue surrounded by green on the floor of Eagle Crater. Surface water would then seep back beneath the surface, or turn to mist or freeze and for which there is documented evidence in the wheel wells of the rover Curiosity; i.e. frozen pure water ice (Joseph et al, 2020b).

Four major reservoirs of water have also been identified, based on data provided by the orbital Atmospheric Chemistry Suite, and the Mars Science Laboratory and it's Environmental Monitoring Station, i.e. in the northern and south poles (Kieffer et al., 1976; Farmer et al., 1977), in Martian clouds (Spinrad et al. 1963; Masursky et al., 1972; Whiteway et al., 2009; Moores et al. 2015) which likely consist largely of water as do the clouds of Earth (Pruppacher and Klett 2010; Hu et al. 2010); within atmospheric vapors (Farmer et al., 1977; Korablev et al., 2001; Smith et al., 2001) and in the upper atmosphere which is subject to "large, rapid seasonal intrusions of water" (Fedorova et al. 2020).

For example, Fedorova and colleagues (2020)--employing three infrared spectrometers which are part of the Atmospheric Chemistry Suite on the ExoMars Trace Gas Orbiter spacecraft-- examined atmospheric spectra between 15 to 100 km above the surface to examine water vapor profiles. During the Spring and Summer, in association with dust storms, water levels increased to supersaturation and with thick ice clouds forming 15 to 40 km above and supersaturated layers 80 to 100 km and intermittently 50 to 60 km above the southern and northern hemisphere. They determined that "large portions of the atmosphere are in a state of supersaturation" thus replicating the findings of other scientists (Maltagliati et al. 2011; Todd et al. 2017)

Columns of water vapor have been observed every spring and summer from orbit (Read and Lewis, 2004; Smith, 2004; Todd et al. 2017), and which are transported from the north toward the equator (and thus to Eagle Crater) by southerly winds (Harri et al. 2014). Moreover, these vapors have a precipitable water content of at least 10–15 pr μm (Smith, 2004), and depending on humidity

(Harri et al. 2014) appear to reach saturation in the early morning hours thereby inducing a mist-like precipitation which may provide moisture to organisms dwelling in Eagle Crater.

Photosynthesis and Seasonal Fluctuations in Martian Oxygen

Cyanobacteria (algae) produce oxygen via photosynthesis (Graham et al. 2016). It is believed that early in the course of evolution, Earthly eukaryotes acquired, via horizontal gene transfer, cyanobacterial genes, which triggered the development of pigmented plastids and organells that made it possible for plants and algae-symbiotes to evolve, engage in photosynthesis and secrete oxygen as a waste product (Buick 1992; Holland 2006), thereby fashioning Earth's oxygen atmosphere.

Molecular oxygen in the atmosphere of Mars was first detected by the Herschel Space Observatory in 2010 (Hartogh et al. 2010). Franz and colleagues (2017) have estimated that the mean volume of Martian atmospheric oxygen is 0.174%. This is similar to the levels of oxygen, on Earth, during the Paleoproterozoic "Great Oxidation Event" ~2.2 to 2.0 bya (Bekker et al. 2004; Farquhar et al. 2011). It was during this "Event" when atmospheric oxygen rose to >1% of modern levels on Earth, an accumulative byproduct of oxygenic photosynthesis and the respiration of oxygen by cyanobacteria (blue-green algae) -- and related species (Buick 2008; Nisbett and Nisbett 2008; Olson 2006)-- which may have first appeared on Earth 3.8 bya (Uyeda et al. 2016).

Initially, however, photosynthesis was anoxygenic, with H2 and iron being employed as oxygen acceptors (Eigenbrode and Freeman 2006; Olson 2006; Sleep and Bird 2008). Lichens have high concentrations of iron (Bajpai et al. 2009; Hauck et al. 2007) and iron is abundant on Mars. Iron and H2, therefore, may have and may still serve as receptors for Martian organisms engaged in photosynthesis.

As cyanobacteria and related photosynthesizing Earthly species proliferated, they formed symbiotic relationships and respired-oxygen slowly built up in the oceans, soil and atmosphere culminating in the "Great Oxidation Event" and reaching levels comparable to and then surpassing those on modern day Mars.

Lichens produce oxygen via photosynthesis (Vinyard et al. 2018; ted Veldhuis et al. 2020). As noted, lichens are a symbiotic organism consisting of at least one green alga or cyanobacterium (photobiont) which makes possible oxygen photosynthesis, and at least one fungus (mycobiont), the latter of which is largely responsible for the lichens' thallus, fruiting bodies, mushroom shape, and bulbous cap (Armstrong 2017, 2019; Brodo et al. 2001; Tehler & Wedin, 2008). Molecular analyses, however, indicate that the lichen consortia also include a wide range of bacterial communities within the photobiont zone and on the lichen-surface such as Sphingomonas, Methylobacterium, and Nostoc, as well as a variety of eukaryotic Rhizaria, Amoebozoa, and Metazoa (Graham et al. 2018). Squires et al (2004) has argued that eukaryotes may have evolved on Mars and features resembling fossilized metazoa have been identified in Gale Crater, though if they are abiotic is unknown (Joseph et al. 2020a). These symbiotic relationships have proven vital in the ability of the lichen to survive life neutralizing and water stressed environments (Armstrong 2017; Margulis and Fester, 1991; Kranner et al. 2008) and which promote mutual metabolism, energy conversion and enhance the respiration of oxygen via photosynthesis.

Measurements of lichen electron transport have demonstrated that O_2 is generated by the alga and consumed internally and any excess is respired, whereas CO_2 is produced by respiration of photosynthetically generated sugars which along with fungal CO_2 are consumed by the alga (ted Veldhuis et al. 2020). More specifically, photosynthesis-produced-oxygen involves the absorption of CO_2, the transfer of multiple electrons, monotonic-OO-bond formation, OH bond cleavage, and the splitting of water molecules (H / O_2) with all excess oxygen released into the soil or atmosphere.

Martian specimens resembling lichens have been previously identified in the Gale Crater and Eagle Crater (Joseph et al. 2019, 2020a), including--and as reported here--vast colonies which are collectively oriented and angled skyward similar to terrestrial lichens and plants engaged in photosynthesis. In addition, apertures resembling open-cone-like gas-bubble vents were identified adjacent to the lichen-like specimens observed in Gale Crater (see Joseph et al. 2020a Figures 16, 17) and these apertures are associated with photosynthesis-oxygen respiration (Bengtson et al. 2009; Sallstedt et al. 2018). It is reasonable to assume that these and related species are respiring oxygen; and this would account for oxygen in the atmosphere and soil of Mars (Leshin et al. 2013; Ming et al. 2014; Rahmati et al. 2015; Sutter et al. 2017; Valeille et al. 2010). Almost all oxygen on Earth is produced biologically and the presence of oxygen is an obvious biomarker for life. The same reasoning should apply to Mars.

The amount of oxygen in the Martian atmosphere also shows seasonal variations, increasing by 30% in the Spring and Summer (Trainer et al. 2019). Levels of oxygen in the soil, oceans, and atmosphere of Earth, also vary according to the season and increase during the Spring and Summer due largely to fluctuations in the biological activity of photosynthesizing organisms (Keeling and Shertz, 1992; Kim et al. 2019) and as related to increases in temperature and the availability of water and water vapor condensation and precipitation (Buenning et al. 2012; Keeling and Shertz, 1992). These seasonal fluctuations on Earth parallel the Spring/Summer increases in oxygen, temperature and water availability on Mars (Fedorova et al. 2020; Read and Lewis, 2004; Smith, 2004; Todd et al. 2017; Trainer et al. 2019); and these variations on Mars in turn parallel seasonable increases in biological activity and the respiration of oxygen on Earth.

Given the evidence indicative of photosynthesis-oxygen-gas vents and what appear to be algae and lichens in the Gale Crater (Joseph et al. 2020a), coupled with what appear to be colonies of lichens in the Eagle Crater which are morphologically oriented in a manner similar to photosynthesizing lichens on Earth, it is thus reasonable to that they, and other photosynthesizing organisms dwelling on Mars, contribute to the seasonal variations in Martian oxygen, which in turn is regulated by increases in temperature and water availability. Increases in oxygen by as much as 30% during Spring and Summer is an obvious biomarker.

Although water may be stored within the lichen (depending on species), water content equilibrates with atmospheric conditions such that their photosynthetic activity, respiration and growth is determined by water availability and decreases or increases accordingly (ted Veldhuis et al. 2020). However, lichens easily survive long-term desiccated states. In consequence, lichens can be repeatedly dehydrated without any loss in their ability to engage in photosynthesis and to release oxygen into the atmosphere and surrounding soils once sufficient water is available (Vinyard et al. 2018). Lichens are well adapted to survive and engage in photosynthesis and oxygen production, on Mars.

For example, despite long term exposure to space and Mars-like analog conditions, over 70% of lichen photobionts and 84% of lichen mycobionts showed average viability rates of 71% to 84% respectively (Brandt et al. 2015; Meesen et al. 2014). Additionally, 50-80% of alga and 60-90% of the fungi symbiote demonstrating normal functioning (Brandt et al. 2015) including the ability to engage in photosynthetic activity with minimal impairment (Meesen et al. 2014). The angled-skyward orientation of the mushroom-shaped lichen-like specimens in this report should also be viewed as an indication of viability and evidence of photosynthetic activity thereby account for oxygen in the Martian atmosphere and soil.

Furthermore, Martian atmospheric oxygen and other gasses are believed to continually bleed into space (Jakosky et al. 2019). Oxygen, therefore, not only increases dramatically when the Martian environment is most conducive to biological activity, but oxygen is continually replenished; otherwise there would be no oxygen in the atmosphere or soil.

On Earth O2 production via biological photosynthesis (Canfield 2014; Hall and Rao, 1986; Vinyard et al. 2018; ted Veldhuis et al. 2020) is the major and primary source of and which constantly replenishes soil, oceanic, and atmospheric oxygen. Therefore, it is reasonable to deduce that cyanobacteria, and the Martian lichen-like organisms identified in Gale and Eagle Crater, also produce and replenish atmospheric and surface oxygen and are responsible for the seasonal variations in oxygen on Mars.

The Spherical Hematite Hypothesis is Refuted

It's been inferred that the spheres of Eagle Crater consist of hematite (Squires et al. 2004). However, a number of investigators have recognized that the hematite hypotheses is not supported by the evidence and is incompatible with the nature of these spheres (Burt, et al. 2005; Dass 2017; Joseph 2014; Knauth et al. 2005; Rabb 2018; Small 2015). As will be detailed: there is no evidence that these spheres consist of hematite. Nor is there any evidence of large bodies of water, ancient hot springs or volcanic activity at any time in the past history of Eagle Crater and thus there was no means of producing hematite which requires a boiling liquid or volcanic source at temperatures of at least 900°C in order to form (Anthony et al. 2003; Morel 2013).

In an attempt to circumvent and explain away the fact that the environment of Eagle Crater has never been conducive to the creation of hematite, it's been claimed that under "dry laboratory conditions" "goethite" can be "converted to hematite" at 300°C (Christensen and Ruff, 2004). However, Eagle Crater is not a laboratory and equatorial temperatures, as reported by NASA, seldom exceed of 20°C (see http://quest.nasa.gov/aero/planetary/mars.html). The last time surface temperatures in Eagle Crater reached or exceeded 300°C may have been hundreds of millions if not billions of years ago when struck by the meteor which cratered the surface (Knauth et al. 2005).

Although Knauth and colleagues (2005; Burt et al. 2005) did not address the mushroom-shaped formation of Eagle Crater, they proposed that the "blue berries" upon the Martian surface may have been created upon meteor impact. However, obviously, these top-heavy mushroom-shaped specimens attached to rocks by thin stems, could not have been formed millions or even thousands of years ago by any catastrophic or geological-weathering process, as the stems would have long ago crumbled, broken, and, along with their bulbous caps, toppled over due to weathering, wind, and frequent Mars-quakes as the planet is seismically quite active Banerdt et al. 2020).

The problems with the hematite hypothesis are legion (Burt et al. 2005; Knauth et al. 2005; Joseph et al. 2019). For example: The Opportunity's instruments were not calibrated to selectively detect hematite and the cameras were not capable of taking true color photos. Therefore, the actual and true color of the landscape, rocks, outcrops, sand, dust, dirt, is unknown. Instead, composite false color images were generated by the Opportunity's panoramic camera's 750-, 530- and 480-nanometer filters (Soderblom et al. 2004). Based on these "color composites" blues and greens were detected throughout the lower landscape. With the exception of the dull gray stones embedded inside cement-like outcrop matrix (Bell et al. 2004; Squires et al. 2004) the spherules of Eagle Crater were judged to be yellow, orange, and purple (Soderblom et al. 2004) whereas the stemmed Martian specimens with mushroom features appear to be purple in color (Figures 1-3).

As is well known, a variety of terrestrial organisms including lichens and fungi may appear green, purple, orange or yellow. The Eagle Crater spheres upon the ground ("blue berries") and those jutting skyward attached to rocks ("Martian mushrooms") were judged to be purple, orange and yellow (Soderblom et al. 2004). By contrast, terrestrial hematite is variably colored black, silver-gray, brown, reddish-brown, or red (Anthony et al. 2003; Morel 2013). Thus, based on these composite colors from the Eagle Crater, the "blue berries" and purple Martian mushrooms could not be hematite.

Although investigators observed, via photographs, specimens with lichen-like mushroom features jutting up from rocks within Eagle Crater (Bell et al. 2004; Squires et al. 2004), there was no selective, focused attempt to determine if they were biological or consisted of hematite or other minerals. Further, despite recognizing that the spheres (blue berries) upon the surface were a different color than hematite (Soderblom et al. 2004) and much smaller than terrestrial hematite, ranging in size from 0.6 to 6 mm in diameter (Herkenhoff et al., 2004) it was assumed they must be hematite based on inference and the interpretation of results generalized from panoramic images that included sand, soil, dust, and outcrops, and as based on generalized all inclusive spectra recorded by the Opportunity's Mössbauer Spectrometer, Alpha Particle x-ray Spectrometer and Miniature Thermal Emission Spectrometer (Bell et al. 2004; Christensen et al. 2004; Klingelhöfer et al. 2004; Rieder et al. 2004; Squires et al. 2004). These instruments were not even mineral specific. The response functioning of these instrument also continually changed "over the course of the mission" and did not correspond to pre-mission "instrument calibration" thereby requiring ad hoc calibration adjustments (Glotch and Bandford, 2006).

Hematite was never directly or positively detected in any of the spheres by the spectrometers. Instead, it's possible presence was inferred based, for example, on the averaging of what were assumed to be high and low temperatures derived from outcrops and plains (Klingelhöfer et al. 2004), and the elimination of spectral signals until arriving at spectral signatures that could be interpreted as similar to hematite in a controlled laboratory setting (Christensen et al. 2004).

As is well established, different colors have different spectra and differentially absorb and reflect light and heat. Likewise, biological organisms generate heat and their pigments reflect and absorb different spectra. Moreover, Martian sand, dust, dirt, rocks, outcrops, all appeared to and would be expected to have different albedos, colors (Bell et al. 2004; Klingelhöfer et al. 2004; Soderblom et al. 2004) and heat signatures. However, the temperature and true color of the Eagle Crater landscape are unknown.

Nevertheless, the spectra from false colors created by the camera filters were analyzed and compared to "test spectrum" obtained from a "magnetite-derived hematite" laboratory sample; and as admitted by Glotch and Bandfied (2006), the data was nevertheless a "poor fit" and did not match laboratory samples.

Moreover, all obtained spectral signatures were confounded and contaminated by numerous uncontrolled and unknown variables, the properties of which could not be accurately and precisely determined. Hence, due to depth of field, reflected light from the Opportunity and the differential angles of the surrounding objects, and layers of obscuring dust and sand, and as the temperature sensors had failed, it was impossible for the Opportunity's suite of spectral sampling instruments to obtain accurate and selective spectral signatures. All data collected included dust, dirt, sand, outcrops, large flat oblong rocks, surrounding matrix and soil, and were affected by reflected light and atmospheric temperatures and solar radiance; and then the data was combined, adjusted, averaged, and then attributed to the spheres which were falsely claimed to contain hematite (Christensen et al. 2004; Klingelhöfer et al. 2004; Rieder et al. 2004). As admitted by Grotzinger et al. (2005): the spectra from rocks lying on the surface were "indistinguishable from that of the average spectral character of dust." And as acknowledged by Klingelhöfer and colleagues (2004): "images obtained by the Microscopic Imager sampled only outcrop matrix." And yet, the Opportunity's team of investigators, claimed that the spheres consisted of hematite despite having no accurate data to support this interpretation.

Despite "the failure of temperature sensors on the on-board calibration targets" (Glotch and Bandfield 2006) Klingelhöfer and colleagues (2004) nevertheless claimed to have averaged high and low temperatures from multiple sources, and inferred the existence of hematite within the spheres based on these generalized averages. Klingelhöfer et al. (2004), admitted that spectra was believed to "imply" hematite and were therefore "assigned to hematite" (Klingelhöfer et al. 2004).

Furthermore, the data that was claimed to have been obtained from single spheres were obtained from panoramic views of the landscape (Christensen et al. 2004; Klingelhöfer et al. 2004) and from flat oblong rocks which were then inexplicably mischaracterized as spheres (see Figure 6 in Bell et al. 2004). Not only did Bell et al (2004) utilize multiple images and spectra from a single large oblong rock, but their data was also based on panoramic photographs depicting multiple features, and was contaminated by solar radiance, atmospheric irradiance, surface temperature and albedo which could only be guessed at and estimated, thus confounding the data. Furthermore, Bell et al (2004) admitted the data is "not consistent" with solid hematite but jarosite and ferric iron and "exhibit crystalline ferric iron spectral signatures." And yet, Bell (2004) and the others claimed that the spheres contained hematite (Christensen et al. 2004; Klingelhöfer et al. 2004; Squires et al. 2004) when this confounded data was actually derived from a single oblong rock and there was no data based on or selectively derived from these spheres to substantiate this assumption which was based on inference and speculation.

Christensen and colleagues (2004), also claimed to have directly examined these spheres but instead relied on panoramic images to determine "the mineral abundances and compositions of outcrops, rocks, and soils" via the "Miniature Thermal Emission Spectrometer (Mini-TES)." According to Christensen et al. (2004) the Mini-TES "collects infrared spectra and were combined with panoramic images and as based on thermophysical properties, atmospheric temperature profiles and atmospheric dust and ice opacities." Thus, the Mini-Tes acquired its data not by examining a single sphere, but from composites obtained from "ice opacities" when no ice was observed, atmospheric temperatures when the temperature sensors had failed and the atmospheric temperature was (and is) unknown, and from large panoramas via "long-integration single-point stares at 14 locations along the out-crops... and vertical scans of the plains." Christensen et al. (2004) also acknowledged that their data was affected by "reduced spectral contrast" and was "likely contaminated" by sand, dust, and other materials, and which led them to "overestimate the hematite."

Christensen et al. (2004) also admit that they removed spectra "by first deconvolving each spectrum with an end member library of 47 laboratory minerals and four scene spectra...and then subtracting" spectra until this "derived spectrum" could be interpreted to resemble "a laboratory hematite sample." Thus, the inferred presence of hematite was based on the selective elimination of spectral signals until arriving at a spectrum that was interpreted to be similar to the spectral signature of a sample of hematite examined in a laboratory setting; the lighting and controlled conditions of which, of course, would be completely different from Eagle Crater or a natural terrestrial environment.

Although soils and outcrop matrixes of Eagle Crater likely contain considerable iron and jarosite (Bell et al. 2004; Christensen et al. (2004; Herkenhoff et al., 2004; Klingelhöfer et al. 2004; Rieder et al. 2004; Squyres et al. 2004), the fact is: no evidence of hematite was found, the presence of iron was inferred to indicate hematite, and not one of the thousands of mushroom-shaped lichen-like specimens (Martian mushrooms) photographed in Eagle crater were individually or selectively examined by Opportunity's suite of spectral sampling instruments for any evidence of hematite. Likewise, the spheres (blue berries) upon the soil were never directly, selectively, individually, and specifically examined for the presence of hematite or spectra that might imply hematite. Oblong rocks are not spheres. The spectrometers employed were not even mineral specific, there were problems with calibration, temperature sensors had failed, and the data obtained was generalized from multiple sources then combined, manipulated, averaged, or selectively deleted, until what remained was still a "poor fit" for spectra obtained from laboratory samples. Claiming that these spheres, especially those with stalks and caps, are hematite, has no factual scientific basis or validity.

Three Types of Spheres: Martian Mushrooms, Cement Concretions, Blue Berry Puffballs

Martian Mushrooms: As reviewed in this document: the "Martian mushrooms" presented here are up to 8 mm in length, have thin stems up to 5 mm in length and less than 1 mm in diameter and topped with bulbous caps up to 6 mm in diameter. These specimens have a different morphology, color, and are smaller than hematite; there is no evidence to support the belief these are hematite; their caps and stalks appear uniform in shape which is a biological and not an abiogenic trait as well as being characteristic of living lichens and mushrooms. As to those on the top-sides of rocks, their collective, flexible upward angled orientation is exactly what would be expected of photosynthesizing organisms.

By contrast, those on the sides of rocks, and being top-heavy with bulbous caps, are pulled downward as if by gravity. Furthermore, those on the tops and sides of rocks have an obvious and completely different structural organization and composition from the outcrops and rocks from which they jut out skyward or downward, and distinctly different visible and infrared properties as compared to these outcrops (Bell et al., 2004) indicating they were not sculpted from rock. In addition, several specimens on the top-sides of rocks appeared to change their angle of orientation during a single day, such that arched downward, thus suggesting that their stems (top-heavy with bulbous caps) are flexible (Figure 10). In all respects these Martian mushrooms appear biological and distinct from surface substrates.

There are, however, two other types of "spheres" that have been observed and photographed in Eagle Crater and which differ significantly from the thin stemmed "Martian mushrooms" and each other, in morphology, location, color and attached substrate: A) "yellow, orange and purple" spheres upon the soil (Soderblom et al. 2004) which have been referred to as "blue berries" and resemble fungal "puffballs" (Dass 2017, Joseph 2016; Joseph et al. 2019; Rabb 2018); and B) gray spheroidal cement-like concretions (Bell et al. 2004; Squyres et al. 2004; Herkenhoff et al. 2004).

Cement-like Concretions and Fossilization: In contrast to Martian mushrooms and "blue berries" the gray spheroidal-cement-like concretions are embedded in a cement-like matrix (Bell et al. 2004; Squyres et al. 2004; Herkenhoff et al. 2004) and have been described as "harder than surrounding rock" (Squyres et al. 2004) though what they consist of was never determined. It is believed that these gray spheroids had undergone "cementation" thereby "cementing" this matrix and the concretions embedded in this cement (Herkenhoff et al. 2004). If these represent a form of fossilization unique to the Martian environment, if they consist of calcium, or were formed secondary to iron metabolism, or if they are completely abiogenic, is unknown.

However, based on terrestrial analogs, the lichen-like specimens growing atop rocks may contain iron which is a lichen characteristic (Bajpai et al. 2009; Hauck et al. 2007). Many species feed on iron. If these specimens, due to the unique Martian and iron-rich environment, have a greater uptake of iron as compared to terrestrial species, is unknown. Moreover, cyanobacteria produce calcium via their secretions. To speculate, could high levels of iron uptake or calcium secretions make these specimens "harder than rock" if fossilized such as following a sudden change in the biosphere due to a catastrophic event-- thus accounting for the embedded gray cement-like concretions observed by Squyres and colleagues (2004; Bell et al. 2004; Herkenhoff et al. 2004)? Fossilization, cannot be ruled out.

Blue Berry Puffballs and Growth vs Wind: Specimens similar to "puffballs" ("blue berries") have been observed on the Martian surface of Meridiani Planum (Dass 2017; Joseph 2014; Rabb, 2015, 2018; Small, 2015). Some of these ground-level specimens have "lemon shapes," others have a short or long thin stalk, but the majority appear to have no discernible stalk. Furthermore, with the exception of those attached to an unknown (fungi-like) substrate and which, via that substrate, are elevated above the ground (Figures 1, 4) most ground level spheres, including those with long and thin stalks lay upon the surface as if they fell over. Even those observed to emerge from beneath the soil do not rise up on their stems as is typical of mushrooms and lichens (Figure 11).

Contrasting the soil mushrooms with the rock mushrooms, could it be that the soil cannot support them so they fall over and lay on the ground? The answer to this is unknown. However, it is possible that those upon the soil (vs those on rocks) consist of many different species, assuming they are biological.

The biological interpretation is supported by the previously reported observation of fifteen spherical specimens which increased in size and emerged from beneath the coarse grained surface (Joseph et al. 2019). Here we present pictorial evidence of twenty-three puffball-shaped specimens, photographed on Sol 1148 which increased in size over a three-day period, twelve of which were not visible three days earlier on Sol 1145 (Figure 11). We have determined that wind was not a factor in the emergence and size increase of these specimens.

It's been estimated that the movement of coarse-grained Meridiani Planum soil requires wind velocities of 70 m/s at least one m above the surface, but that velocities of 40 m/s may "occasionally" displace coarse-grained sand and soil (Jerolmack et al. 2006). All surrounding soil in Sol 1145 and 1148 is coarse (vs fine) grained. Soil crusts, including and especially those infiltrated by micro-organisms, are relatively resilient to wind erosion. In a two year study of soil and wind in the Negev Desert--considered to be a Mar-like analog environment (Kidron 2019)--it was reported that only exceptionally strong and prolonged winds were capable of crust rupture, disintegration or flaking and the removal or erosion of buried crust (Kidron et al. 2017). No strong winds or dust storms in Meridiani Planum were reported by NASA on Sols 1145, 1146, 1146, or 1148. Likewise, there is no evidence or comparative evidence of wind, dust storms, or dust devils or the accumulation of dirt, sand, ripples, lines, or dust as based on a visual examination of all photos between Sol 1145 and 1148. Nor is there any evidence of soil or sand displacement, soil or sand buildup or "filling in" or that soil is higher or lower on one side of any of these specimens as might be expected if subject to powerful directional winds (Kidron et al. 2017). It is reasonable to deduce that these puffball-shaped spheres grew up out of the ground and expanded in size over a three-day period.

Figure 7. Opportunity - Sol: 85. Lichen-like specimens approximately 2 to 6 mm in length. Note upward orientation. Specimens appear to change their angle and direction of orientation depending on gravity and their length and time of day.

Figure 8. Opportunity - Sol: 85. Lichen-like specimens approximately 2 to 8 mm in length. Orientation of specimens on tops of rocks appears to be affected by gravity and may differ depending on if they are or are not sheltered from the wind; i.e. those on opposite sides of rocks vs tops of rocks may face different directions.

Figure 9. Opportunity - Sol: 85. Lichen-like specimens approximately 2 to 8 mm in length. Note similar orientation of specimens on tops of rocks and which may be affected by gravity due to the top-heavy bulbous caps.

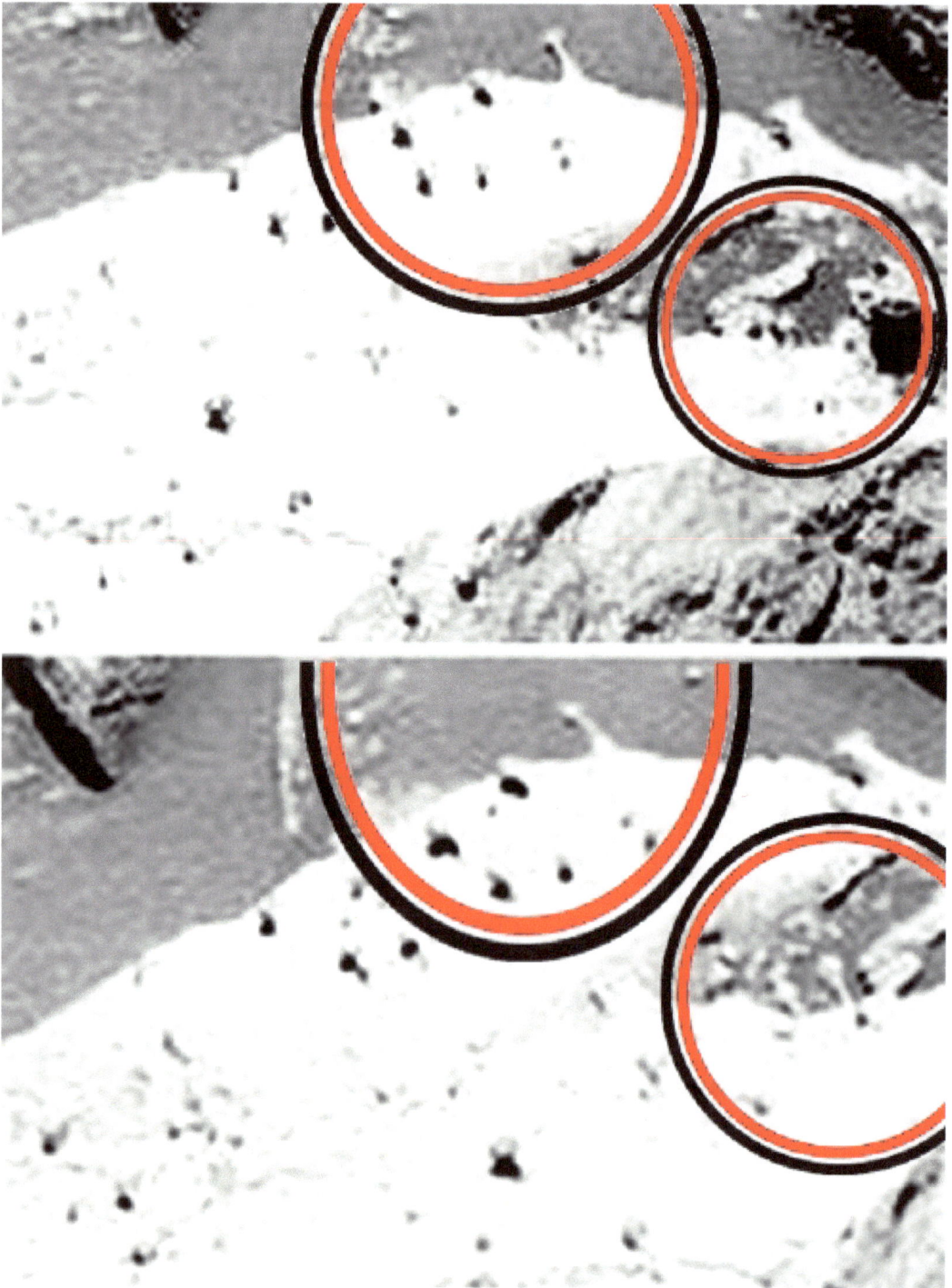

Figure 10. Opportunity - Sol: 85. Two photos of the same specimen at different times on the same day. Note that seven of the "mushrooms" within the red circles, arch and bend down in a leftward direction (bottom vs top photo). This downward arching cannot be explained by the slight change in the angle of the photographs, as several of the specimens have become bent downward. This supports the hypothesis that the stems, top heavy with mushroom (fruiting body) caps, are flexible.

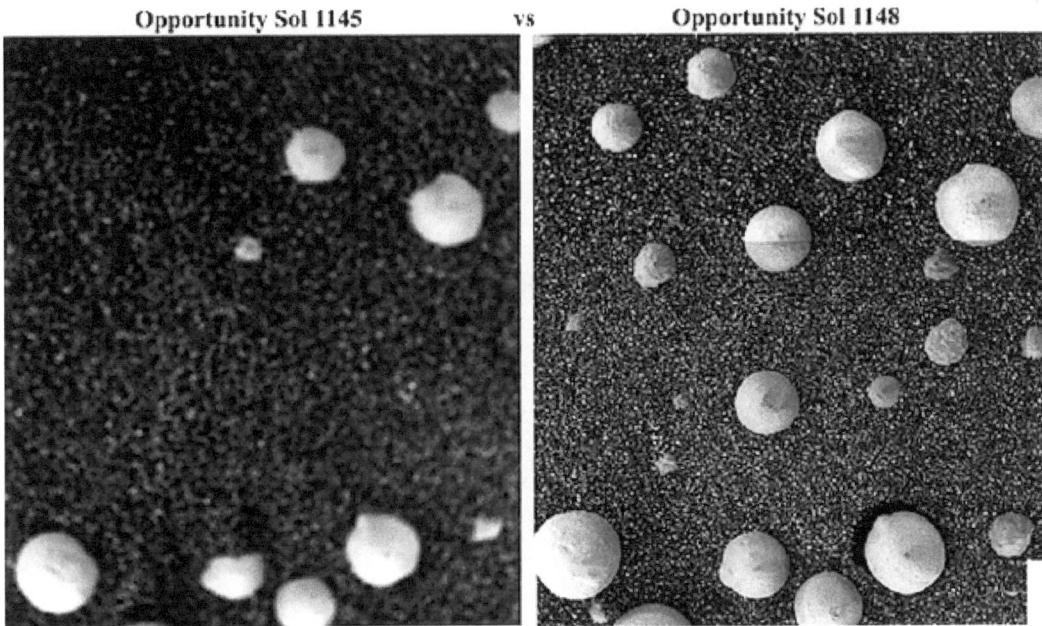

Figure 11. Opportunity - Sol 1145-left v Sol 1148-right). Comparing Sol 1145-left vs Sol 1148-right. Growth of twenty-three Martian specimens over three days, twelve of which emerged from beneath the soil and all of which increased in size.Wind speeds up 70 m/h are required to move coarse grained soil, and no strong winds, dust clouds, dust devils, or other indications of strong winds were observed, photographed, or reported during those three days in this vicinity of Mars. Nor does the Sol 1148 photograph show any evidence that the surface has been disturbed by wind, as there are no ripples, waves, crests, or build up of soil on one side of the specimens as would be expected of a directional wind (Kidron et al. 2017). Photographed by the Rover Opportunity, NASA/JPL. Differences in photo quality are secondary to changes in camera-closeup-focus by NASA.

Martian Mushrooms in Eagle Crater. Lichenised vs Non-Lichenized Fungi

If the mushroom-like specimens presented in this report are fungi or composite algae-fungal organisms, is unknown. Most of these specimens resemble the lichen, Dibaeis baeomyces in morphology, shape, growth patterns, and size including possessing stalk/thallus and a bulbous apothecia (Joseph et al. 2019). Dibaeis baeomyces are similar in a number of respects to these Martian specimens which appear to depict the gradual development of ascomata from small globulars which become stalked structures capped with a fruiting body (Figures 2,3). Like their putative terrestrial counterparts, there appear to be thallus granules and nodules on the Martian substrate surface.

Dibaeis baeomyces are well adapted for life on Mars, and have colonized the most extreme environments and been found growing in desert sand, dry clay, on rocks, and in the arctic (Brodo et al. 2001; Jonsson et al. 2008; Platt & Spatafora 2000; Ryan et al. 2002; U.S. Department of the Interior 2010). Because of their stress-tolerance, slow growth rates, low demands for water and nutrients, longevity, and adaptations to stressful conditions, lichens might easily colonize Mars.

The amount and availability of water within Eagle Crater is unknown. Lichens can tolerate long periods of drought and dehydration, a function, in part, of their slow growth rate and metabolism (Armstrong and Bradwell 2010). Some species spend considerable time in a dehydrated state in which there is little physiological activity and no demand for nutrients (Armstrong 2017, 2019). These organisms survive long periods without water and in nutrient-poor habitats.

Extremes in cold temperature would not be a limiting factor. Lichens flourish on the Antarctic continent and its adjacent islands (Llano 1965; Ahmadjian 1970; Longton 1979; Lindsay 1978; Smith 1984) and despite subzero temperatures for prolonged periods. Dark and nPS respiration is maintained even at subzero temperatures (Schroeter & Scheidegger 1995).

Hundreds of these Martian mushrooms form colonies which are selectively oriented skyward. The orientation of the presumed fruiting-bodies with respect to light appears to be behavioral and indicative of phototropism during fruiting body development.

Many of the specimens presented here lack an obvious crustose thallus which is a Dibaeis baeomyces' characteristic (Armstrong, 2019; Armstrong and Bradwell 2010; Kidron 2019). If these specimens were produced by a lichen-forming fungus, then it might be expected that the symbiotic lichen thallus would be endolithic and inside the substrate, and for which there is evidence as depicted in Figures 2-3. These endolithic attachments could be individual as well as collective fungal hyphae. However, terrestrial hyphae are usually 2-5 microns in diameter (Armstrong 2017).

If these are lichens, then perhaps they have adapted and evolved in response to the Martian environment and its high levels of ground radiation. Hence, the absence of an obvious crustose thallus may be an evolved adaptation. For example, in response to heightened radiation exposure--well beyond the "0.67 millisieverts per day" on Mars (Hassler et al. 2013)--lichens as well as fungi have developed adaptive features (reviewed by Joseph et al. 2019)--a property described as "radiostimulation," "radiation hormesis," and "adiotropism" (Levin 2003; Tugay et al. 2006; Zhdanova et al 2004). These radiation-induced adaptations include tissue and cellular regeneration and new growth (Basset 1993; Becker 1984; Occhipinti et al. 2014; Levin 2003; Maffei 2014; Moment, 1949). The varying levels of radiation on Mars may have contributed to the evolution of unique features making Martian species distinct from, albeit remaining similar to terrestrial organisms.

The specimens presented here could also be non-lichenised fungi similar to Leotia lubrica, Cordyceps capitata, Tulostoma brumale. Given their size and morphology, they could also represent stalked reproductive fungal components, such as ascomata or basidiomata (complex fruiting-bodies producing sexual spores) or stilbelloid synnemata (complex conidiophores producing asexual spores). If these Martian mushrooms are fungi then it can be assumed they produce these fruiting bodies to facilitate spore dispersal. If they are lichens, they are likely engaged in photosynthesis.

Conclusions

Edgar, Grotzinger, Hayes, and colleagues (2012) have argued that "the stratigraphic architecture of sedimentary rocks on Mars is similar (though not identical) to that of Earth, indicating that the processes that govern facies deposition and alteration on Mars can be reasonably inferred through reference to analogous terrestrial depositional systems." The same reasoning should apply to biology and oxygen and rule out any possibility these Martian mushrooms are abiogenic.

We have provided evidence of over 200 specimens with thin stalks and spherical-caps which appear to be purple in color, and have a mushroom-shape and resemble lichens in size and morphology and jut upward toward the sky. Thousands of similar specimens have been observed (see Methods). There are no terrestrial abiogenic processes that can sculpt high density colonies of mushroom-shapes, attached by thin stalks to rocks and which form colonies that selectively orient their bulbous caps skyward in the same general direction exactly as what might be expected of photosynthesizing organisms. This interpretation is supported by the seasonal fluctuations and dramatic (30%) increases in Martian atmospheric oxygen during the Spring and Summer and which parallel seasonal fluctuations in the biological production of oxygen on Earth. It is reasonable to deduce that photosynthesizing organisms on Mars are responsible for the production and replenishment of oxygen which is an obvious biomarker.

In addition, twenty-three sphere-puffball-shaped specimens were photographed by the rover Opportunity either increasing in size or emerging from beneath the soil, over three days in the absence of any contributing wind or other abiogenic process. These behavioral indices (i.e. growth, skyward orientation, flexibility of the stems), coupled with morphology and oxygen production, are indicative of biology, the first evidence of which was detected by the Viking experiments (Levin & Straat 1976, Levin et al. 1978).

These "Martian mushrooms" are uniform in appearance and do not resemble and are smaller and a different color than hematite; and they were never directly or selectively examined by any means for evidence of hematite--despite misleading claims to the otherwise. The Eagle Crater environment was not and is not conducive to producing hematite; there were significant problems with calibration and target sensors such that the instrumentation and methodology employed to detect spectral signatures that could be interpreted as hematite were dubious at best; and the claim that any of the spheres observed in Eagle Crater contain hematite has been described as "inappropriate" (Burt et al. 2005; Knauth et al. 2005) and a "poor fit" when compared to laboratory samples (Glotch and Bandfield 2006). The spherical hematite hypothesis is based on speculation and inference and devoid of any scientific or factual foundation.

It is not probable that these specimens consist of salt, sand, or other abiogenic substances. Consider: these mushroom-shaped specimens look identical to mushrooms and lichens, and are attached to rocks by thin stalks, and top heavy with spherical caps that weigh some of these specimens so they arch upward then downward. If abiotic, these thin stems, top-heavy with skyward orientated bulbous caps, would have long ago broken apart and shattered in response to powerful winds, Mars-quakes, meteor strikes, or, more recently, by turbulence created by the rover Opportunity. They did not.

There are no "analogous terrestrial" processes which can explain the unique and uniform morphology, size, color, thin hollow stems, and collective skyward orientation of these mushroom-shaped specimens, or the seasonal fluctuations and increases and replenishment of Martian oxygen, other than biology. That these are living organisms "can be reasonably inferred through reference to analogous terrestrial" organisms; although their precise identity is unknown.

In conclusion, coupled with reports of seasonal variations and increases in oxygen in the Martian atmosphere during the spring and summer, and based on size, color, morphology, flexibility, and what appears to be indications of photosynthesis and growth, the evidence presented in this report strongly supports the hypotheses that mushroom-shaped, lichen-like organisms have colonized Eagle Crater and that there is life on Mars.

References

Ahmadjian, V. (1970). Adaptation of Antarctic terrestrial plants. In: Antarctic Ecology, Vol 2. MW Holdgate (ed.), Academic Press, New York, pp 801-811.

Anthony et al. (2003). Handbook of Mineralogy. V (Borates, Carbonates, Sulfates). Chantilly, VA, US: Mineralogical Society of America. ISBN 978-0962209741.

Armstrong, R.A. (1981). Field experiments on the dispersal, establishment and colonization of lichens on a slate rock surface. Environmental and Experimental Botany 21: 116-120.

Armstrong R.A. (2017) Adaptation of Lichens to Extreme Conditions. In: Shukla V., Kumar S., Kumar N. (eds) Plant Adaptation Strategies in Changing Environment. Springer, Singapore.

Armstrong, R. A. (2019). The Lichen Symbiosis: Lichen "Extremophiles" and Survival on Mars Journal of Astrobiology and Space Science Reviews, 1, 378-397.

Armstrong, R.A., Bradwell, T. (2010). Growth of crustose lichens: A review. Geografiska Ann, Series A, Phys Geog 92A: 3-17. Armstrong, R.A., Bradwell, T. (2011). Growth of foliose lichens: a review. Symbiosis 53: 1-16.

Ayupova, N., Maslennikov, V. V., Tessalina, S., Statsenko, E. O. (2016). Tube fossils from gossanites of the Urals VHMS deposits, Russia: Authigenic mineral assemblages and trace element distributions. Ore Geology Reviews 85, DOI: 10.1016/j.oregeorev.2016.08.003

Ayupova, N. R., Valeriy V. Maslennikov, Sergei A. Sadykov, Svetlana P. Maslennikova and Leonid V. Danyushevsky (2006) Evidence of Biogenic Activity in Quartz-Hematite Rocks of the Urals VMS Deposits, Frank-Kamenetskaya et al. (eds.), Biogenic—Abiogenic Interactions in Natural and Anthropogenic Systems, Lecture Notes in Earth System Sciences, DOI 10.1007/978-3-319-24987-2_10

Bajpai, R., UpretiD K., Dwivedi, S.K. (2009) Passive monitoring of atmospheric heavy metals in a historical city of central India by Lepraria lobificans Nyl, Environmental Monitoring and Assessment 166(1-4):477-84.

Banerdt, W.B., Smrekar, S.E., Banfield, D. et al. Initial results from the InSight mission on Mars. Nat. Geosci. (2020). https://doi.org/10.1038/s41561-020-0544-y

Baque, M., de Vera, J.P., Rettberg, P. and Billi, D. (2013). The BOSS and BIOMEX space experiments on the EXPOSE-R2 mission: Endurance of the desert cyanobacterium Chroococcidiopsis under stimulated space vacuum, Martian atmosphere, UVC radiation and temperature extremes. Acta Astronautica 91:180-186.

Baque, Mickael and Böttger, Ute and Leya, Thomas and de Vera, Jean Pierre Paul (2017) Preservation of carotenoids in cyanobacteria and green algae after space exposure: a potential biosignature detectable by Raman instruments on Mars. EANA17, 14-18 August 2017, Aarhus, Denmark.

Basset C.AL. (1993). Beneficial effects of electromagnetic fields. J Cell Biochem 31:387-393.

Baucon, A., De Carvlho, C. N., Felletti, F., Cabella, R. (2020). Ichnofossils, Cracks or Crystals? A Test for Biogenicity of Stick-Like Structures from Vera Rubin Ridge, Mars, Geosciences 2020, 10(2), 39; https://doi.org/10.3390/geosciences10020039

Bauer, H., A. Kasper-Giebl, M. Lo" flund, H. Giebl, R. Hitzenberger, F. Zibuschka, and H. Puxbaum, The contribution of bacteria and fungal spores to the organic carbon content of cloud water, precipitation and aerosols, Atmos. Res., 64, 109 – 119, 2002.

Becker RO. (1984). Electromagnetic controls over biological growth processes. Journal of Bioelectricity 3:105-118.

Becket, K. et al. (2008). Stress Tolerance in Lichens. In Lichen Biology (T, H. Nash III Ed) Cambridge University Press.

Bell, J. F., et al.,Pancam Multispectral Imaging Results from the Opportunity Rover at Merid

Biemann, K., J. Oro, P. Toulmin III, L. E. Orgel, A. O. Nier, D. M. Anderson, D. Flory, A. V. Diaz, D. R. Rushneck, and P. G. Simmonds (1977), The search for organic substances and inorganic volatile compounds in the surface of Mars, J. Geophys. Res., 82, 4641–4658, doi:10.1029/JS082i028p04641.

Bosea, S., HochellaJr., M. F., .Gorby, Y.A. Kennedy, D. W., McCready, D. E., Madden, A. S., Lower, B. H. (2009) Bioreduction of hematite nanoparticles by the dissimilatory iron reducing bacterium Shewanella oneidensis MR-1, Geochimica et Cosmochimica Acta, 73, Issue 4, 962-976.

Brandt, A., et al. (2015). Viability of the lichen Xanthoria elegans and its symbionts after 18 months of space exposure and simulated Mars conditions on the ISS--International

Journal of Astrobiology, 14, 411-425.

Brodo, I.M. et al. (2001). Lichens of North America. Yale University Press. pp. 50, 55, 173-4.

Burt, D.M., Knauth, L.P., Woletz, K. H. (2005). Origin Of Layered Rocks, Salts, And Spherules At The Opportunity Landing Site On Mars: No Flowing Or Standing Water Evident Or Required. Lunar and Planetary Science XXXVI (2005).

Claeys. P. (2006). Experimental Observations of the Patterns of Fungi-Mineral Surfaces Interactions with Muscovite, Biotite, Bauxite, Chromite, Hematite, Galena, Malachite, Manganite and Carbonate Substrates. SAO/NASA ADS Physics Abstract Service. http://adsabs.harvard.edu/abs/2006AGUFM.B11A1005C

Christensen, P. R. et al., Mineralogy at Meridiani Planum from the Mini-TES Experiment on the Opportunity RoverScience 306, 1733-1739 (2004).

Christensen, P. R., Ruff, S. W., (2004), Formation of the hematite□bearing unit in Meridiani Planum: Evidence for deposition in standing water, JGR Planets, 109, E8 2004

Dadachova E., Bryan RA, Huang X, Moadel T, Schweitzer AD, Aisen P, et al. (2007) Ionizing Radiation Changes the Electronic Properties of Melanin PLoS One, doi:10.1371/journal.pone.0000457.

Dass, R. S. (2017) The High Probability of Life on Mars: A Brief Review of the Evidence, Cosmology, Vol 27, April 15, 2017.

De la Torre Noetzel, R. et al. (2017). Survival of lichens on the ISS-II: ultrastructural and morphological changes of Circinaria gyrosa after space and Mars-like conditions EANA2017: 17th European Astrobiology Conference, 14-17 August, 2017 in Aarhus, Denmark.

De Vera, J. -P. (2012). Lichens as survivors in space and on Mars. Fungal Ecology, 5, 472-479.

De Vera, J. -P. et al. (2014). Results on the survival of cryptobiotic cyanobacteria samples after exposure to Mars-like environmental conditions, International Journal of Astrobiology, 13, 35-44.

de Vera, J-P, M., Backhaus, T., et al. (2019). Limits of Life and the Habitability of Mars: The ESA Space Experiment BIOMEX on the ISS, 19, Astrobiology, https://doi.org/10.1089/ast.2018.1897.

Dighton, J, Tatyana Tugay, T., ZhdanovaN., (2008) Fungi and ionizing radiation from radionuclides, FEMS Microbiol Lett 281, 109-120.

Edgar, L. A. John P. Grotzinger, Andalex G. Hayes, David M. Rubin, Steve W. Squyresandjames F. Bell, Ken E. Herkenhoff (2012) Stratigraphic Architecture Of Bedrock Reference Section,Victoria Crater, Meridiani Planum, Mars, Sedimentary Geology of Mars, ISBN 978-1-56576-312-8, CD/DVD ISBN 978-1-56576-313-5, p. 195–209.

Farmer, C.B., D.W. Davies, A.L. Holland, D.D. Laporte, and P.E. Doms (1977), Mars—Water vapor observations from the Viking orbiters, J. Geophys. Res., 82, 4225–4248, doi:10.1029/JS082i028p04225.

Farmer, C.B. Liquid water on Mars. Icarus 28(2), 279–289 (1976).

Fedorova, A. A. Montmessin, F., Korablev, O. et al. (2020) Stormy water on Mars: The distribution and saturation of atmospheric water during the dusty season, Science 09 Jan 2020: DOI: 10.1126/science.aay9522

Fredrickson, J, et al. (2008). Towards environmental systems biology of Shewanella." Nature Reviews in Microbiology. Volume 6:592-603.

Glotch, T. D., Bandfield, J. L. (2006). Determination and interpretation of surface and atmosphericMiniature Thermal Emission Spectrometer spectralend-members at the Meridiani Planum landing site, Journal of Geophysical Research, VOL. 111, E12S06, doi:10.1029/2005JE002671, 2006

Gralnick, R., & Hau, S. (2007). Ecology and biotechnology of genus Shewanella." Annu Rev Microbiol. 61:237-58.

Grotzinger, J. P., Arvidson, R. E., Bell, J .F. et al. (2005), Stratigraphy and sedimentology of a dry to wet eolian depositional system, Burns Formation, Meridiani Planum, Mars. Earth and Planetary Science Letter, 240, 11-72.

Harri, A.-M., et al (2014). Mars Science Laboratory relative humidity observations: Initial results, JGR Planets, 119, 2132-2147.

Hassler, D, M., et al. (2013) Mars' Surface Radiation Environment Measured with the Mars Rover. Science, doi: 10.1126/science.1244797.

Hauck, M., Huneck, S., .Elixc, J. A., Paul, A. (2007) Does secondary chemistry enable lichens to grow on iron-rich substrates? Flora - Morphology, Distribution, Functional Ecology of Plants\202, 471-478.

Herkenhoff, K. E. et al., Evidence from Opportunity's Microscopic Imager for Water on Meridiani Planum Science 306, 1727-1730 (2004)

Hu, Y., et al. (2010). Occurrence, liquid water content, and fraction of supercooled water clouds from combined CALIOP/IIR/MODIS measurements, JGR Atmospheres, https://doi.org/10.1029/2009JD012384.

Jerolmack, D. J., Mohrig, D., Grotzinger, J.P., Fike, D.A., Watters, W. A. (2006). Spatial grain size sorting in eolian ripples and estimation of wind conditions on planetary surfaces: Application to Meridiani Planum, Mars, JGR Planets, Volume111, IssueE12

Jonsson, A.V., Moen, J., Palmqvist, K. (2008). Predicting lichen hydration using biophysical models. Journal of the Royal Society Interface, Oecologia, 156:259-273.

Joseph, R. (2014) Life on Mars: Lichens, Fungi, Algae, Cosmology, 22, 40-62.

Joseph, R. (2016) A High Probability of Life on Mars, The Consensus of 70 Experts, Cosmology, 25, 1-25.

Joseph, R (2019). Life on Venus and the Interplanetary Transfer of Biota from Earth. Astrophysics and Space Science, 364(11), DOI: 10.1007/s10509-019-3678-x

Joseph, R. G, Dass, R. S., Rizzo, V., Cantasano, N., Bianciardi, G. (2019), Evidence of Life on Mars? Journal of Astrobiology and Space Science Reviews, 1, 40–81.

Joseph, R. Gibson, C., Schild, R. (2020). Water, Ice and Mud in the Gale Crater: Implications for Life on Mars. Submitted and Under Review.

Joseph, R. Graham, L., Budel, B., Jung, P., Kidron, G. J., Latif, K., Armstrong, R. A., Mansour, H. A., Ray, J. G., Ramos, G.J.P., Consorti, L., et al. (2020). Specimens Resembling Algae (Cyanobacteria), Lichens, Fungi, Microbial Mats, Stromatolites, and Trace Fossils in Gale Crater. Submitted and Under Review.

Korablev, O.I., J.L. Bertaux, and J.P. Dubois (2001), Occultation of stars in the UV: Study of the atmosphere of Mars, J. Geophys. Res., 106, 7597–7610, doi:10.1029/2000JE001298.

Kieffer, H.H., B.M. Jakosky, and C.W. Snyder (1992), The planet Mars: From antiquity to present, in Mars, edited by H. H. Kieffer et al., pp. 1–33, Univ. of Ariz. Press,

Tucson, Ariz.

Klein, H.P., Horowitz, N.H., Levin, G.V., Oyama, V.I., Lederberg, J., Rich, A., Hubbard, J.S., Hobby, G.L., Straat, P.A., Berdahl, B.J., Carle, G.C., Brown, F.S., and Johnson, R.D. (1976) The Viking Biology Investigation: Preliminary Results. Science. 194, 4260, p. 92-105.

Kidron, G. J. (2019). Cyanobacteria and Lichens May Not Survive on Mars. The Negev Desert Analogue Journal of Astrobiology and Space Science Reviews, 1, 369-377, 2019.

Kidron, G. J., Ying, W., Starinksy, A., Herzberg, M. (2017). Drought effect on biocrust resilience: High-speed winds result in crust burial and crust rupture and flaking, Science of The Total Environmment, 579. 848-859, Doi: 10.1016/j.scitotenv.2016.11.016.

Klingelho"fer, G. Jarosite and Hematite at Meridiani Planum from Opportunity's Mössbauer Spectrometer, Science 306, 1740-1745, (2004).

Knauth, L., Burt, D. & Wohletz, K. Impact origin of sediments at the Opportunity landing site on Mars. Nature 438, 1123–1128 (2005). https://doi.org/10.1038/nature04383

Ksanfomality, L. W., (2013). An Object of Assumed Venusian Flora Doklady Physics, 2013, Vol. 58, No. 5, pp. 204–206.

Levin, G.V., Straat, P.A., and Benton, W.D. (1978) Color and Feature Changes at Mars Viking Lander Site. J. Theor. Biol., 75: 381-390.

Levin, G., Straat, P. A. (1976) Viking Labeled Release Biology Experiment: Interim Results, Science, 194, 1322-1329.

Levin, G. V., Straat, P. A. (1977) Life on Mars? The Viking labeled release experiment, Biosystems 9 :2-3, pp. 165-174.

Levin, G. V., Straat , P. A. (1979) Completion of the Viking Labeled Release Experiment on Mars, J. Mol. Evol., 14, 167-183.

Levin, M. (2003). Review: Bioelectromagnetics in Morphogenesis. Bioelectromagnetics 24:295-315.

Lindsay, D.C. (1978). The role of lichens in Antarctic ecosystems. Bryologist 81: 268-276.

Llano, G.A. (1965). The flora of Antarctica. In: Antarctica Ed. T Hatherton, Methuen & Co, London, pp 331-350.

Longton, R.E. (1979). Vegetation ecology and classification in the Antarctic zone. Canaadian Journal of Botany 57: 2264-2278.

Lowy, D.A. et al. (2006) "Harvesting energy from the marine sediment- water interface II - kinetic activity of anode materials." Biosens. Bioelectron. 21, 2058-2063.

Maffei, M. E. (2014). Magnetic field effects on plant growth, development, and evolution (2014). Front. Plant Sci., 04.

Malin, M. C., Edgett, K. S., (1999). Oceans or Seas in the Martian Northern Lowlands: High Resolution Imaging Tests of Proposed Coastlines, Geophys. Res. Letters, V. 26, No. 19, p. 3049-3052.

Malin, M.C. K.S. Edgett, Evidence for recent groundwater seepage and surface runoff on Mars. Science 288(5475), 2330 (2000).

Maltagliati, L., F. Montmessin, A. Fedorova, O. Korablev, F. Forget, J.-L. Bertaux, Evidence of water vapor in excess of saturation in the atmosphere of Mars. Science 333, 1868–1871 (2011). doi:10.1126/science.1207957pmid:21960630

Masursky, H., et al. (1972), Mariner 9 Mars television experiment, Bull. Am.

Astron. Soc., 4, 356.

Mellon, M.T. R.J. Phillips, Recent gullies on Mars and the source of liquid water. J. Geophys. Res. 106(E10), 23165–23180 (2001).

Meessen, J., Backhaus, T., Sadowsky, A., Mrkalj, M., Sanchez, F.J., de la Torre, R., Ott, S. (2014). Effects of UVC254 nm on the photosynthetic activity of photobionts from the astrobiologically relevant lichens Buellia frigida and Circinaria gyrosa. Int J Astrobiol 13: 340-352.

Moment GB. 1949. On the relation between growth in length, the formation of new segments, and electric potential in an earthworm. J Exp Zool 112:1-12.

Moores, J. E., et al. (2015). Atmospheric movies acquired at the Mars Science Laboratory landing site: Cloud morphology, frequency and significance to the Gale Crater water cycle and Phoenix mission results, Advances in Space Research, 55, 2217-2238.

Morel, D. (2013). Hematite: Sources, Properties and Applications, Nova Biomedical

Mustard, J. F., F. Poulet, B. L. Ehlmann, R. Milliken, and A. Fraeman (2012), Sequestration of volatiles in the Martian crust through hydrated minerals: A significant planetary reservoir of water, in 43rd Lunar and Planetary Sci. Conf., Abstract No. 1539, Lunar and Planetary Institute (LPI), Houston, Tex.

Novikova, N (2009) Mirobiological research on board the ISS, Planetary Protection. The Microbiological Factor of Space Flight. Institute for Biomedical Problems, Moscow, Russia.

Novikova, N et al. (2016) Long-term spaceflight and microbiological safety issues. Space Journal, https://roomeu.com/article/long-term-spaceflight-and-microbiological-safety-issues.

Occhipinti, A., De Santis, A., and Maffei, M. E. (2014). Magnetoreception: an unavoidable step for plant evolution? Trends Plant Sci. 19, 1-4. doi: 10.1016/j.tplants.2013.10.007

Onofri, S., et al (2018). Survival, DNA, and Ultrastructural Integrity of a Cryptoendolithic Antarctic Fungus in Mars and Lunar Rock Analogues Exposed Outside the International Space. Astrobiology, 19, 2.

Onofri, S., Selbman, L., Pacelli, C. et al. (2019) Survival, DNA and ultrastructural integrity of a cryptoendolithic Antarctic fungus on Mars and lunar rock analogues exposed outside the International Space Station. Astrobiology, 19, 2, DOI:10.1089/ast.2017.1728.

Owocki. K., Kremer, B., Wrzosek, B., Królikowska, A., Kaźmierczak J. (2016). Fungal Ferromanganese Mineralisation in Cretaceous Dinosaur Bones from the Gobi Desert, Mongolia. Plos One. https://doi.org/10.1371/journal.pone.0146293

Perron, J. Taylor; Jerry X. Mitrovica; Michael Manga; Isamu Matsuyama & Mark A. Richards (2007). Evidence for an ancient Martian ocean in the topography of deformed shorelines Nature. 447: 840-843.

Platt, J. L., Spatafora, J.W. (2000). Evolutionary Relationships of Lichenized Fungi: Molecular Phylogenetic Hypotheses for Genera Siphula, Thamnolia from SSU and LSU rDNA. Mycologia. 92, 475-487.

Plaut, J. J., et al. (2007), Subsurface radar sounding of the south polar layered deposits of Mars, Science, 316(5821), 92– 95.

Pruppacher H., Klett J. (2010) Microstructure of Atmospheric Clouds and Precipitation. In: Microphysics of Clouds and Precipitation. Atmospheric and Oceanographic Sciences Library, vol 18. Springer, Dordrecht.

Rabb, H. (2015) Life on Mars - Visual Investigation. https://www.scribd.com/doc/288486718/Life-on-Mars-Visual-Investigation. Scrib D. publishers.

Rabb, H. (2018). Life on Mars, Astrobiology Society, SoCIA, University of Nevada, Reno, USA. April 14, 2018.

Raggio J, Pintado A, Ascaso C, De La Torre R, De Los Ríos A, Wierzchos J, Horneck G, Sancho LG (2011). Whole lichen thalli survive exposure to space conditions: results of Lithopanspermia experiment with Aspicilia fruticulosa. Astrobiology. 2011 May;11(4):281-92. doi: 10.1089/ast.2010.0588.

Read, P.L., and S.R. Lewis (Eds.) (2004), The Martian Climate Revisited—Atmosphere and Environment of a Desert Planet, 326 pp., Springer Verlag, Berlin.

Rieder, R. et al., Chemistry of Rocks and Soils at Meridiani Planum from the Alpha Particle X-ray SpectrometerScience 306, 1746-1749 (2004)

Rennó, N.O., B.J. Bos, D. Catling, B.C. Clark, L. Drube, D. Fisher, W. Goetz, S.F. Hviid, H.U. Keller, J.F. Kok, et al., Possible physical and thermodynamical evidence for liquid water at the Phoenix landing site. J. Geophys. Res. 114(E1), 0003 (2009).

Roffman, D. A. (2019). Meteorological Implications: Evidence of Life on Mars? Journal of Astrobiology and Space Science Reviews, 1, 329-337, 2019

Ryan, BD, Bungartz F, Nash TH (2002). Morphology and anatomy of the lichen thallus. In Lichen Flora of the greater Sonoran Desert region (eds Nash TH, Ryan BD, Gries C, Bungartz F), pp. 8-23. Tempe, AZ.

Saleh, Y. G., M. S. Mayo, and D. G, Ahearn (1988) Resistance of some common fungi to gamma irradiation." Appl. Environm. Microbiol. 1988, 54: 2134-2135

Sanchez, F. J., E. et al. (2012) The resistance of the lichen Circinaria gyrosa (nom. provis.) towards simulated Mars conditions-a model test for the survival capacity of an eukaryotic extremophile." Planetary and Space Science, 2012, 72(1), 102-110.

Sancho L. G., de la Torre, R., Horneck, G., Ascaso, C., de los Rios, A. Pintado,A., Wierzchos, J.,Schuster, M. (2007). Lichens Survive in Space: Results from the 2005 LICHENS Experiment Astrobiology. 7, 443-454.

Schroeter, B., Scheidegger, C. (1995) Water relations in lichens at subzero temperatures - Structural changes and carbon-dioxide exchange in the lichen Umbilicaria aprina from continental Antarctica. New Phytologist 131: 273-285.

Small, L. W, (2015) On Debris Flows and Mineral Veins - Where surface life resides on Mars. https://www.scribd.com/doc/284247475/On-Debris-Flows-eBook

Smith, M.D. (2004), Interannual variability in TES atmospheric observations of Mars during 1999–2003, Icarus, 167, 148– 165.

Smith, M.D., J.C. Pearl, B.J. Conrath, and P.R. Christensen (2001), One Martian year of atmospheric observations by the Thermal Emission Spectrometer, Geophys. Res. Lett., 28, 4263–4266, doi:10.1029/2001GL013608.

Soderblomet. L. A. al.,Soils of Eagle Crater and Meridian Planum at the Opportunity Rover Landing Site" (PDF). Science. 306 (5702): 1723–1726. Bibcode:2004Sci...306.1723S. doi:10.1126/science.1105127. PMID 15576606.

Spinrad, H., G. Münch, and L.D. Kaplan (1963), Letter to the Editor: The detection of water vapor on Mars, Astrophys. J., 137, 1319, doi:10.1086/147613.

Squires, S. W. , Grotzinger, J. P. R. E. Arvidson,3J. F. Bell III,1W. Calvin,4P. R. Christensen,5B. C. Clark,6J. A. Crisp,7W. H. Farrand,8K. E. Herkenhoff,9J. R. Johnson,9G. Klingelho Ter,10A. H. Knoll,11S. M. McLennan,12H. Y. McSween Jr.,13R.

V. Morris,14J. W. Rice Jr.,5R. Rieder,15L. A. Soderblom9 (2004). "In Situ Evidence for an Ancient Aqueous Environment at Meridiani Planum, Mars" (PDF). Science. 306 (5702): 1709–1714. Bibcode:2004Sci...306.1709S. doi:10.1126/science.1104559. PMID 15576604.

Todd, R. Clancy, M. D. Smith, F. Lefèvre, T. H. McConnochie, B. J. Sandor, M. J. Wolff, S. W. Lee, S. L. Murchie, A. D. Toigo, H. Nair, T. Navarro, (2017) Vertical profiles of Mars 1.27 μm O2 dayglow from MRO CRISM limb spectra: Seasonal/global behaviors, comparisons to LMD-GCM simulations, and a global definition for Mars water vapor profiles. Icarus 293, 132–156 (2017). doi:10.1016/j.icarus.2017.04.011

Tugay, T. Zhdanova, N.N., Zheltonozhsky, V., Sadovnikov, L., Dighton, J. (2006). The influence of ionizing radiation on spore germination and emergent hyphal growth response reactions of microfungi, Mycologia, 98(4), 521-527.

U.S. Department of the Interior (2010) Lichen Inventory Synthesis Western Arctic National Parklands and Arctic Network, Alaska. Natural Resource Technical Report NPS/AKR/ARCN/NRTR--2010/385.

Vesper, S.J., W. Wong, C.M. Kuo and D.L. Pierson. (2008) Mold species in dust from the ISS identified and quantified by mold-specific quantitative PCR. Research in Microbiology. 159: 432-435.

Villanueva, G. Mumma, M. Novak, R. Kufl, H. Hartogh, P., Encrenaz, T., Tokunaga, A., Khayat, A., Smith, M. (2015) Strong water isotopic anomalies in the Martian atmosphere: Probing current and ancient reservoirs". Science. 348: 218-21.

Wember, V. V., Zhdanova, N. N. (2001) Peculiarities of linear growth of the melanin-containing fungi Cladosporium sphaerospermum Penz. and Alternaria alternata (Fr.) Keissler. Mikrobiol. Z. 63: 3-12. Whalen, S.C. (2005). Biogeochemistry of Methane Exchange between Natural Wetlands and the Atmosphere, Environmental and Atmospheric Science, 22, 1093-1096.

White, O. et al. (1999) Genome Sequence of the Radioresistant Bacterium Deinococcus radiodurans R1, Science, 286, 1571-1577.

Whiteway, J. A., et al. (2009), Mars water ice clouds and precipitation, Science, 325(5936), 68–70, doi:10.1126/science.1172344.

Zakharova,K., et al. (2014). Protein patterns of black fungi under simulated Mars-like conditions. Scientific Reports, 4, 5114.

Zhdanova NN, Lashko TN, Vasiliveskaya AI, Bosisyuk LG, Sinyavskaya OI, Gavrilyuk VI, Muzalev PN. (1991). Interaction of soil micromycetes with 'hot' particles in the model system. Microbiol J 53:9-17.

Zhdanova, N. N., T. Tugay, J. Dighton, V. Zheltonozhsky and P McDermott, (2004) Ionizing radiation attracts soil fungi." Mycol Res. 2004, 108: 1089-1096.

V. IS MARS A FAILED EARLY? THE ORIGINS AND EVOLUTION OF MARTIAN LIFE
Rhawn Joseph

It is now established that Mars at one time, had a magnetic field, as well as rivers, streams, and oceans of water; whereas reports from the European Space Agency and other investigators, indicates that liquid water continues to percolate and pool upon the surface of Mar. Where there is water, there is life.

1. Interplanetary Origins and the Earliest Evidence of Life on Earth and Mars

Life on Mars, Venus, and Earth, may have come from other planets, including those in other solar systems (Joseph 2000, 2009, 2019). A number of scientists have also argued that Mars and Earth may have repeatedly exchanged life beginning billions of years ago via microbial-laden atmospheric dust and bolides ejected into space (Arrhenius, 1914; Beech et al. 2018; Davies, 2007; Joseph 2019; Melosh 2003; Schulze-Makuch, et al. 2005). As the ultimate source and origins of life are unknown, its also been argued that during the heavy bombardment phase which came to a close around 3.8 bya, that Earth, and thus Mars, were seeded with cyanobacteria, fungi and other forms of life which originated in solar systems external to our own (Arrhenius, 1908; Joseph and Schild, 2010; Gibson et al. 2010a,b; Wickramasinghe et al. 2018).

The ultimate origins and source for life are unknown. All Miller-Urey life-creation-experiments, initially conducted by Nobel Laureate Harold Urey and his student Stanley Miller, have failed to produce life (Miller & Urey, 1959; Lazcano & Bada, 2004). However, Urey and colleagues (Claus & Nagy 1961; Nagy et al. 1962; Nagy et al. 1963) did find evidence of fossilized cyanobacteria in the Orgueil meteor--discoveries which were dismissed as due to terrestrial contamination. Subsequently a number of independent investigators also reported what they believed to be extraterrestrial fossilized cyanobacteria and other microbes, including fungi, in five different meteorites: the Murchison, Ivuna, Orgueil, Allende, and Efremovka (Folk and Lynch 1997; Hoover, 2011; Pflug, 1984; Zhmur and Gerasimenko 1999; Zhmur et al. (1997). Although the likelihood of contamination should be taken seriously, it is noteworthy that all are infested with the same "contaminants" i.e. fossilized cyanobacteria (algae) and fungi; species which may have colonized Earth and Mars at around the same time, during a period in which both planets were subject to a heavy bombardment by meteors, asteroids, comets, and moon-size stellar objects until around 3.8 bya (Schoenberg et al. 2002).

Coupled with repeated failures to produce life from non-life, a number of scientists, beginning with the Greek Atomists, have proposed that life on Earth (and thus Mars) must have come from space and other planets much older than our own; a view championed by notable scientists including von Helmholtz (1872), Lord Kelvin (Thomson 1881), Nobel Laureates Svante Arrhenius (1908) and Francis Crick (1981), Sir Fred Hoyle (1978) and Carl Sagan (Sagan and Shklovskii, 1966). Arrhenius (1908) proposed that spores journey through space propelled by photons and galactic winds. Hoyle and Wickramasinghe (2000) have identified comets and interstellar dust as a delivery source for viruses and bacteria (Wickramasinghe et al. 2018; Joseph and Wickramasinghe 2010; Gibson et al. 2011a,b). Joseph and Schild (2010) have argued that microbe-infested bolides and solar winds deposited extraterrestrial life on Earth and other planets during the heavy bombardment phase which came to a close around 3.8 bya. If these theories are correct, then the same species (i.e. algae and fungi) should have taken root on Earth and Mars at around the same time; a hypothesis which is supported by considerable evidence.

There is no substantial evidence of life in Eagle Crater and Gale Crater. Perhaps life was deposited on Mars and in the vicinity of these craters following impact; with the survivors going forth and multiplying.

The Gale Crater is believed to have formed 3.7bya when Mars was struck by a large asteroid, creating a massive crater which filled with water. Noffke (2015) has reported evidence of structures resembling stromatolites which were fashioned 3.7 bya in water. Could those stromatolites been fashioned by algae that originated from space?If extraterrestrial debris harbored living organisms, then life should have taken root on Mars and Earth well before 3.7 bya; and this hypothesis is supported by evidence of biological residue, including PAHs, in Martian meteor ALH84001.

McKay et al. (1996) and Thomas-Keprta et al. (2002, 2009), as based on a detailed analyses of fossilized evidence of biological activity in Martian meteor ALH 84001, have determined that 4 billion years ago Mars was flush with microbial life, including bacterial magnetosomes, such as those found in fresh pond water. Bacterial magnetosomes (Magnetotactic bacteria) fashion magnetic crystals which enable them to orient along the magnetic lines of a planet's magnetic field.

Although Thomas-Keprta et al. (2002, 2009) claim their findings "constitute evidence of the oldest life yet found" the fact is, around the same time if not earlier, life had also colonized Earth. Specifically, microprobe analyses of the carbon isotope composition of metasediments in Western Australia formed 4.2 BY revealed very high concentrations of carbon 12, or "light carbon" which is typically associated with microbial life (Nemchin et al. 2008). Moreover, biological evidence of Earthly magnetotactic bacteria was found in banded iron formations in northern Quebec, Canada, consisting of alternating magnetite and quartz dated to 4.28 BY (O'Neil et al. 2008). Similar life forms may have appeared on Mars and Earth around the same time.

Thomas-Keprta and colleagues (2002, 2009) have argued that life appeared on Mars between 4.2 to 3.8 bya. Sedimentary rocks in the Gale Crater have been dated from 4.2 bya to 2.7bya (Grotzinger et al. 2015). On Earth, when terrestrial rocks first began to reform 4.2 bya, there is evidence of biological activity (Nemchin et al. 2008; O'Neil et al. 2008), as suggested by the very high concentrations of carbon 12, or "light carbon" detected in those rocks and which is typically associated with microbial life (Nemchin et al. 2008). More definitive evidence of terrestrial life has been dated from 3.8 to 3.9 bya, and includes carbon-isotopes discovered in quartz-pyroxene rocks on Akilia, West Greenland dated to 3.8 vya, and within a phosphate mineral, apatite, which includes tiny grains of calcium and high levels of organic carbon; the residue of photosynthesis, oxygen secretion, and thus biological activity (Manning et al. 2006; Mojzsis et al. 1996). In addition, microfossils resembling fungi were discovered in 3.8 billion year old quartz, recovered from Isua, S. W. Greenland (Pflug 1978). Evidence of terrestrial biological activity including photosynthesis was also discovered in this area dated from the same time period (Rosing 1999, Rosing and Frei 2004). By 2.7 bya ago, a variety of eukaryotes and prokaryotes were flourishing on Earth.

2. Speculation: Evolution of Life on Mars?

Estimates, based on an analyses of 25 binary traits indicate that cyanobacteria, appeared on Earth between 2.7 to 3.8 bya (Uyeda et al. 2016) which corresponds to estimates as to the construction of stromatolites (Allwood et al. 2009; Gardwood 2012; Graham et al., 2016; Lepot et al. 2008). On Earth it is believed by some scientists that the earliest cyanobacterial lineages dwelled in watery habitats and that their evolution shifted, diversified and accelerated in response to major geological events, i.e. the "Great Oxidation Event" and the world-wide "Snowball Earth" glaciation 2.5 to 2.3 bya (Bekker et al. 2004; Farquhar et al. 2011). Evolutionary innovations included filamentous and branched morphology, increases in size and motility and the appearance or complex planktonic-like forms (Uyeda et al. 2016) followed by fuculinida, and treptichnids (Sen Gupta, 1999; Vannier et al. 2010; Zakrys et al. 2017)

Likewise, there is evidence of numerous major, perhaps cataclysmic environmental transitions on Mars, as indicated by the what appears to be repeated alternating cycles of water filling then disappearing from the Gale Crater. The repeated loss then replenishment of a watery habitat, in-itself, could have triggered evolutionary innovations that led to what appears to be trace fossils of fusulinida which is an extinct order within the Foraminifera marine heterotrophic protists (Sen Gupta, 1999).

On Earth it's been estimated that the radiation of early foraminifera may have had its onset 1,150 bya (Pawlowski et al. 2003). Foraminifera are believed to have descended from Platysolenites antiquissimus around 1.15 bya, and have left a rich biomineralized fossilized assembly of their multiple forms and lineages; evolving from simple soft-walled unilocular forms into multilocular (Hensen 1979) and which include globular, spiral, spherical, and tubular species (Tappen and Loeblich, 1998; McIlroy et al. 2001) with all these diverse forms appearing together in a single species (Pawlowski et al. 2002).

Joseph et al., (2020a) have presented evidence of what may be fossilized sponges, corals, and a variety of metazoans on Mars, within Gale Crater. It could be argued that these specimens resembling siliceous filamentous sponges, bioherms, biostromes and other porifera-like and coral-like organisms evolved on and were fossilized on Mars perhaps at the same time similar species on Earth evolved or became extinct during the Cambrian Odovician transition (Lee et al. 2014, 2016). Moreover, trace fossils similar to tunneling cyanobacteria and (to speculate) worms circa 500 mya (Vannier et al. 2010) have been presented, which, coupled with the sponge-like fossils, supports the speculation that some these putative Martian fossils may have been formed around 500 mya on Mars.

The peer reviewed evidence compiled by a number of independent investigators, coupled with the evidence reported and reviewed by Joseph et al. (2019, 2020a,b), supports the hypothesis that algae (cyanobacteria) and other organisms may have been flourishing on Mars 3.7 bya. If Martian life did not become completely extinct, then those survivors might be expected to evolve in response to the changing environment, in parallel with evolutionary events on Earth. In fact, the evolutionary atmosphere-geology of Mars may have progressed more rapidly than events on Earth, perhaps "to the point of allowing low-volume partial melts before 4.2 Ga, while the Earth only cooled this far at ~2.7 Ga" (Grotzinger et al. 2015); a time period in which there is firm evidence terrestrial cyanobacteria were building stromatolites (Lepot et al. 2008; Graham et al. 2016), though others date the first Earthly stromatolites to around 3.7 bya (Allwood et al. 2009; Garwood 2012) at around the same time Martian stromatolites may have first been fashioned (Noffke 2015).

If life on Mars and Earth appeared around the same time, then the pace of biological evolutionary innovation may have also occurred in parallel on Mars and Earth during the first few billions years in response to major environmental-geological transitions; which on Earth, are associated with major extinction events and great leaps in evolutionary change (Eldredge and Gould, 1972; Grey et al., 2003). The Gale Crater (and presumably the entire planet) was repeatedly subject to repeated cycles of major environmental upheavals involving the loss and replenishment of water resources. Hence, it can be predicted that those species which survived these cataclysmic events on Mars, adapted and evolved; and that some of these organisms left trace fossils or became fossilized after they died.

3. Evolution and Adaptation of Algae On Mars

Algae and other organisms may have been deposited on Mars, Earth, Venus, and other planets encased in meteors, asteroids and other debris cast into space following bolide impacts—and some of that living-cargo may have contaminated not only planets in this solar system, but other solar systems—perhaps even arising from other solar systems.

Terrestrial algae have colonized the most extreme and life-neutralizing environments on Earth (Hoham and Ling 2000; Leya 2013; Scalzi et al., 2012; Sharma and Shukla, 2019; Holzinger and Pichrtová 2016), including frigid climates and hot deserts surface crusts (Büdel et al. 2004; Lewis and Flechtner, 2002 ; Flechtner, 2007) Antarctic rocks (Büdel et al. 2008; Broady, 1996; Graham et al., 2016) and despite inadequate or a paucity of available nutrients (Rindi and Guiry, 2002 ; Häubner et al., 2006 ; Rindi, 2007). As reviewed in this report, algae which evolved on Earth can also survive Mars-simulated environments. Certainly, algae which evolved on Mars would be adapted to living in the Martian environment.

The Martian algae-like specimens presented by Joseph and colleagues (2020a) were photographed by the rover Curiosity in the Gale Crater which is located along Mars' highlands-lowlands boundary. It's been estimated that the crater was formed around 3.7 bya and then filled with water (Le Deit et al., 2013; Thomson et al., 2011). The overall pattern of evidence currently available, based also on hydrated minerals, indicates that the Gale Crater has been periodically flush with rivers, streams, and lakes, and that water continues to periodically moisten the surface. As also based on the evidence reviewed in this paper, there was sufficient water to induce mineral formation, the possible mineralization of fossils, and to sustain what appears to be stromatolite- and ooid-building algae, beginning around 3.7 bya and continuing to the present. When those large bodies of Martian surface waters first disappeared only to return and how often this cycle has been repeated is unknown, but these drastically changing conditions would have required any surviving organisms to evolve and adapt.

Many scientists believe that blue-green algae first flourished in low-salinity aquatic systems (Uyeda et al. 2016), and only later migrated to land (Rindi 2010). If this scenario is correct, then just as water-dwelling green algae migrated to dry land on Earth (Lewis and Lewis, 2005; Lewis, 2007; López-Bautista et al., 2007; Cardon et al., 2008), the same phenomena may have taken place on Mars with water-dwelling algae becoming land-dwelling algae which evolved and adapted to these wet-dry conditions and exposure to high levels of radiation.

That algae can rapidly evolve to drastic changes in environmental and climatic conditions is well documented (Collins et al., 2006ab; Miller et al., 2005; Finkel et al., 2005, 2007) including from wet to dry desert conditions (Lewis and Lewis, 2005 ; Lewis, 2007) where they are subject to high levels of UV radiation, near complete desiccation, and extremes and faster variations of temperature (Fagliarone et al. 2017). If the transition is rapid, and/or if exposed to high levels of CO_2, profound evolutionary changes may ensue (Rindi 2010). This includes, in just two years after only 1000 generations (Collins and Bell, 2004; Collins et al., 2006ab), dramatic alterations in morphology and phenotype (Rindi and Guiry, 2002; Luo et al., 2006; Lewis and McCourt, 2004).

Algae and other species including fungi, have also been found to evolve in response to heightened radiation exposure, and rapidly develop adaptive features--a property described as "radiostimulation," "radiation hormesis," and "adiotropism" (Levin 2003; Tugay et al. 2006; Zhdanova et al 2004). These radiation-induced adaptations include tissue and cellular regeneration and growth and enhanced spore production (Basset 1993; Becker 1984; Becker & Sparado 1972; Occhipinti et al. 2014; Levin 2003; Maffei 2014; Moment, 1949). This may account for why the algae depicted in this report maintained their green coloration, even when they appear to be dried and dehydrated.

Terrestrial algae are naturally adapted to high levels of UV radiation (Rindi 2010) which is employed for photosynthesis (Karsten et al., 2007b). Also making them well adapted for life on Mars: low vs high levels of direct sunlight enhance algae growth and photosynthesis (Ong et al., 1992; Hughes, 2006; Gray et al., 2007; Pierangelini, et al. 2017). Likewise, exposure to high CO_2, environments can even increase growth and photosynthetic activity in some algae taxa (Miyachi et al., 2003; Huang et al. 2018). For example, an Antarctic green alga, Chlorella, when exposed to elevated CO_2 levels, was found to manipulate levels of metabolites associated with energy, amino acid, fatty acid and carbohydrate production, thereby maintaining its growth and photosynthetic abilities under acidified conditions (Tan et al. 2019). Algae are well adapted for life on the Red Planet.

Cyanobacteria/Chroococcidiopsis Are Adapted For Life On Mars

A number of research teams have also documented that algae/cyanobacteria (as well as fungi and lichens) survive in Mars-simulated environments and even direct exposure to space outside the ISS (de Vera et al. 2014; Nicholson et al. 2012; Scalzi et al., 2012). For example, de Vera and colleagues (2014) reported that cyanobacteria collected from cold and hot deserts survived "Mars-like conditions such as atmospheric composition, pressure, variable humidity (saturated and dry conditions) and strong UV irradiation." Olsson-Francis and colleagues (2009) exposed akinetes (dormant cells formed by Nostocales of filamentous cyanobacteria) to extraterrestrial conditions, periods of desiccation, temperature extremes (-80 to 80°C), and UV radiation (325-400 nm). These organisms all displayed high levels of viability

A number of scientists have argued that the blue-green algae, such as Chroococcidiopsis are well adapted for life on the Red Planet (Bothe, 2019; reviewed by de Vera et al. 2019). Chroococcidiopsis are believed to be among the oldest and most ancient of life forms to have colonized Earth. Others argue, however, that Chroococcidiopsis evolved in parallel to the filamentous, heterocyte bearing cyanobacteria (Clade 5). Data based on genetic, molecular clocks suggest they began to evolve about 2 billion years ago, while the origin of clade 1 cyanobacteria dates back at least 2.600 billion years (Schirrmeister et al 2015). Nevertheless, evidence from molecular clocks and the fossil record indicates that algae or algae-like species were already flourishing in the Archean (Hayes 1996; Mojzis et al.1996), between 2.7 to 3.7 bya (Uyeda et al. 2016), and building stromatolites (Allwood et al. 2009; Garwood 2012).

Many species of cyanobacteria/blue-green algae, of the order Nostocales, such as Anabaena cylindrica, form resting cells (akinetes), which are well suited for a Mars-like and other extreme environments (Olsson-Francis et al. 2009; Sharma and Shukla, 2019). Likewise non-akinetes forming species, such genus Chroococcidiopsis, are resistant to drought, extremes in temperature and can withstand harsh environmental conditions including exposure to high levels of radiation (Fagliarone et al 2017; Olsson-Francis et al., 2009). The genus Chroococcidiopsis, in particular, are adapted for life on Mars and extraterrestrial survival (e.g., Billi et al. 2019; Cockell et al., 2005; Olsson-Francis and Cockell, 2010; Baque et al., 2013, 2016; Verseux et al., 2017) and can survive prolonged exposure, in a desiccated state, and maintain genetic stability (Billi 2009).

Chroococcidiopsis includes both salt-tolerant and salt-sensitive species, some of which have a growth requirement for salt (Cumbers and Rothschild, 2014) for which there is evidence in the Gale crater where all the algae-like specimens were photographed. Chroococcidiopsis also survives near the surface of salt (halite) deposits (Stivaletta, Barbieri and Billi, 2012; Finstad et al., 2017), occupying crevices between salt crystals in the Atacama Desert (Wierzchos, Ascaso and McKay, 2006).

A particular binding of Chroococcidiopsis to gypsum soils has repeatedly been reported, e.g. in the Atacama and Mojave deserts, in the Al-Jafr Basin of Jordan (Dong et al., 2007) and in mound evaporite deposits in Tunisia (Stivaletta and Barbieri, 2009). Evidence of gypsum has been detected on Mars (Vaniman et al., 2018) and may form a substrate for Chroococcidiopsis-like algae. A marked feature of Chroococcidiopsis is its resistance to desiccation (Billi 2009; Fagliarone et al. 2017) which is why it is found in the driest places world-wide, such as the deserts of Atacama Chile (Azua-Bustos et al., 2012; Gomez-Silva, 2018; Jung et al. 2019), the Mojave in the United States (Smith et al., 2014) and the Antarctic (Friedmann, 1982). These species can also survive extreme cold and freezing and can warm the soil surface by as much as 10 °C through the production and accumulation of scytonemin which absorbs photons thereby generating heat. Among the ecological strategies of some cyanobacteria, when subjected to drier environments, is the production of akinetes and presence of mucilage thereby minimizing the effects of cell desiccation (Fletcher 2007, Ramos et al. 2019). These latter structures may remain inactive and withstand for long periods until environmental conditions become favorable for cell development.

The formation of pigments, such as carotenoids and scytonemin also protects this genus of algae against UV radiation (Vitek et al., 2014, 2017). The latter compound, an aromatic indole alkaloid, is specifically synthesized by cyanobacteria as an effective sunscreen; and its synthesis can be enhanced upon periodic desiccation (Fleming and Castenholz, 2007). Tolerance to UV radiation is also related to the damage and repair constants in the Photosystem II reaction centre. For example, an Antarctic strain of Chlorella exhibited higher repair ability at elevated temperature when exposed to increased UVR, suggesting algae living in cold environments may be stressed by high UVR (Wong et al. 2015). This may be countered by the protective activity of the antioxidative compounds like carotenoids and the photoprotective mucosporine amino acids (MAA).

Algae would not be negatively impacted even by direct exposure to high levels of radiation coupled with periodic loss of water on Mars, as also documented in space- and -Mars-like-environmental studies (Billi et al., 2019; Cockell et al. 2005; de Vera et al. 2019; Fagliarone et al. 2017). Typically, algae/cyanobacteria on Earth, protect themselves against irradiation by living in shadows, or on the sides or a few millimeters within rocks (Jung et al. 2019). On the other hand, there is substantial evidence that photosynthesis is promoted by UV exposure coupled with low doses of sunlight (Bothe 2019). Therefore, these and other species of algae/cyanobacteria would be well adapted to surviving on the surface of rocks and sands of Mars; especially if provided with water; and for which there is now evidence.

Although it is impossible to identify what species of algal are depicted in the photographs presented in this report, it is notable that the blue-green Chroococcidiopsis and other algae species also respond to long periods of drought by turning a yellow or a very faint green (Pattanaiki et al. 2007), similar to some the specimens depicted in this report.

Martian Water and Desiccation

It is believed there are large amounts of Martian water beneath the surface, in the atmosphere, and sequestered in rocks, hydrated minerals, and frozen ground and icecaps (Plaut et al., 2007; Mustard et al., 2012; Steele et al. 2017; Titus et al. 2003). Moisture may be available in limited quantities depending on the season, as melt water, run off, or by percolating to the surface (Castro et al. 2015; Steele et al. 2017). Water ice appears to have formed in the Curiosity rover wheel wells. Hence, there appears to be sufficient water to maintain these algae-like species, which can easily adapt to rapidly changing and increasingly hostile environmental conditions.

Although Martian moisture may be available only in the mornings, or just beneath the surface, some species can survive extreme desiccation, and up to 35 years without water (Trainor and Gladych, 1995) and at temperatures of 1°C in the absence of moisture (Häubner et al., 2006). Moreover, some species are highly resistant to cellular water loss and dehydration under low light (Pierangelini, et al. 2017) and may inactivate photosynthesis during dehydration in low light and then reactivate photosynthesis after rehydration (Herburger and Holzinger 2015; Karsten et al. 2016). For example, the cyanobacterium Nostoc flagelliforme, has evolved extreme desiccation tolerance and survives in the semi-arid deserts of Spain and Australia by forming colonies protected by mucilage (Aboal et al. 2016) and calcium encrusted "Nostoc balls" which protects the akinetes which are resting cells--and similar structures have been presented by Joseph et al. (2020a) Algae, therefore, are well adapted for a life on Mars, despite lower levels of direct sunlight, extremes in temperature, high levels of radiation and atmospheric CO_2, and diminished water supplies and can quickly evolve and adapt to the most drastic, life neutralizing conditions. The evidence presented here indicates that algae have colonized Mars.

Algae and Water

How much water is available in the Gale Crater is unknown. However, unless the putative algae-like specimens depicted here evolved the ability to flourish with a minimal water supply--the availability of which may decrease exponentially with decreases in temperatures (De Vera et al., 2014)--then the ability to grow and flourish on Mars would be severely impacted.

For example, it is well established that the thickness of water film is crucial for cell functioning (Kidron 1999; Kidron et al. 2014). According to Lange et al. (1992), the threshold for liquid water and hence the minimum water thickness required for cyanobacterial growth is 0.1 mm--and the same appears true for algae-fungal symbiotes (Armstrong, 1976, 1981, 2017). Yet, some have argued that water thickness on Mars may fall to a low of 0.06 mm depending on temperature and other variables (Fouchet et al., 2007) but which may increase to 2 mm or more which is sufficient to facilitate cyanobacterial growth on Earth and would not negatively impact the ability to engage in photosynthesis (see Lange et al., 1992). However, exactly how much moisture may be available on a daily and seasonal basis is unknown.

Consider, by way of analogy, the biota of the Negev Desert which is characterized by an abundance of dew (200 daily dew events a year amounting to 33 mm). Average daily amounts of 0.1-0.3 mm dew have been documented across the Negev Highlands (Kidron, 1999), with a precipitable water of circa 25 mm (Tuller, 1968), i.e., at least two orders of magnitude higher than the lowest estimates for Mars. In the moderately high altitudes of the Negev Highlands (~500 m above msl), endolithic cyanobacteria were found to grow only on south-facing rock outcrops, the high surface temperatures of which prevent vapor condensation during the night and yields 0.02-0.03 mm of moisture (Kidron et al., 2014), substantially lower than the 0.1 mm threshold required for net photosynthesis of cyanobacteria (Lange et al., 1992). This indicates that these rock-desert-dwelling cyanobacteria rely on rainwater for growth (Kidron et al., 2014) and for which there is no evidence on Mars.

Massive biocrusts associated with the photosynthetic activity of lichens, cyanobacteria, microfungi, green algae along with the equivalent of 0.25 mm of water secondary to fog and dew have been identified in the Atacama Desert (Jung et al. 2019). Therefore, it can be predicted that much less water, secondary to dews or vapors, would be sufficient to promote biological activity and the fashioning of biocrusts on Mars which are only a few centimeters (or larger) in size as presented in this report.

As detailed by Joseph and colleagues (2020c) water appears to have formed by condensation and then froze in the wheel wells of the Curiosity rover.

Moreover, several of the algae-like specimens presented by Joseph et al. (2020a) appear to be frozen, or moist. If moisture builds up, freezes, then melts, these organisms would be supplied with sufficient water to maintain and promote biological activity.

Water beneath the surface may also keep these organisms continually supplied with moisture. Consider, for example, the hyper arid desert regions of Kuwait which is characterized by temperature extremes and intense solar radiation (Al-Temeemi & Harris, 2001). There is little precipitation and soil water content is 2% (Al-Sanad & Ismael, 1992; Al-Temeemi & Harris, 2001). And yet, these hyper-arid desert conditions have also been found to draw moisture and water up from the subterranean depths (Al-Sanad & Ismael, 1992). If there are underground pools of water on Mars, then as water is drawn up and pools upon the surface, Martian organisms would be continually moisturized before this water completely evaporates. And the same would be true if water is causing these organisms to freeze, only to thaw.

It is believed that Martian water pools beneath the surface, and is sequestered in rocks, hydrated minerals, and frozen ground and icecaps (Plaut et al., 2007; Mustard et al., 2012; Steele et al. 2017), some of which may melt, forms streams, run off, or percolate to the surface (Castro et al. 2015; Steele et al. 2017). For example, it's been argued that during "the evening and night, local downslope flows transport water vapour down the walls of Gale crater. Upslope winds during the day transport vapour desorbing and mixing out of the regolith up crater walls, where it can then be transported a few hundred metres into the atmosphere" (Steele et al. 2017). Martian water, therefore, may be continually recycled, becoming vapor and forming ice which melts thereby regularly providing water to the surface which may freeze, only to melt, thereby continuing the cycle.

Hence, if water was available in sufficient quantities over 3.7 bya and is still present and available on modern day Mars then that water could have, and would still, support life, including the putative algae/cyanobacteria which are believed to have constructed the stromatolites on Mars, dated to 3.7 bya (Noffke 2015; Rizzo and Cantasano 2016). The daily and seasonal replenishment of surface moisture would also support what appears to be the Martian algae-symbiotes described in this report.

Moreover, the presence of large bodies of water in the ancient past of Mars, would also account for what appears to be trace fossils of sponges, worms, and other marine organisms which may have lived upon, or burrowed beneath the sea floor (Joseph et al. 2020a). When those organisms died, or their tunnels filed with sediment, trace fossils appear to have been formed secondary to mineralization.

4. Cyanobacteria, Calcium, Ooids, and Gas Domes

Joseph and colleagues (2020a) have presented specimens resembling fossilized stromatolites and microbial mats. External and internal morphology of these

Martian specimens are remarkably similar to those in Shark's Bay, Lake Thesis, and other terrestrial locations. These shared features include gas domes and fossilized open cone-like oxygen gas bubbles typically produced by cyanobacteria and which are associated with photosynthesis-oxygen respiration. These Martian mats and stromatolites are also adjacent to calcium and gypsum deposits, and are sometimes enmeshed with what appears to be the fossilized tubular impressions of filamentous cyanobacteria engaged in mat building.

There is also evidence of specimens almost identical to calcium-carbonate encrusted cyanobacteria consisting of interconnected spherical "Nostoc balls;" similar to Nostoc flagelliforme (Aboal et al. 2016). These have been photographed in association with species similar to fungi-lichen, and enmeshed with surrounding surfaces consisting of calcium incrustations.

Calcium carbonate is a byproduct of cyanobacteria mucous secretions and forms the cementing matrix for the construction of stromatolites and ooids. Ooid-like specimens, along with subsurface hyphae and gas-dome like protrusions, along with calcium biosignatures and specimens resembling lichen-fungi and algae, appear in many of the photos we've presented.

5. Cyanobacteria, Calcium, Oxygen, Stromatolites

On Earth, cyanobacteria (blue-green algae) engage in photosynthesis, produce oxygen and calcium and build stromatolites. Stromatolites have also been discovered on Mars (Rizzo & Cantasano 2009); and, on Earth, the first stromatolites appeared around 3.7 bya in shallow seas. Cyanobacteria (blue-green algae) likely colonized Mars early in its history.

Photosynthesizing, calcium-secreting, oxygen-releasing Cyanobacteria were among the first to take root on Earth. They formed thick cyanobacteria mats, contributed to the eukaryotic gene pool, gave rise to plants, and secreted oxygen and calcium carbonate into the oceans and seas (Alois 2008; Kazmierczak & Stal 2008). The buildup and liberation of vast quantities of oxygen and calcium over four billion Earthly years, acted on gene selection and resulted in skeletal metamorphosis and the evolution of oxygen-breathing animals.

On Earth, Cyanobacteria (algae) provided numerous genes to the eukaryotic gene pool, as well as substances, such as calcium, which promote adhesion and multicellularity as well as shell and bone development. CA buildup in the sea led to two main eukaryotic lineages, one with cell walls rich in polysaccharides (which led to plants), the other containing collagen (metazoans). Multicellularity required calcium and the synthesis of collagen, leading to biocalcification, and then plants and animals were able to stand upright and leave the ocean and migrate to land.

6. Martian Stromatolites

By 3.7 billion years ago, Earthly Cyanobacteria began building stromatolites in shallow waters (Nutman et al. 2016). Cyanobacteria are the only known prokaryotes capable of oxygenic photosynthesis (DesMarais 2000) whereas stromatolites contain high concentrations of calcium. Thus, Cyanobacteria were producing the calcium necessary to build bones as well as oxygen so that oxygen-breathing creatures could evolve and flourish.

The stromatolites of Mars (Rizzo & Cantasano 2009) are fossilized and ancient. The algae which glued them together with calcium, were also likely capable of photosynthesis and were releasing oxygen into the Martian ocean and atmosphere By 3.46 bya these photosynthesizing microbes had released significant amounts of oxygen into Earth's atmosphere and oceans (Hoashi et al., 2009). In fact, they were performing the same functions from deep beneath the sea and were congregating near undersea volcanoes and thermal vents and reducing metals, minerals and carbon dioxide. As based on an analysis of microfossils, stromatolites, and chemical biomarkers in Australia and South Africa, chlorophyll containing cyanobacteria had switched to oxygenic photosynthesis by 2.8 Ga (Olson 2006).

The age of the stromatolites discovered on Mars (Rizzo & Cantasano 2009) is unknown. However, using Earth as a model, it can be surmised that Martian cyanobacteria also began building stromatolites as early as 3.7 billion years ago. And like their cousins on Earth, Martian cyanobacteria began secreting calcium and releasing oxygen into the Martian biosphere. If so, then the biological foundations for the evolution of metazoans and later, oxygen breathing creatures equipped with shells or bones, may have laid upon both planets at around the same time.

7. Fossils of Martian Metazoans

It can be argued that Martian minerals have been subject to considerable biological activity, bio-weathering and tunneling, and serve as an energy source for a variety of chemolithotrophic and biomineralizing organisms including those which engage in photosynthesis, i.e. algae, fungi, and lichens (Cecchi et al. 2019; Rawlings et al. 2003; Wilson, 2004).

A variety of organisms accumulate minerals internally and externally (Polgari et al. 2006; Gadd et al. 2011; Bender et al. 2002; Ayupova and Maslennikov, 2012), including chlorite (Ayupova et al. 2016; Haggart & Bustin, 1999). Many Martian minerals have been formed, hydrated, in recessive bodies of water (Xue & Jin, 2013; Schwenzer & Kring, 2009; Carter et al. 2010; Michalski & Niles, 2010; Lin et al. 2016), which, on Earth, are host to innumerable organisms.

Terrestrial microorganisms not only feed on minerals but bind metallic ions and act as nucleation sites for initiating and mediating biomineralization. This causes minerals to accumulate within and on the exterior of these organisms (Polgari et al. 2006; Gadd et al. 2011; Bender et al. 2002; Ayupova and Maslennikov, 2012). In consequence, when fossilized in regressive bodies of water, only the accumulated minerals may be detected (see Ayupova et al. 2016; Haggart & Bustin, 1999; Ran et al. 1999; Sanz-Montero and Rodriguez-Aranda, 2009; Sakakibara et al. 2014).

Gale Crater is believed to have been a water-rich environment that underwent early mineralization, conditions which would have preserved the denizens of this ancient habitat (Grotzinger et al. 2012, 2015). Therefore, the presence of various minerals on Mars may serve as a food and energy source and byproduct of biological activity and could represent the biomineralized, fossilized remains of these organisms as is true on Earth (Ayupova et al. 2016; Haggart and Bustin, 1999; Ran et al. 1999; Sanz-Montero and Rodriguez-Aranda, 2009; Sakakibara et al. 2014). In addition, Eigenbrode and colleagues (2018) reported organic matter preserved in 3-billion-year-old mudstones at Gale crater, which may represent the residue of Martian organisms. This hypothesis is supported by evidence of fossil-like specimens, many similar to metazoans and which may have formed in regressive bodies of water (Joseph et al. 2020a).

8. Speculations: Evolution, Trace Fossils, Tube Worms, Metazoans?

The Gale Crater is marked by potassium-rich fluvial valleys and water pathways (Grotzinger et al. 2015b). The Gale Crater appears to have repeatedly filled with water (Bibring et al. 2006; Cabrol et al. 1999; Fairen et al. 2014; Murchie et al. 2009; Siebach and Grotzinger, 2014; Buz et al. 2017) and could have sustained a variety of species and promoted their evolutionary development due to major changes in their watery environment and availability of liquids. Wet followed by dry spells may also account for what could be fossilized impressions of water dwelling and subsurface Martian organisms including metazoans.

To speculate, some of the fossil-like specimens are similar to complex metazoans and burrowing worms, which evolved on Earth 500 million years ago. These include formations which resemble Cambrian fauna that first appeared during the Ordovician (e.g. Calymene callicephala, Flexicalymene meeki, Homotelus bromidensis, Isotelus sp., Pseudogygites canadensis, Streptelasma sp.). Our speculative interpretation of the organisms represented (pro and con) are detailed in Joseph et al. (2020a) Figures 33-37. For example, Foraminfera live under the effects of water currents and if they accumulate, and die, they are disposed isoriented. In fact, what appears to be "fossilized" Foraminfera-like specimens could also be explained as an accumulation of plagioclase crystals.

Unfortunately, it is impossible to make precise determinations as to the identity or exact nature of these mineralized, fossil-like specimens, and this might be true if these same "fossils" were discovered on Earth (Graham 2019). Microfossil-like shapes are ubiquitous in Earth's geological record and despite careful extraction, preparation and microscopic examination there is great debate as to their authenticity (Knoll 2015, Brazier et al. 2003, De Gregorio et al. 2011; Marshall et al. 2011; Wacey et al. 2016). Samples have been repeatedly challenged as abiotic and a consequence of mineralization.

However, as to the controversy over the authenticity of the earliest terrestrial microfossils, it may also be that accumulated minerals are all that remains of these fossilized organisms. This would explain their fossil-like biogenic shape and why all that remains are the accumulated minerals.

Clockwise from top) Sol 890, 298, 1905: Specimens resembling colonies of micro-organisms and tube worms

Sol 1905: Specimens resembling colonies of tube worms and mushroom-lichens (bottom right). Contrast with Figure 28 the colors of which have been desaturated by NASA, thereby giving these same specimens (top photo) a gray-rock-like fossilized appearance (see Figure 28).

Sol 1905: Specimens resembling mineralized trace fossils. These specimens are approximately 1 to 5 mm in length on average. In Figures 26, 27, and 28, one of these "worm-like" specimens has an open aperture at one end, suggesting either an orifice (Figures 26, 27) or that they are hollow (Figure 28). Whereas the specimens depicted in Figure 27 have colorization suggesting biology, the gray specimens depicted here are similar to trace fossils of burrowing tube worms and the tunneling and burrowing systems made by worms in Cambrian sedimentary rocks (see Vannier, 2010), with the tunnels filling with sediment then mineralizing and fossilizing after waters receded in the Gale Crater. As depicted in this NASA desaturated photo, these specimens also resemble the trace fossils of treptichnids, priapulid worms, and their burrowing curved branching beneath and along the mud of the seafloor on Earth, circa 545 mya and which filled with sediment and mineralized (Vannier ,2010). As the NASA image curator did not reply to inquiries, it is unknown if NASA/JPL desaturated the colors of these "worm-like" specimens and their surroundings, or if NASA applied false colors to these same specimens which are much darker in color in Figure 27 and have other biological-features as also depicted in other photos taken on Sol 1905.

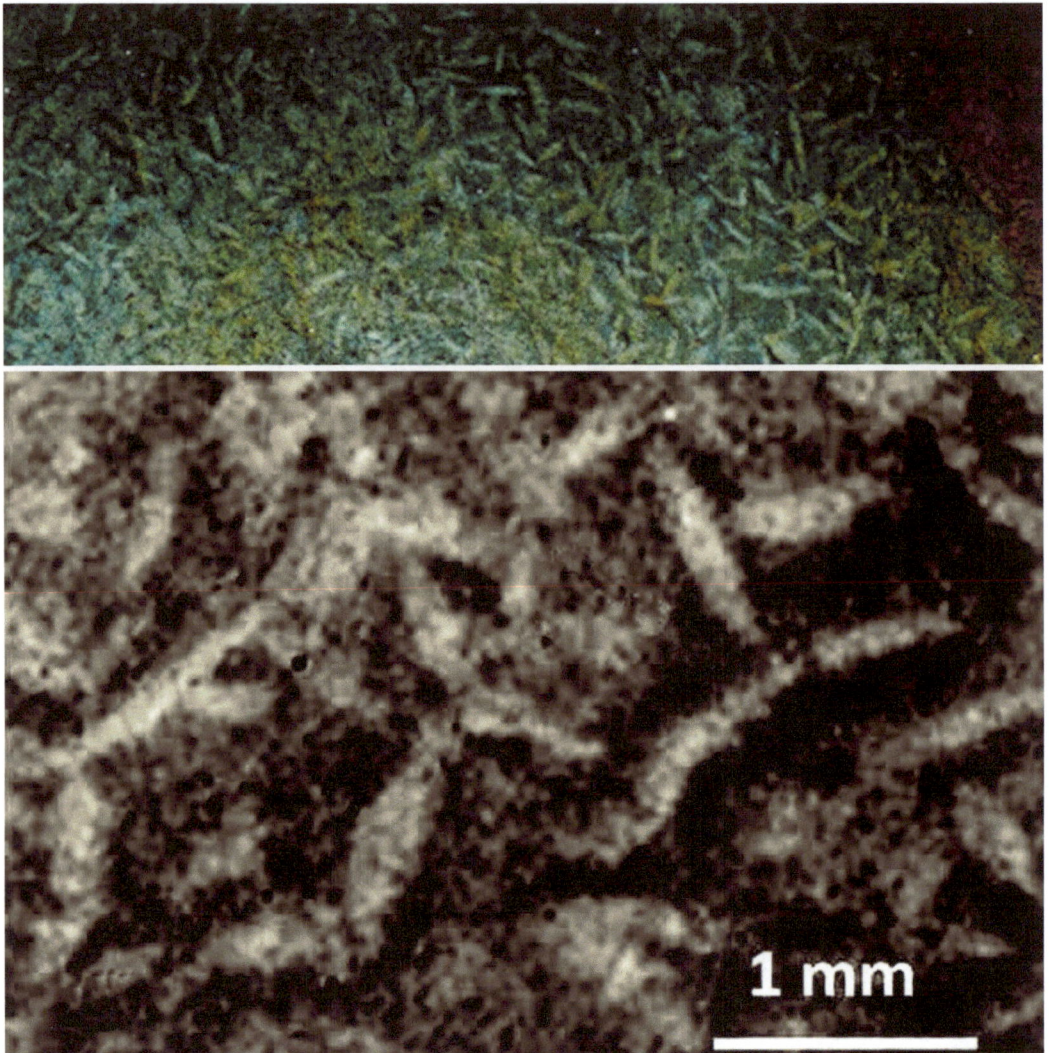

Sol 809: (Bottom enlarged, desaturated). These specimens resemble mineralized trace fossils formed, on Earth, during the Cambrian Ordovician transition, by microbes (Girvanella) and siliceous filamentous sponges, with lime, mud and biocasts between them (see Lee et al. 2014). Specimens similar to these have been found on the sediment surface of a relatively flat seashore in northern China during the late Cambrian extinction and the Great Odovician Biodiversification Event (Lee et al. 2014, 2016; Lou and Reitner 2015). Specimens similar to a variety of metazoans can also be viewed in the following figures. That these represent a pure mineralogical argumentation appears implausible as the orientation of the spindle-like, whitish structures are not organized like mineral structures (Graham, 2019; Graham et al 2016).

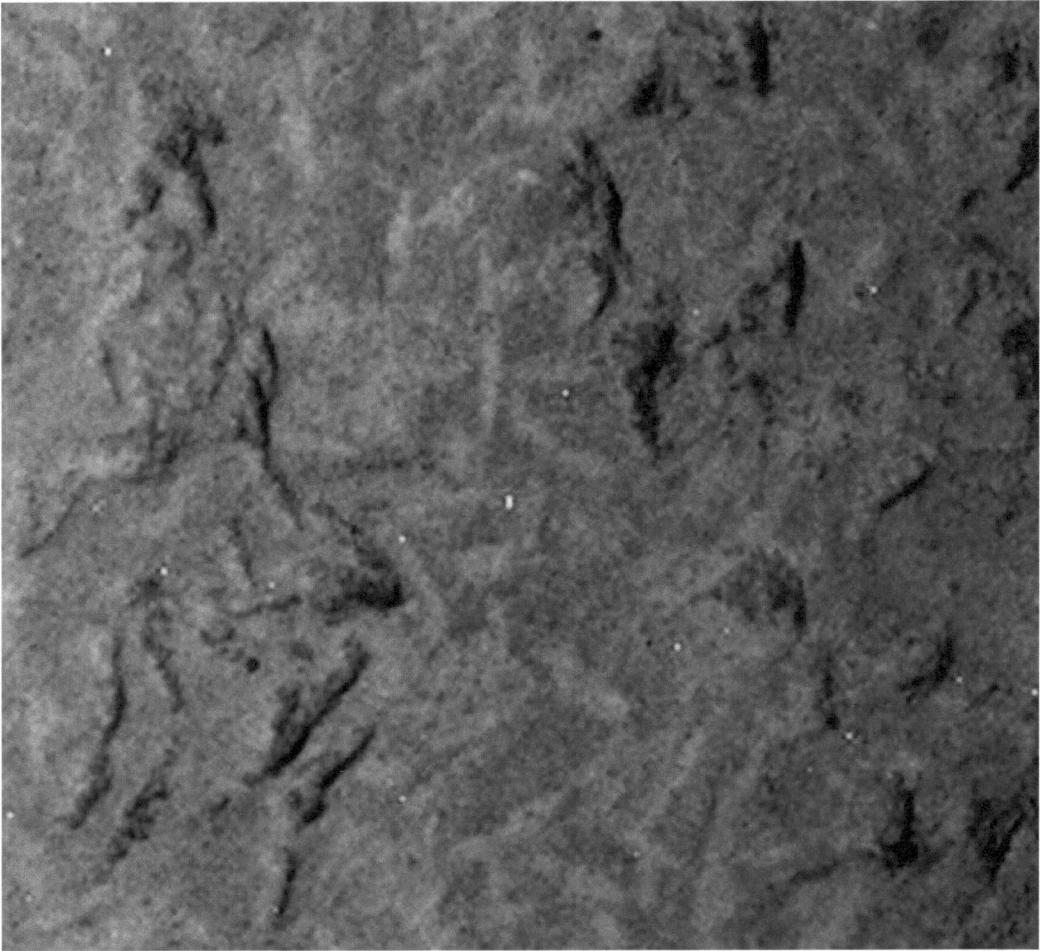

Sol 880: Specimens resembling trace fossils of metazoans (e.g. Calymene callicephala, Flexicalymene meeki, Homotelus bromidensis, Isotelus sp., Pseudogygites canadensis, Streptelasma sp.) including filamentous silica sponges upon and impressed into the surface of this Martian rock lying face-up on the floor of the Gale Crater. These formations are approximately 2 -3 mm in length on average. They are similar to metazoan fossils formed at the end of Cambrian (Lee et al. 2016). Although some resemble foraminifera in shape, their patterns of organizational preservation do not. On Earth, dead foraminifera rain down on the sea floor and their mineralized tests often become preserved as fossils in the accumulating sediment. Foraminifera are believed to have evolved around 1.15 bya, and have left a rich biomineralized, fossilized assembly of their multiple forms which include globular, spiral, spherical, and tubular species (Tappen and Loeblich, 1998; McIlroy et al. 2001) with all these diverse forms appearing together in a single species (Pawlowski et al. 2002). It is noteworthy that nearly identical complex metazoan-like forms have been photographed in other nearby locations within Gale Crater, including and especially the "ice-cream-cone" specimen at center right.

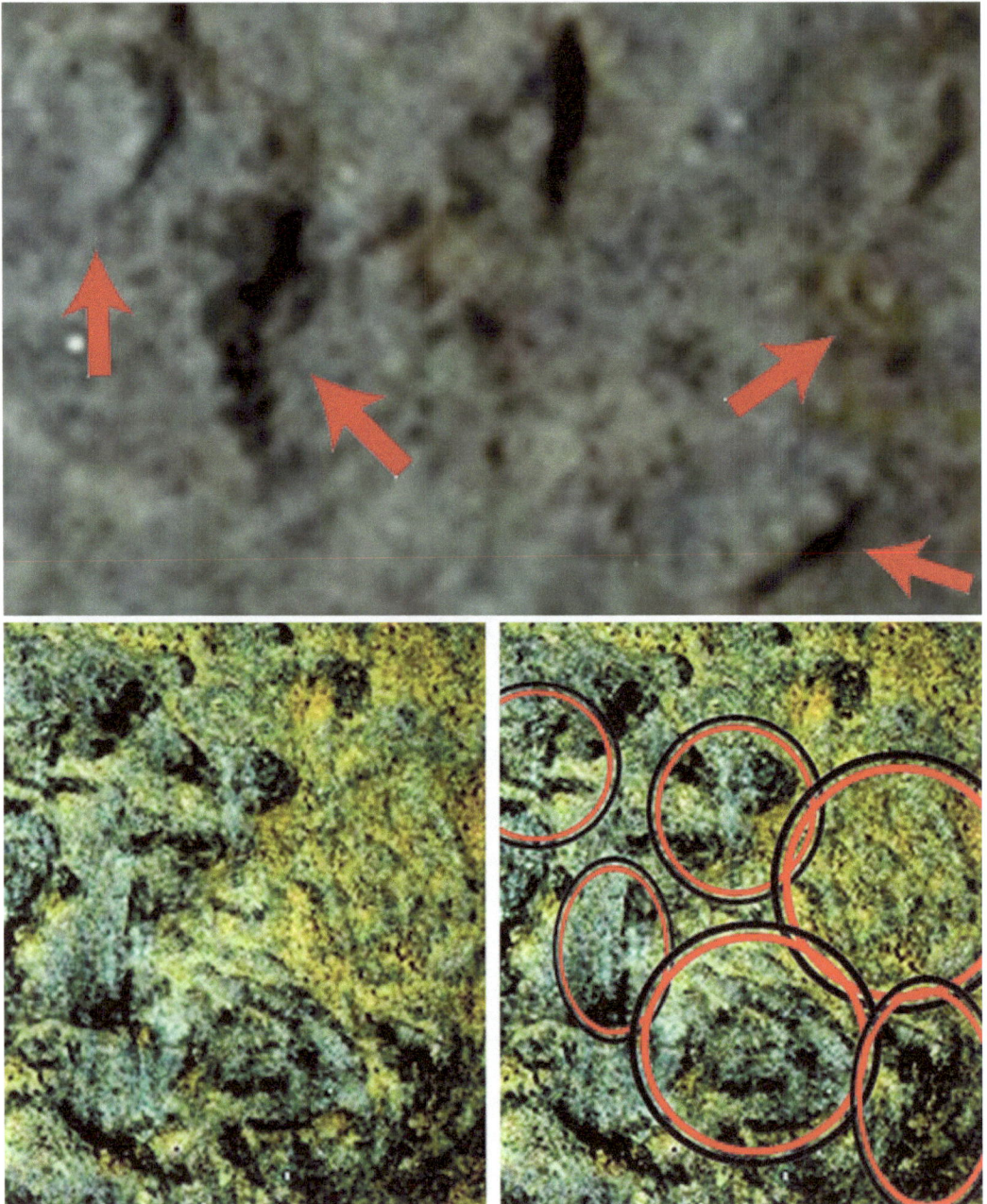

(Top) Sol 809. (Bottom) Sol 809. Similar specimens in two different locations, photographed alongside tubular, curved, and other fossil-like structures which resemble a variety of metazoans.

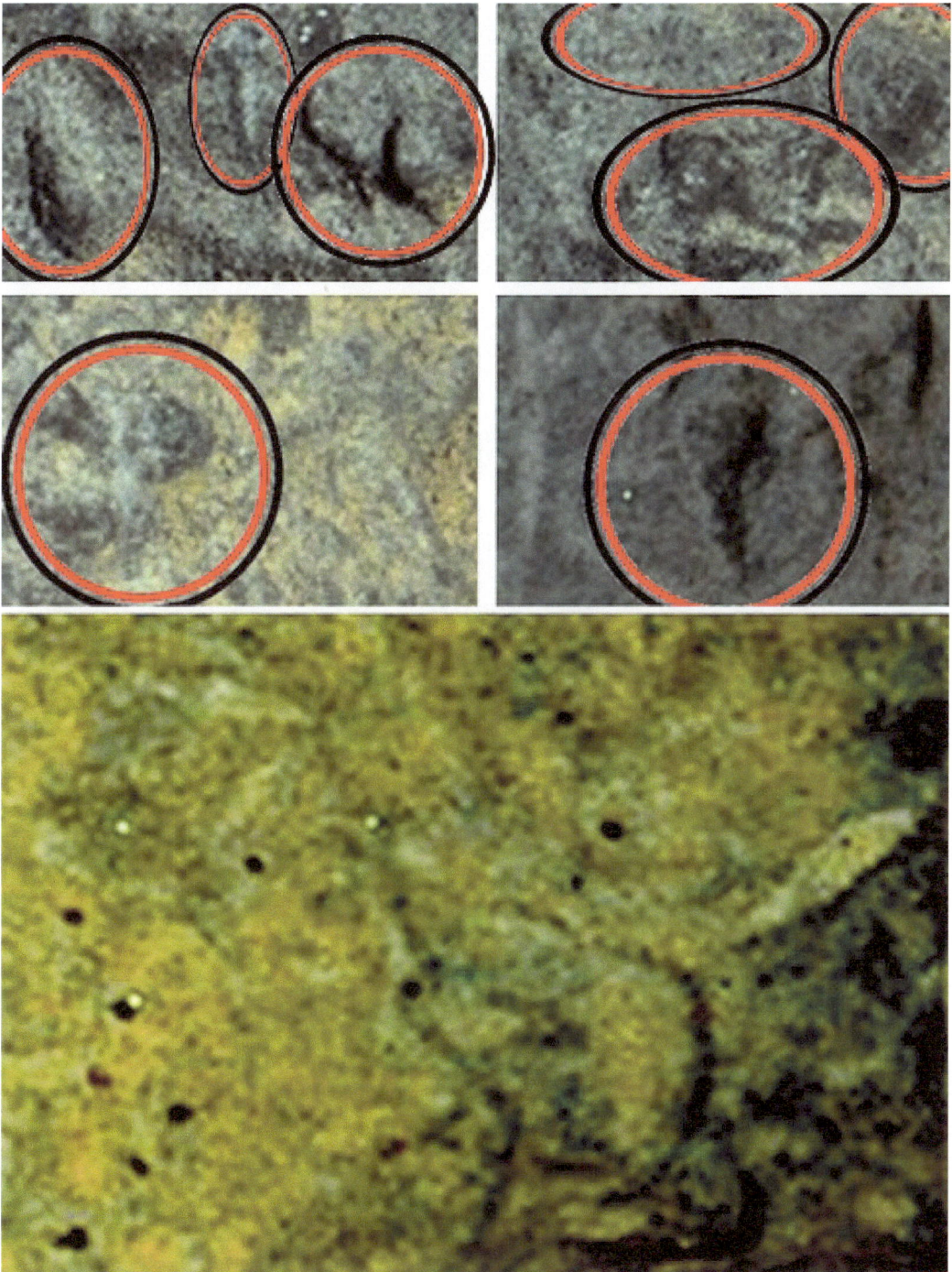

Sol 809, 880, 869: A variety of specimens, resembling the fossilized remains of metazoans (Calymene callicephala, Flexicalymene meeki, Homotelus bromidensis, Isotelus sp., Pseudogygites canadensis, Streptelasma sp.), photographed on rocks on the floor of the Gale Crater.

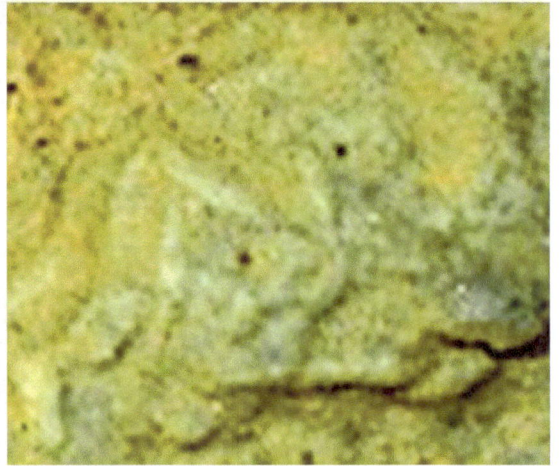

(Top) Sol 869: Specimens resembling mineralized fossils of worms and metazoans, approximately 1 to 2 mm in length. (Bottom): Sol 1905 compared with Sol 869.

9. Calcium, Bones, Brains, Oxygen, Ozone

Calcium is a major component of the skeletal-muscular system and promotes nerve cell development, neural generation and the functioning of the synapse, as well as acting on DNA which codes for neural functional organization and expression, and thus the development and functional integrity of the brain (Hong et al., 2000; Llinás et al., 2007; Perez-Reyes 2003). In fact, Ca2+ ions also have a special affinity for genes which code for functions mediated by the central nervous system.

Calcium is the most ubiquitous metal ion in the cellular system and also plays a universal role as messenger and regulator of protein activities. Calcium acts directly on gene expression and the regulation of programmed cell death (apoptosis), cellular proliferation and differentiation, and cell to cell adhesion and fusion, and the creation of bones and brains. In the absence of CA cells stop aggregating, embryos fail to adhere, cells disintegrate, and bones become soft and easily break. Therefore, until sufficient quantities of calcium had been biologically produced and then liberated into the oceans, embryos and bones were an impossibility for the first four billion years of Earth's history.

Cyanobacteria secrete calcium carbonate within their mucous. These secretions are used to glue and cement stromatolites together, and to create thick cyanobacterial mats, allowing vast colonies of cyanobacteria to adhere to one another. The lithification of these marine cyanobacterial mats, to create rock-like sediments, is thought to be driven by metabolically-induced increases in calcium carbonate saturation (Alois 2008). The secretion of calcium carbonate to form cyanobacteria mats, also infiltrated carbonate rocks and accelerated the mineralogy of reef-building (Porter 2006). On Earth vast stores of calcium began to build up in these mats, stromatolites, carbonate rocks and reefs. Thus, by 600 mya, vast ocean preserves of calcium carbonate had been established.

Increased levels of calcium carbonate potentiates photosynthesis in eukaryotes and prokaryotes. Increased photosynthesis also increases the production and secretion of calcium carbonate by eukaryotes and prokaryotes. Thus, a feedback mechanism is maintained where calcium carbonate potentiates photosynthesis which results in the release of more calcium carbonate as well as more oxygen.

On Earth, over the course of the last 4 billion years, and even under global glacial conditions, Cyanobacteria living upon the ice, those living beneath the surface of frozen seas, and those receiving only a limited amount of light, were able to engage in photosynthesis and calcium and oxygen production--and the same can be said of Mars. Yet others could engage in heterotrophic activity, and produce oxygen or high levels of C resulting in large pools of C and then the oxidation of this C upon the release of molecular oxygen via enhanced Corg burial (Kelly et al., 2007). The ultimate result was the creation and buildup of massive amounts of calcium carbonate which had been biologically produced and which acted on gene selection and turning on of silent genes; all of which paralleled significantly increased oxygenation of the ocean and atmosphere and dramatic alterations in temperature.

Cyanobacteria secret calcium as a means of fashioning stromatolites. However, once stromatolites begin to melt or deteriorate, they release all this calcium into the oceans--and the same can be assumed to have taken place on Mars, depending on its magnetic field.

Following the end of the Marinoan/Gaskiers glaciation, and beginning around 580 mya, Earth began to significantly warm. The increase in temperatures triggered bacterial mat evaporation and cyanobacterial mucous decomposition. The seas became saturated with calcium carbonate.

10. A Martian Cambrian Explosion? Calcium, Multicellularity, Brains, Bones, Skeletons

Six hundred million years ago, Mars was still flush with bacterial life, as based on analyses of meteor EETA 79001 (Pillinger et al. 1996). Six hundred million years ago, the vast majority of life forms sojourning on Earth and beneath the seas, were single celled organisms and simple multi-celled creatures composed of less than 11 different cell types such as algae and fungi. Therefore, the same can be predicted of Mars.

On Earth, the flooding of the oceans with (cyanobacteria-produced) calcium coupled with increased oxygen levels, led to the most dramatic explosion of life in the history of Earth, 540 mya., known as the Cambrian Explosion. So much oxygen had been released into the atmosphere by Cyanobacteria that ozone was established which blocked out life-neutralizing UV rays. With the establishment of ozone and the release of calcium, innumerable oxygen--breathing creatures equipped with bones and brains quite suddenly evolved around 540 million years ago, and began to emerge from the sea and conquer the land.

Therefore, we can assume, that Cyanobacteria played a similar role in the terraforming of Mars, and secreted calcium and oxygen, thereby providing the foundation for evolution of oxygen-breathing creatures with bones and brains (Joseph 2009). Martians with bones and brains could have evolved 540 million years ago. Melting Martian stromatolites would have saturated the ocean of Mars triggering an explosive evolutionary events similar to the Cambrian Explosion on Earth --but only if the planet's protective magnetic field was still intact.

Ca2+ ions have a special affinity for genes which code for functions mediated by the central nervous system (Glezer et al., 1999; Hong et al., 2000; Llinás et al., 2007; Köhler et al., 1996; Mori et al., 1991; Perez-Reyes 2003; Weisenhorn 1999; Ubach et al., 1998). The high levels of calcium that flooded the oceans acted on gene selection and triggered the evolution of the nerve cell, the nerve net, and the brain, and the bones of the skull to protect it.

On Earth, until sufficient oxygen, silica, and calcium had been released and the oceans had become oxygenated, body and cell size were restricted and unable to expand or engage in strenuous physical activity. Larger bodies require skeletal support. Internal organs require skeletal protection. Moreover, in the absence of ozone, larger sized bodies would be burnt by UV rays and would pop and explode if the organism was near the surface. However, once calcium levels and other substances built up sufficiently, and as the changing environment acts on gene selection (Joseph 2000, 2009), calcium-sensitive genes were activated giving rise to the first shelled animals and those equipped with exoskeletons (e.g., the trilobites) and thus the Cambrian Explosion.

Therefore, beginning around 640-580 mya, once silica, calcium, and oxygen levels had increased and a protective (oxygen-initiated) ozone layer was established, creatures expanded in size, diversified, and grew spines, silica skeletal compartments, then silica-collagen skeletons, collagen-calcium skeletons, armor plates (sclerites) and small shells like those of brachiopods and snail-like molluscs (Butterfield 2003; Conway Morris 2003; Lin et al., 2006).

In fact, a wide range of increasing complex species appeared following the Gaskiers glaciation and ensuing warming cycle, leading to an explosive burst of evolutionary change and diversification 540 mya when large animals and chordates equipped with bilateral bodies, eyes, bones and brains quite suddenly evolved (Chen et al., 1995, 1999; 2003; Shu et al., 2001; Siveter et al., 2001). It was during this same time period, 540 mya, that the genome duplicated in size and there was an explosive evolutionary burst of complex life forms, including the evolution of every phylum which is in existence today.

Hence, beginning around 540 mya, there was a vast explosion of bilateral metazoan diversity and complexity that appeared multi-regionally throughout the oceans of the Earth within 5 my to 10 millions (Levinton, 1992; Kerr, 1993, 1995). Over 32 phyla rapidly evolved, many with the "modern" body plans seen in modern animals (Conway and Morris 2000; Budd and Jensen 2000; Peterson et al. 2005). Many of these creatures were very strange and bizarre in appearance and immediately died out (Mooi and Bruno,1999). These included the five-eyed Opabinia and weird multi-legged creatures. Some of these organisms were so unusual it has been assumed they must represent phyla that became extinct.

There followed a rapid and progressive diversification, with amphibians evolving around 380 million years ago, then reptiles (320m), repto-mammals / therapsids (250m), dinosaurs (225m), mammals (100m), monkeys then apes (60 then 30 million), hominids (2m), and anatomically "modern" humans (35,000 years ago).

If Mars also experience a similar "Cambrian Explosion" is unknown. Hoever, based on fossilized evidence of metazoans discovered in the Gale Crater (Joseph et al. 2020a) there is evidence that life evolved on Mars. The likelihood that or the ability of complex intelligent life to have evolved on Mars would have been determined by levels of oxygen, calcium, the creation of an ozone layer; and when Mars lost its protective magnetic field, its atmosphere and the oceans which covered over 1/4 of the planet. The loss of the magnetic field was catastrophic. In addition to a Martian mass extinction, whatever creatures survived did so because they could adapt, mutate, and evolve under these harsh new conditions.

11. How and When Mars Lost Its Magnetic Shield and Atmospheric Armour: Illiad and Valles Marineris

If complex, intelligent, or human-like organisms equipped with bones and brains evolved on Mars may depend on when Mars collided with another planet. Indeed, based on surface anomalies and the gigantic Borealis basin, it appears Mars was struck by a moon-sized planet which may have *turned off* the magnetic field and stripped Mars of its oceans and atmosphere, the remainder of which would have been blown into space by solar winds. If this impact (or impacts) was the responsible agent, and it this occurred as recently as 65 million years ago when Earth was also struck by a large asteroid, or more recently or billions of years further back in time, is unknown.

However, curiously, ancient Greek "mythology" and Homer's epic, the Iliad--written 2,800 years ago, around 850 B.C.--make direct mention and refer to a collision between the "god" planet Mars/Ares and the goddess "Athene" who emerged from the head of Jupiter (like an ejected planet-sized moon) only to careen throughout the solar system defying the rules and laws obeyed by all the other planetary gods (Homer, Iliad 5. 699 ff):

"Pallas Athene then took up the whip and the reins, steering first of all straight on against blood-stained Ares who was striping gigantic Periphas, shining son of Okheios, of his shield and armour. But Athene put on the helm of Death that stark Ares might not discern her.

"Now as manslaughtering Ares... made straight against Diomedes... as they in their advance had come close together, Ares lunged with a bronze spear, furious to take the life from him. But the goddess Athene catching the spear in her hand pushed it away... and leaning in on it, drove it into the depth of Ares' belly... she stabbed, driving it deep into his flesh and wrenched the spear out again.

"Then Ares the brazen bellowed with a sound as great as nine thousand men make... when they cry as they carry in to the fighting the fury of the war god. And a shivering seized hold alike on Akhaians and Trojans in their feet at the bellowing of battle-insatiate Ares.

"Ares went up with the clouds into the wide heaven... grieving in his spirit, and showed Zeus the immortal blood dripping from the spear cut and addressed him in winged words : 'Father Zeus, are you not angry looking on these acts of violence? It is your fault we fight, since you brought forth this maniac daughter accursed, whose mind is fixed forever on unjust action. For all the rest, as many as are gods on Olympos, are obedient to you, and we all have rendered ourselves submissive. Yet you say nothing and you do nothing to check this girl, letting her go free, since yourself you begot this child of perdition." -Homer, Iliad 5. 699 ff

In fact, a giant asteroid, comet, or moon-sized planet may have struck the Red

Planet, slicing across the belly of Mars, creating the Valles Marineris: a stabbing, ripping cut that is over 3000 kilometers long, 600 kilometers wide, and 8 kilometers deep.

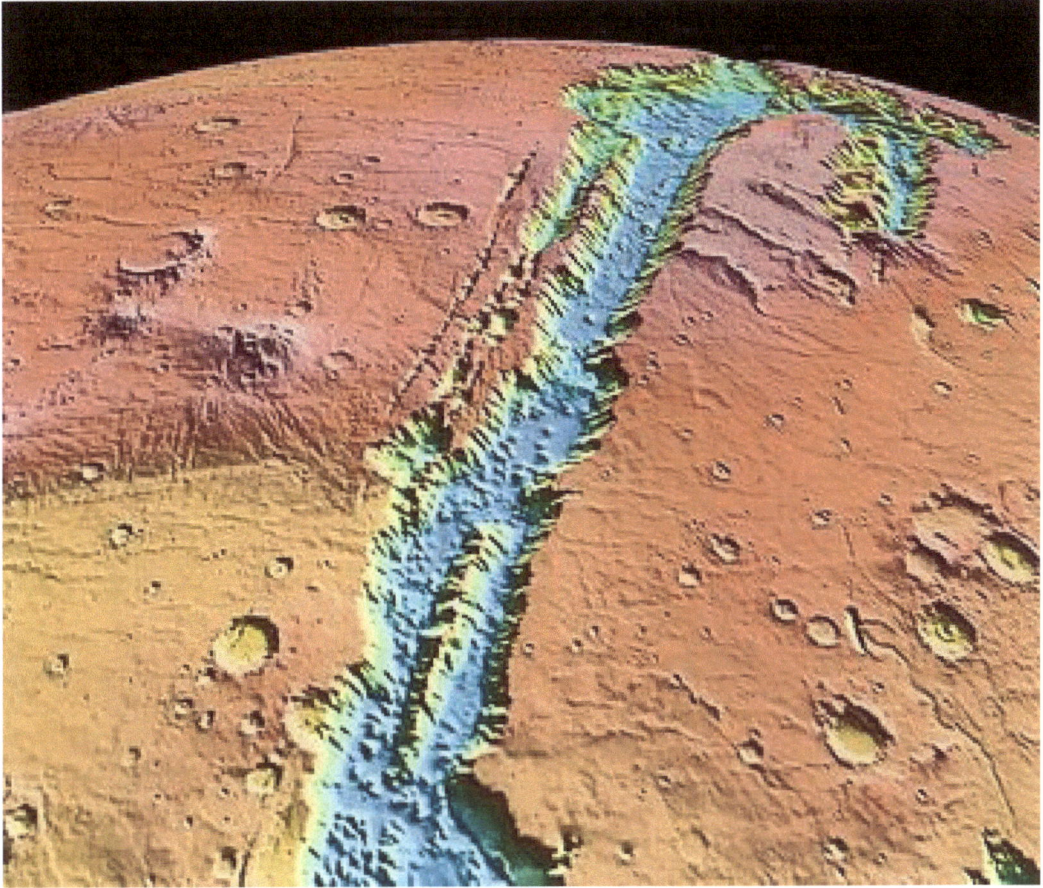

The spear-shaped Valles Marineris: "Athene catching the spear in her hand... and drove it into the depth of Ares' belly... she stabbed and driving it deep into the flesh and wrenched the spear out again..." (Homer, Iliad).

The Iliad is believed to be a retelling of celestial and historical events which took place around 400 years before the time of Homer, over 3000 years ago. Could the collision and "spear cut" suffered by Ares/Mars have been witnessed by the humans of Earth and later reported by Homer and the ancient Greeks as part of the Trojan War epic? It seems implausible as "Valles Marineris" can't be seen by modern telescopes on Earth, and was not discovered until 1972 when filmed and photographed by NASA's Mariner 9 spacecraft, the first *known* satellite to orbit Mars.

The spear-shaped Valles Marineris: "...and Ares... grieving in his spirit, showed Zeus the immortal blood dripping from the spear cut..." (Homer, Iliad).

One can only speculate; but when Mars was struck, and if this caused the Martian magnetic field--which sheilded Mars from powerful solar winds and cosmic radiation--to be lost, then, in consequence the Martian oceans and atmosphere would have boiled away and been blown into space. Indeed, in the days before the collision, Mars would have been convulsed with titanic tidal floods, dozens of erupting volcanoes, mega-magnitude Marsquakes, and buffetted by hundreds of super hurricanes cyclones and tornadoes... commencing in a catastrophic convulsing Mars-shaking stabbing slicing collision followed by the loss of the magnetic fields... and then, over the following days, weeks, and months, the irradiated Martian oceans and atmosphere bled away --blown by solar winds deep into space... and in consequence, much of Martian life became extinct... only those who could adapt survived

238

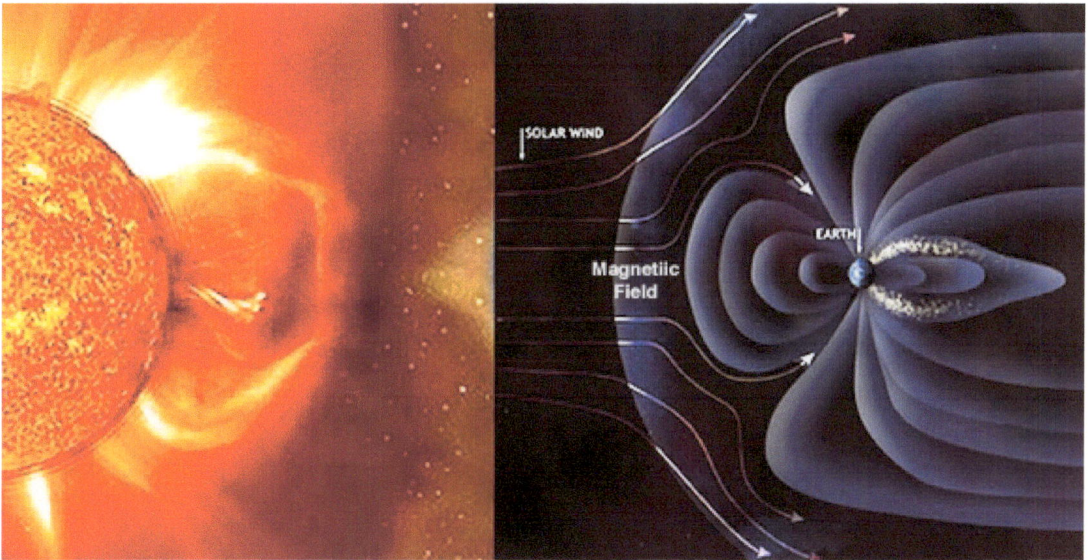

The solar winds vs Earth's Magnetic Field

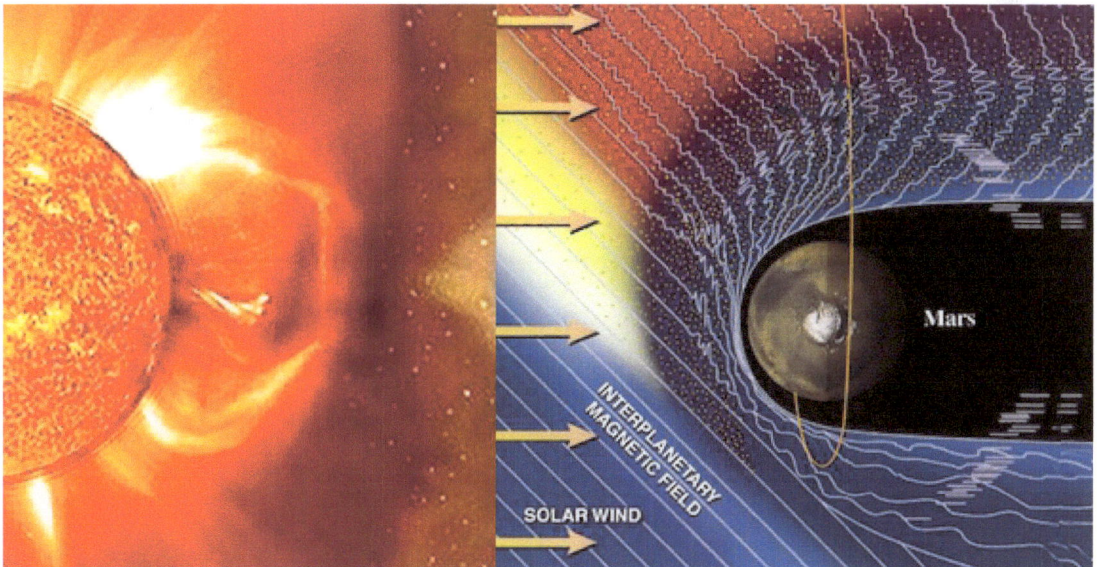

The solar winds vs Mars without a Magnetic Field

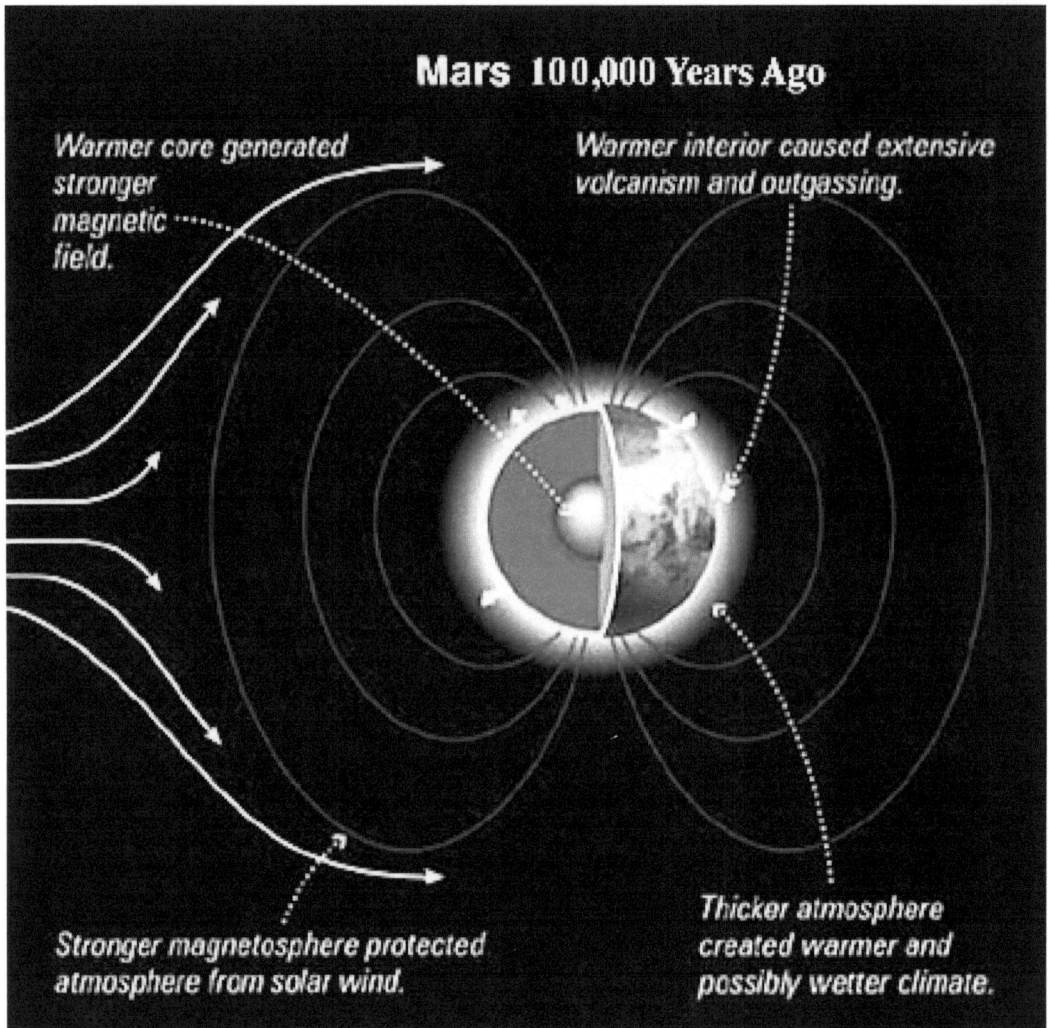

Mars 100,000 Years Ago

Warmer core generated stronger magnetic field.

Warmer interior caused extensive volcanism and outgassing.

Stronger magnetosphere protected atmosphere from solar wind.

Thicker atmosphere created warmer and possibly wetter climate.

The solar winds vs Mars with a Magnetic Field

12. Impact Craters and the Martian Magnetic Fields

When an asteroid or meteor crashes into a planet like Earth or Mars, the kinetic energy from the melting and vaporization, and the shockwaves that travel throughout the planet, alter the magnetization of the crust at the site of impact and which may become demagnetized or remagnetized to varying degrees. Thus, many geoscientists believe that by measuring the magnetization of the craters of Mars it is possible to make estimates as to the ambient Martian magnetic field at the estimated time it was struck.

As measured by the Mars Global Surveyor, there are still strong crustal magnetic fields on Mars (Acuña et al., 1999, 2001; Connerney et al., 2001), which are believed to be evidence of a dynamo-driven global magnetic field comparable in strength to that at the Earth's surface (i.e., ~50 µT). H owever, as measured above

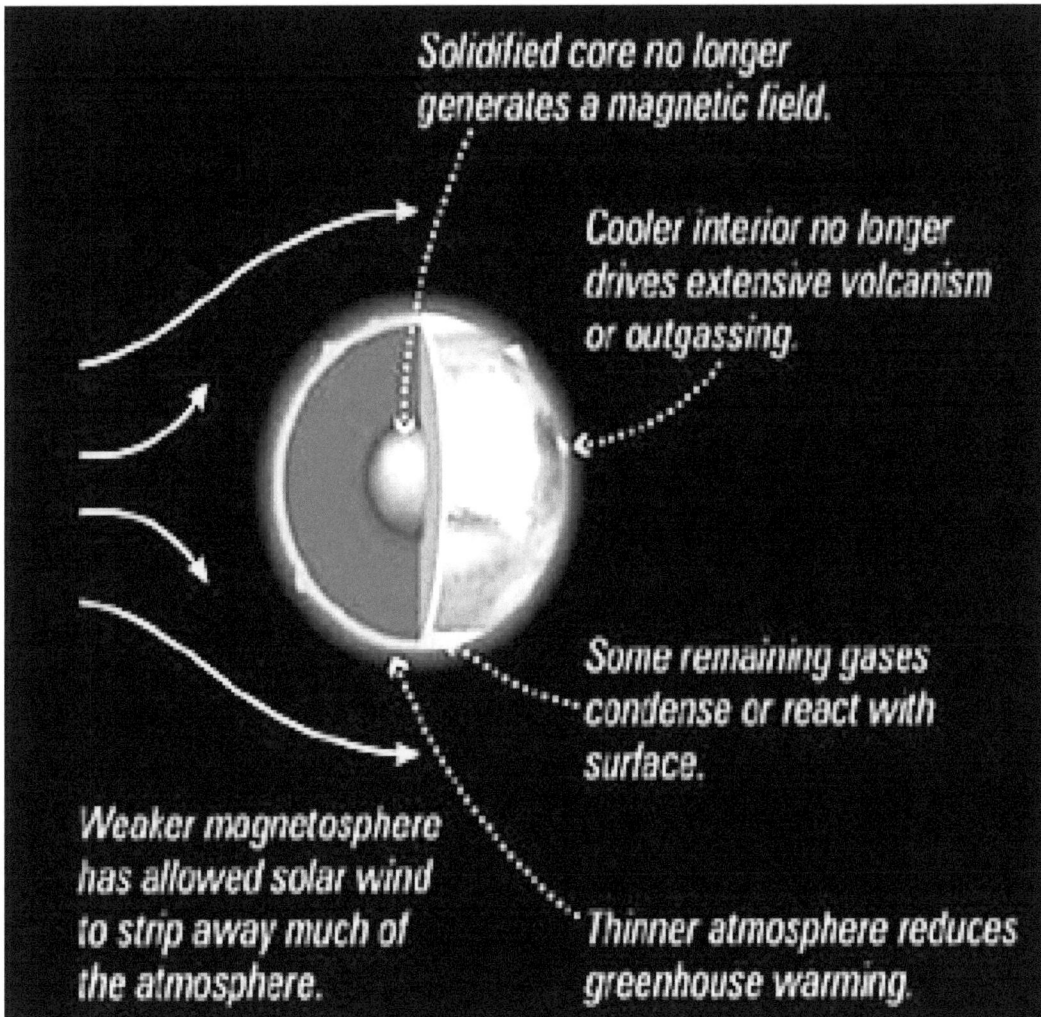

Solidified core no longer generates a magnetic field.

Cooler interior no longer drives extensive volcanism or outgassing.

Some remaining gases condense or react with surface.

Weaker magnetosphere has allowed solar wind to strip away much of the atmosphere.

Thinner atmosphere reduces greenhouse warming.

The solar winds vs Mars without a Magnetic Field

the large impact basins Hellas, Argyre, Utopia, Isidis, and North Polar impact basins, the magnetic fields appear to be very weak, which suggests to some investigators that the basins were demagnetized by the impact of meteors and asteroids (Acuña et al., 1999; Mitchell et al., 2007; Lillis et al., 2008a). However, as the degree of demagnetization is not uniform, and as some are still magnetized and others unmagnetized, it was hypothesized that the differences were due to when the impacts occurred and when Mars lost its magnetic fields. Thus, based on these theories, assumptions and hypotheses, and proxy data obtained not at ground level, but in orbit, it has been suggested that Mars may have lost its magnetic field between 3.2 billion years ago (Hood et al. 2010) to 4.1 billion years ago (Lillis et al., 2008b; Hartmann & Neukum, 2001).However, it turns out that craters on

Earth also show varying levels of magnetization and demagnetization; and Earth still has its magnetic field. So, the fact remains: when (and how) Mars lost its protective magnetic field, is unknown.

13. Evolution of Life From Space

As there is no evidence to support the belief in an "organic soup" it is likely that Mars and Earth were contaminated with a wide range and variety of archae, bacteria, algae, and fungi at the same time, via comets, asteroids, and meteors, and even moon-size stellar objects which crashed into these two worlds early in their history (Joseph 2000, 2009; Joseph & Schild 2010).

For example, fossilized colonies resembling cyanobacteria (blue-green algae) have been discovered in the Orgeuil, Murchison (Hoover 1984, 1997) and Efremovka meteorites (Zhmur and Gerasimenko 1999; Zhmur et al. (1997). Moreover, virus particles and clusters of an extensive array of microfossils similar to methanogens and archae have been found in the Murchison (Pflug, 1984); and organized elements and cell structures that resemble fossilized algae and microscopic fungi have been recovered from within the Orgeuil (Claus & Nagy 1961; Nagy et al. 1962; Nagy et al. 1963a,b,c). If these stellar objects also contained living organisms when they struck Mars and Earth, it can be assumed they went forth and multiplied.

It is also likely, due to solar winds, that bacteria, archae, and fungi have been repeatedly blown into space such that after life took root on Mars and Earth, there was then a continual exchange of living organisms between these (and other) planets; creatures which in turn likely exchanged genes and engaged in horizontal gene transfer once they came into contact with life forms already dwelling on these worlds. (Joseph 2000, 2009; Joseph & Schild 2010).

Thus, based on data from meteors, the oldest Earthly rock formations, and genomic analysis, it can be deduced that the first creatures to take root on Mars and Earth likely included archae, bacteria, and blue-green algae (cyanobacteria), and possibly simple eukaryotes such as yeast and fungi.

If complex, intelligent, life, equipped with bones and brains, evolved on Mars, is unknown and would depend on when Mars lost its magnetic field. However, given the fossilized evidence of fungi, algae, and cyanobacteria in a host of meteors--some older than this solar system--what we can determine is that life, and oxygen-calcium secreting cyanobacteria, have taken root throughout this galaxy and that complex, intelligent life may have evolved on innumerable worlds...

Life in the Radiation Intense Environment of Mars

On Earth, most species consisted of less than 11 cell types until around 540 million years ago with the onset of the Cambrian Explosion (Joseph 2009a, 2010). If life on Mars evolved to or beyond 11 cell types is unknown. However, once Mars lost its magnetic field only simple, radiation-tolerant species, or those dwelling beneath rocks, below the surface, or within caves and crevices would have remained viable. In fact, the growth of many species may have been enhanced by increased radiation exposure.

It has been experimentally demonstrated that sustained exposure of microfungi to gamma irradiation (121 Sn) 200–400 Gy and mixed beta and gamma irradiation (137 Cs) (100–150) Gy (equivalent to an absorbed dose from electrons of 300–500 Gy), has beneficial effects and resulted in the expression of adaptive features, including enhanced spore germination and hyphal growth; a property described as "radiostimulation" (Tugay et al. 2006). Specifically, Tugay, Dighton, Zhdanova and colleagues exposed fungi--isolated from a range of long term background radiation levels in the region around the damaged Chernobyl nuclear reactor--to pure or mixed radiation (137 Cs, 123 Te, 109 Cd, 121 Sn) and found that more than 60% of fungal strains exhibited positive radiotropism (Zhdanova et al 2004). In addition, after repeated irradiation these isolates showed significant growth. In fact, sustained radiation exposure to biota as a whole and to mycobiota in particular has generated a number of formerly unknown adaptive features (Tugay et al. 2003; Zhdanova et al. 1991, 2004).

Moreover, the processes of radiostimulation and radiation hormesis are also known to occur in plants and animals living with increased background radiation (Alshits et al 1981; Zhuravskaya et al 1995; Tverskoy et al 1997; Calabrese & Baldwin 1999, 2000). Likewise, plants and animals have demonstrated adaptive functions following radiation exposure, including tissue and cellular regeneration ((Becker, 1972b, 1984; Becker & Murray1967; Becker & Sparado 1972; Kurtz & Schrank, 1955; Levin 2003; Moment, 1946, 1949).

Fungal growth can therefore be stimulated by repeated irradiation and long term radiation exposure as these and other species appear to utilize radiation as a food source which promotes metabolism. High levels of accumulation of radionuclides in mushrooms have been recorded many times, especially following the Chernobyl accident (Dighton & Horrill, 1988; Dighton et al. 2008). High levels of radiation, and the loss of the Martian magnetic field, do not, therefore preclude survival or evolution even if accompanied by a mass extinction of those species unable to adapt.

On Earth, each mass extinction was followed by explosions of diversity as new species evolved in response to these changed and changing environmental conditions (Bradshaw & Brook, 2009; Elewa 2008; Joseph 2009b; Raup 1992; Ward 2000). Likewise, massive influxes of gamma rays (Melott et al. 2004) and alterations in the polarity of Earth's magnetic field (Maffei 2014) not only resulted in mass extinctions, but the evolution of new species (Elewa & Joseph 2009; Maffei 2014). Therefore, it is not unreasonable to ask if species completely alien to those of Earth may have evolved on Mars before and after the loss of the Martian magnetic fields--creatures adapted to extreme ranges of temperatures, high levels of radiation, minimal oxygen, and a scarcity of surface ground water.

Martian ground level radiation has been estimated to equal "0.67 millisieverts per day" (Hassler et al. 2013). This is significantly and profoundly below the radiation tolerance levels of a variety of prokaryotes (Moseley & Mattingly 1971; Ito et al. 1983) and simple eukaryotes including fungi which can withstand radiation doses up to 1.7×104 Gy (Saleh, et al. 1988). The bacterium, Deinococcus radiodurans which is the most radiation resistant organism so far discovered, can easily survive doses ranging from 5,000 GY to 15 GY (Moseley & Mattingly 1971; Ito et al. 1983).

Moreover, fungi (Wember & Zhdanova 2001; Zhdanova et al. 2004) and radiation-loving bacteria (Moseley & Mattingly 1971; Ito et al. 1983), will seek out and grow towards sources of radiation which serves as an energy source which they metabolize (Dighton et al. 2008; Tugay et al. 2006). Even if their DNA is damaged by radiation levels above their tolerance levels, they can easily repair these genes due to a redundancy of genes with repair functions (White et al. 1999).

Many species of fungi are able to produce melanin (Gupta, et al 2015) which serves a protective function and which is believed to convert radiation into reproductive energy (Dadachova et al. 2007; Drewnowska et al. 2015; Dighton et al. 2008). For example, Dadachova et al (2007) exposed three different species of fungus (Cladosporium sphaerospermum, Cryptococcus neoformans and Wangiella dermatitidis) to gamma radiation from rhenium-188 and tungsten-188 and reported that they thrived and grew rapidly in response to radiation. Dadachova and colleagues theorized that melanin may enhance fungal survival in radiation intense environments: "Many fungi constitutively synthesize melanin which... confers a survival advantage... by protecting against UV and solar radiation. Melanized microorganisms... and... biological pigments play a major role in photosynthesis by converting the energy of light into chemical energy. Chlorophylls and carotenoids absorb light of certain wavelengths and help convert photonic energy into chemical energy during photosynthesis. Given that melanins can absorb visible and UV light of all wavelengths... exposure to ionizing radiation change the electronic properties of melanin and affect the growth of melanized microorganisms... and enhanced growth of melanized fungi under conditions of radiation flux" (Dadachova et al 2007) In other words, high levels of radiation are a food source which can be turned into reproductive energy.

Fungi, Deinococcus radiodurans, and numerous other species could easily flourish on Mars. The loss of the Martian magnetic field would have had no deleterious effect on these organisms other than diminishing access to large volumes of water.

14. Evolution in the Radiation Intense Environment of Mars

When Mars lost its magnetic field and due to subsequent alterations in the biosphere, the result was most likely a mass extinction of innumerable life forms-

-even if consisting of fewer of 11 cell types--except for those already adapted for a high radiation environment, such as fungi and species like Deinococcus radiodurans, Certainly, any complex Earth-like Martian species similar to animals would have died out if such species had evolved prior to this catastrophe. However, the alteration, reduction or loss of the magnetic field may have also triggered the evolution of radiation-loving Martian species quite unlike those of Earth (Joseph 2011).

Earth was devoid of any significant ozone protection for billions of years. However hundreds of time before and after the buildup of an oxygen atmosphere or Cambrian Explosion, our planet was exposed to great fluctuations in electromagnetic radiation and profound alterations in magnetic polarity and geomagnetic reversals (De Santis et al., 2004; Occhipinti et al., 2014; Shcherbakov et al., 2002). Each event appears to have been associated with a mass extinction followed by the evolution of various species adapted to the changed environment.

For example, during the Triassic (250-200 million years ago), the Jurassic (200-145 million years ago), and the Cretaceous (145-65 million years ago) polarity and dipole moment fluctuated widely and was significantly reduced ($\approx 4 \times 1022$ Am2) (Shcherbakov et al., 2002). A variety of species were driven to extinction including entire plant families which never recovered. However, new species evolved (Maffei 2014). Alterations in radiation exposure may have been a major factor in the evolution of increasingly complex species on Earth and Mars.

In fact, numerous laboratory studies have examined the effects of a wide spectrum of electromagnetic activity at the cellular and molecular level and have found a major amplification of beneficial process involving antibodies, hormones, calcium ions, and the binding of neurotransmitters to synaptic receptors thus enhancing cellular communication and even the aggregation of cells involved in tissue formation in higher animals (Adey 1993; Barnes 1992; Basset 1993; Becker, 1972a,b, 1979, 1984; Levin 2003). Moreover, electromagnetic radiation in the range of 0-100 Hz are attracted to cellular pathways, usually not more than 150 A wide, since they provide a low level of electrical impedance. Radiation is then directed along the cell membrane surface thereby producing tissue gradients in the range 10-'-10-1 V/cm, which in turn promotes essential physiological functions, and promoting cell growth, bone-growth, and even immune responses (Adey 1993; Barnes, 1992; Burr 1941a,b, 1952; Levin 2003; Lund 1947).

Therefore, it is possible that species evolving in the radiated environment of Mars--over billions of years of time-- may have evolved physiological functions and cell-bone-and-tissues quite unlike those of Earth. For example, radiation can enhance tissue regeneration depending on the bioelectric properties of the tissue (Becker, 1984; Becker & Murray, 1967; Becker & Sparado 1972). Regenerating vs non-regenerating limbs, for example, demonstrate strong endogenous electromagnetic fields which promote cellular and tissue growth. Moreover experimental application of exogenous fields is able to induce regeneration even in nonregenerating species (Becker, 1972b, 1984; Becker & Murray1967; Becker & Sparado 1972; Kurtz & Schrank, 1955; Levin 2003; Moment, 1946, 1949). This is believed to be due to the interaction of electromagnetic activity with DNA (Barnes 1992; Levin 2003).

In the radiated environment, however, instead of merely regenerating lost tissues, arms, legs, and so on, and given the mutagenic effect of radiation, Martian life forms may have evolved additional arms, eyes, legs, and so on, giving rise to species completely unlike those of Earth.

It should be recognized that for several billion years Earth had no ozone protection and innumerable species of microbes and simple eukaryotes were exposed to high doses of ultraviolet and other forms of radiation. Radiation and diurnal and other fluctuations in radiation levels likely played a significant role in the evolution of photosynthesis and cellular complexity and the evolution of various organisms consisting of up to 11 different cell types long before the development of a significant ozone layer (Joseph 2009a,b).

Even mutagenic effects may have been beneficial, giving rise, for example, to the evolution of diverse species especially adapted to this radioactive environment. Ionizing radiation has in fact been found to greatly increase the rate of mutation by ionizing the bases in the DNA chain thereby effecting the sequences and interactions of these bases and even their structure thus altering DNA synthesis (Bridges 1980; Kimball, 1978; Baumstark-Khan & Facius, 2002). Moreover, exposure to radiation can stimulate growth and promote new adaptive functions (Tugay et al. 2006). In consequence a new species may evolve, or it may sicken and die. The radiation levels on Mars can cause death and deleterious mutations (Straume, Blattnig, Zeitlin, 2011), or promote evolutionary diversity (Joseph 2011; .

Mars lacks an oxygen atmosphere. On Earth, it was not until around 2.45 bya, that atmospheric oxygen levels rose to values between 0.02 and 0.04atm (Holland 2006) and oxygenic photosynthesis became widespread (Buick 2008). By 1.5 BYA, atmospheric oxygen levels had increased to levels ranging from 0.02 and 0.04atm (Holland 2006)--well below current levels. Likewise, the shallow oceans remained mildly oxygenated and the deep oceans continued to be mostly anoxic (Holland 2006). It wasn't until around 500,000 years ago that oxygen levels reach current levels, about the time of the Cambrian Explosion (Joseph 2009a,b).

How much oxygen was present in the Martian atmosphere billions of years ago or even a few thousand years ago, is unknown. However, insufficient or a lack of oxygen in fact protects against bad mutations (Quintiliani 1979). By contrast as oxygen and cellular oxygen levels increase, so too does the severity of tissue damage due to radiation exposure which is enhanced at the cellular level (Baumstark-Khan & Facius, 2002). Radiation and the lack of significant oxygen could have been a driving force in the evolution of life on Mars.

Although we can only speculate, the loss of the Martian magnetic field may have triggered a mass extinction followed by the evolution of truly alien life which utilize radiation for energy; life forms which may dwell beneath the soil, within crevices and caves and thus accounting for the numerous anomalies resembling complex, Earth-like, and frankly quite bizarre shapes and life-like fossil-like forms which have been photographed, by NASA, on Mars.

15. Anomalies Resembling Skulls, Bones, Skeletal Remains and Complex and Intelligent Life on Mars

The following photos, taken by NASA and the ESA from Mars orbit, and at ground level by the rovers Spirit, Opportunity, and Curiosity, are "anomalies" which are suggestive of and which resemble fossilized, skeletal, and complex and intelligent life, including the remains of dwellings, space craft and debris fields which appear to be strewn with wreckage, tools, and the remains of Martians or other aliens. What they are, is unknown, and thus, it is up to the reader to decide if these are illusions or evidence of life.

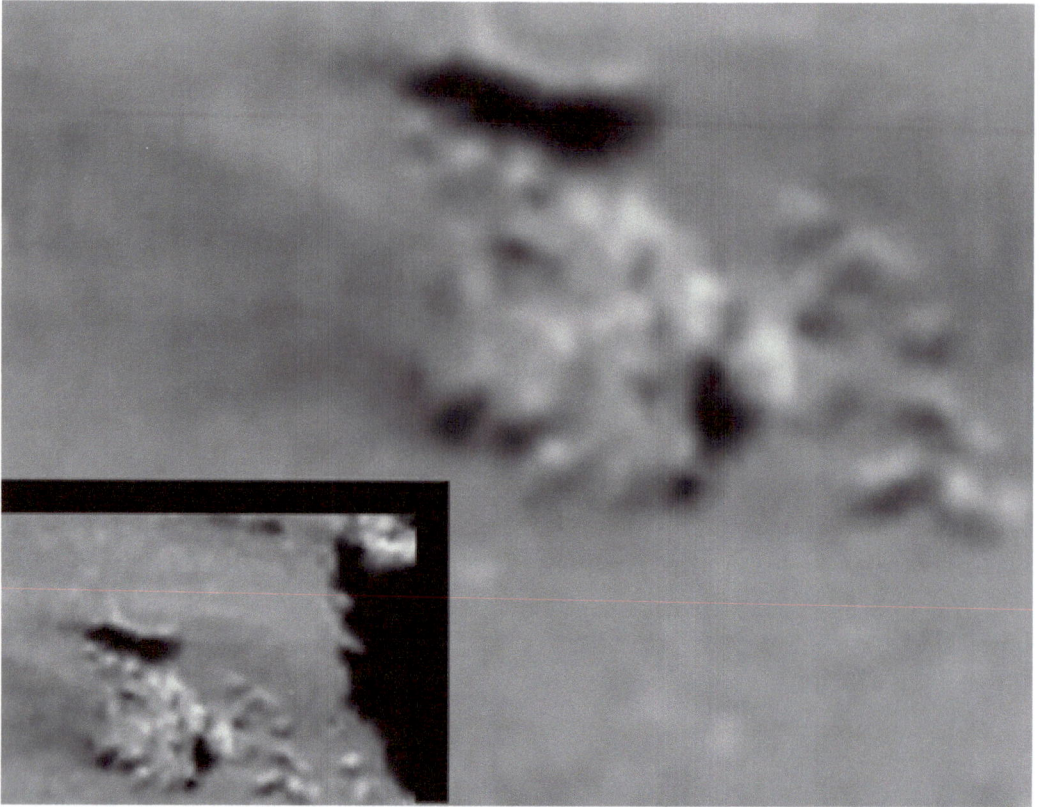

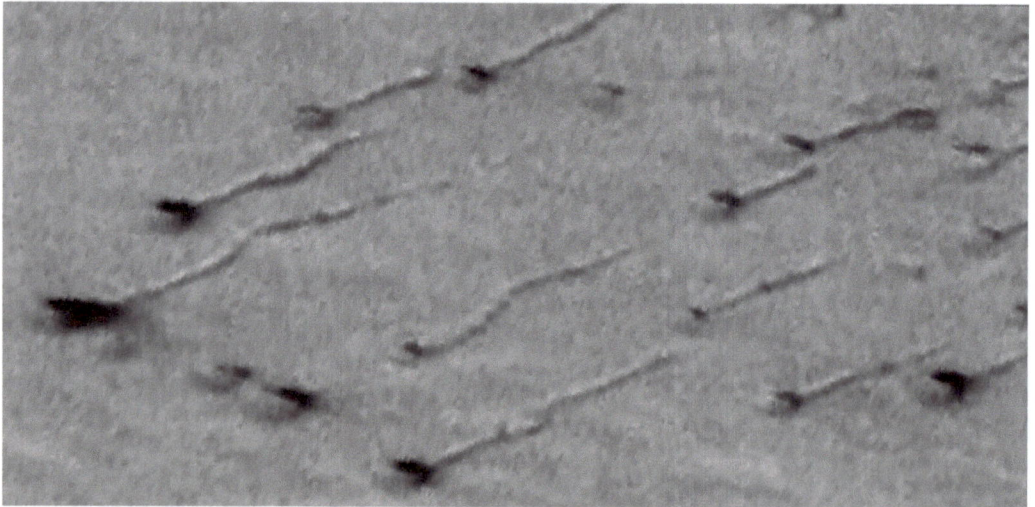

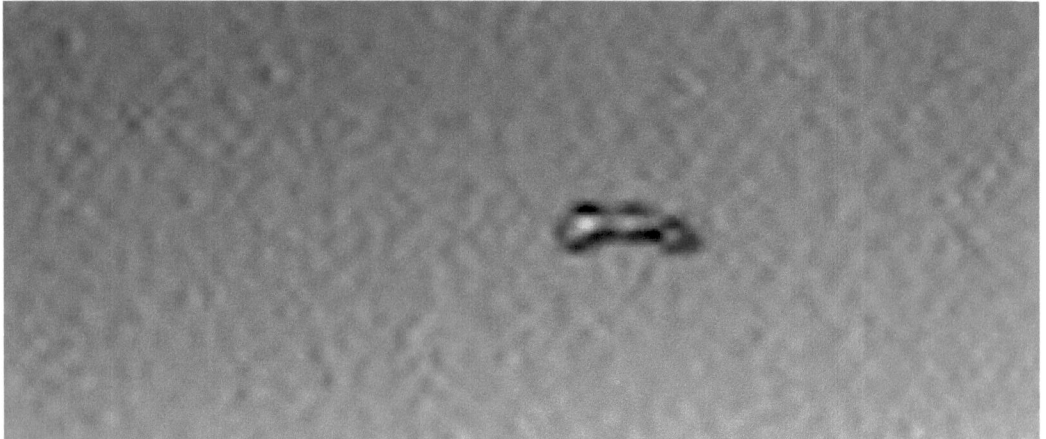

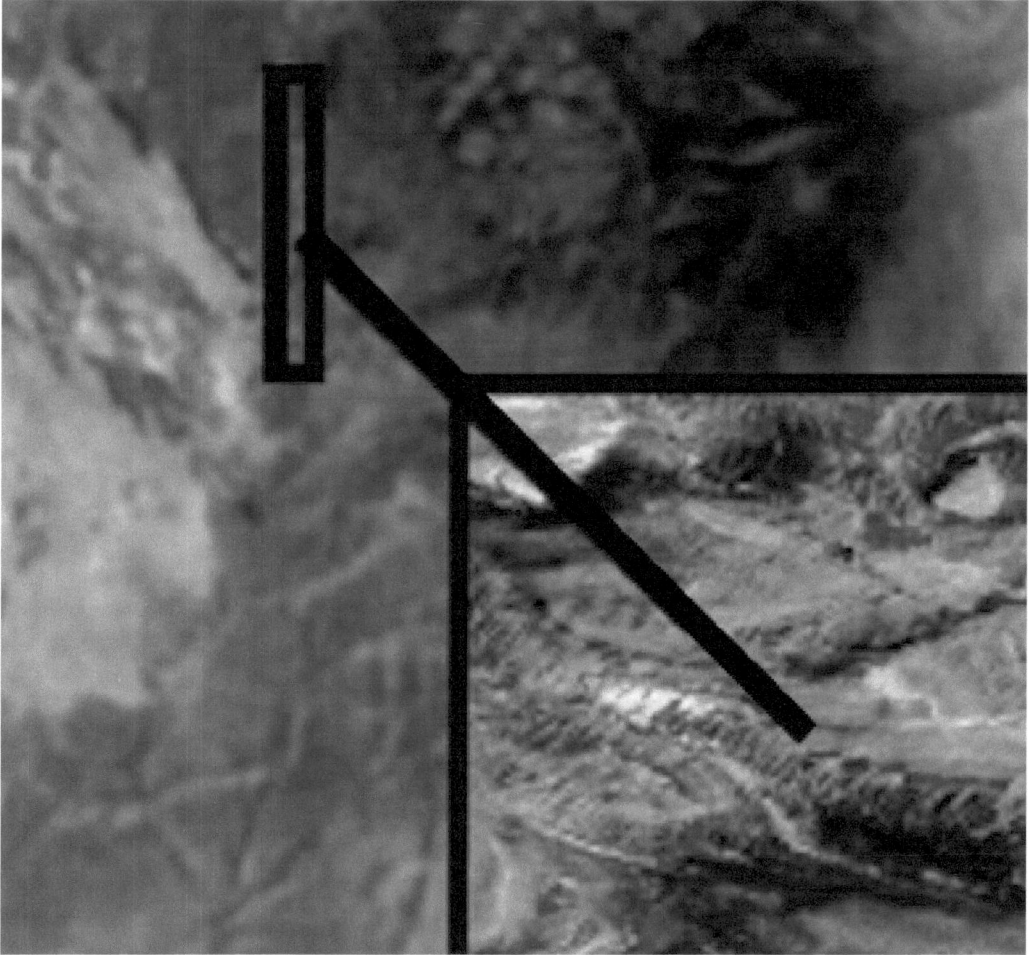

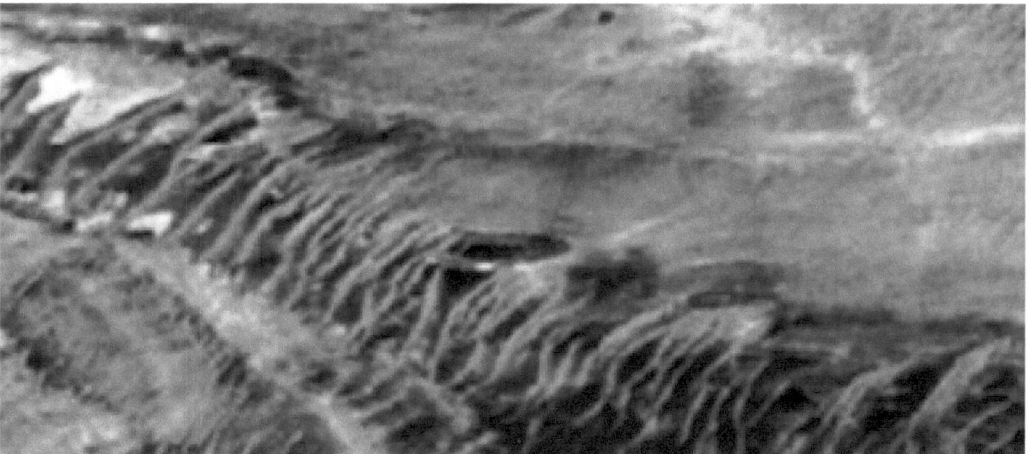

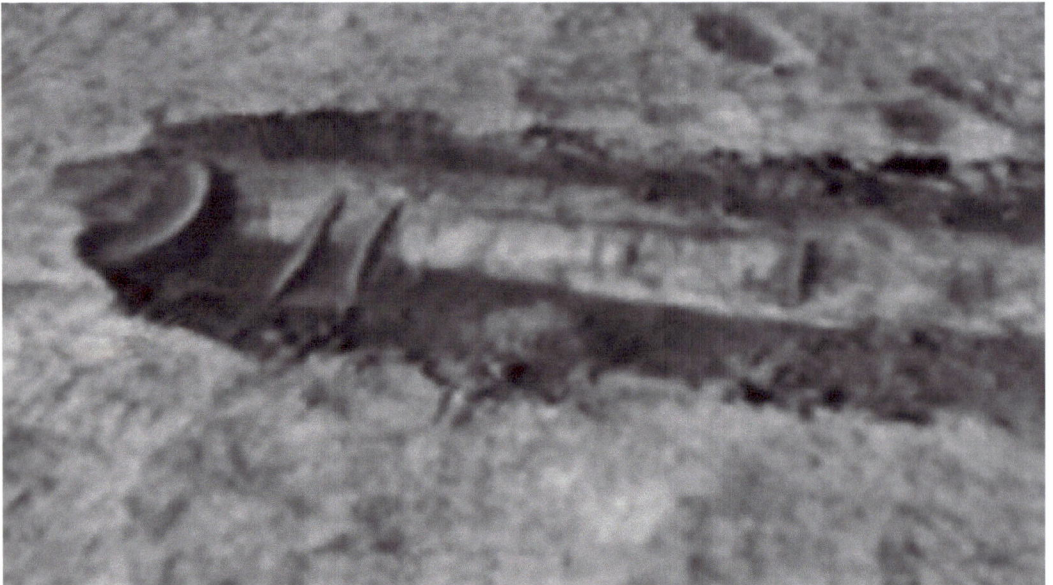

Life on Venus and Mars

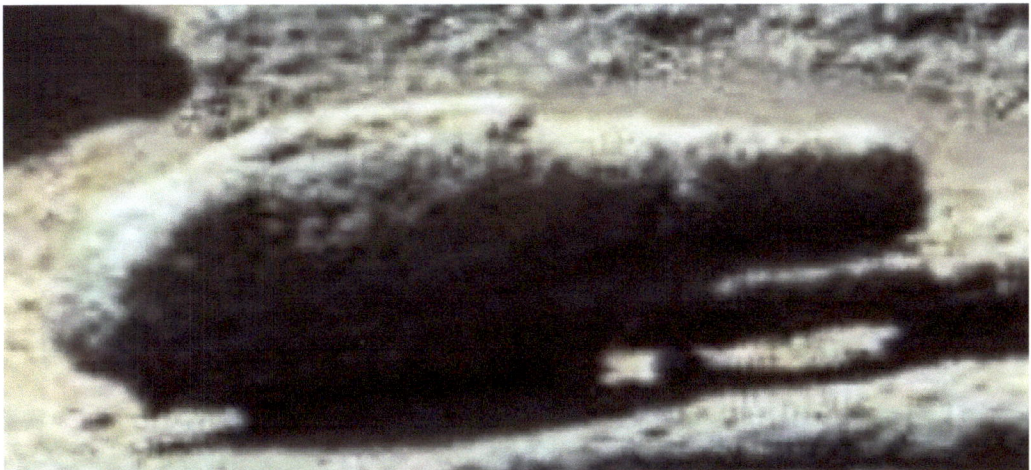

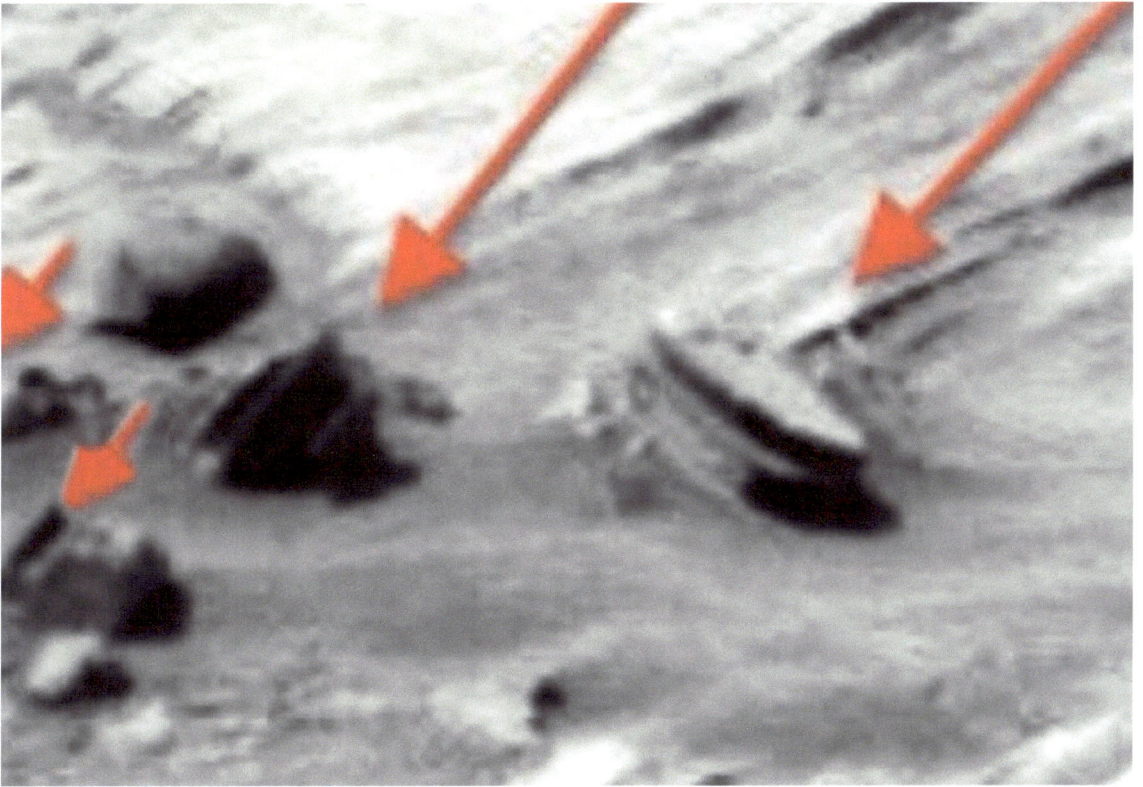

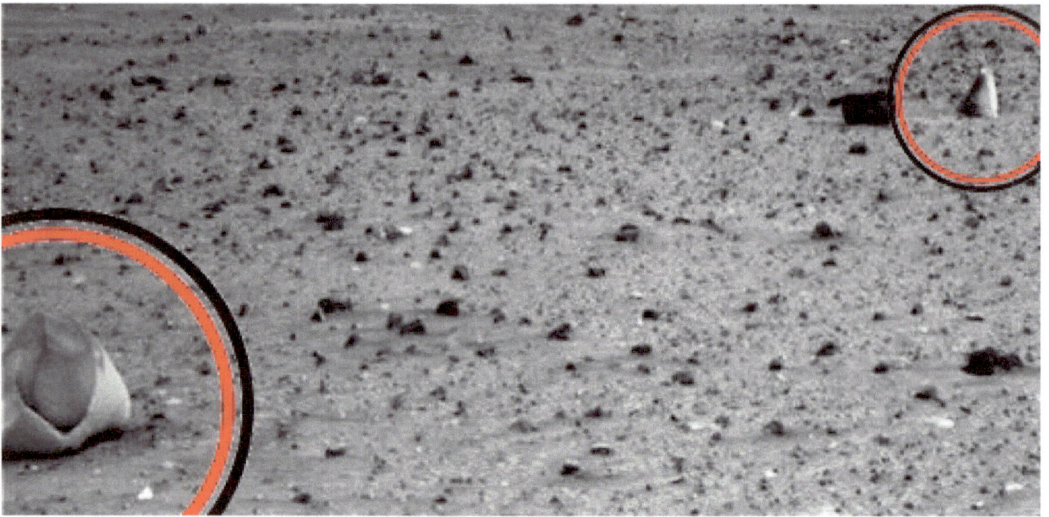

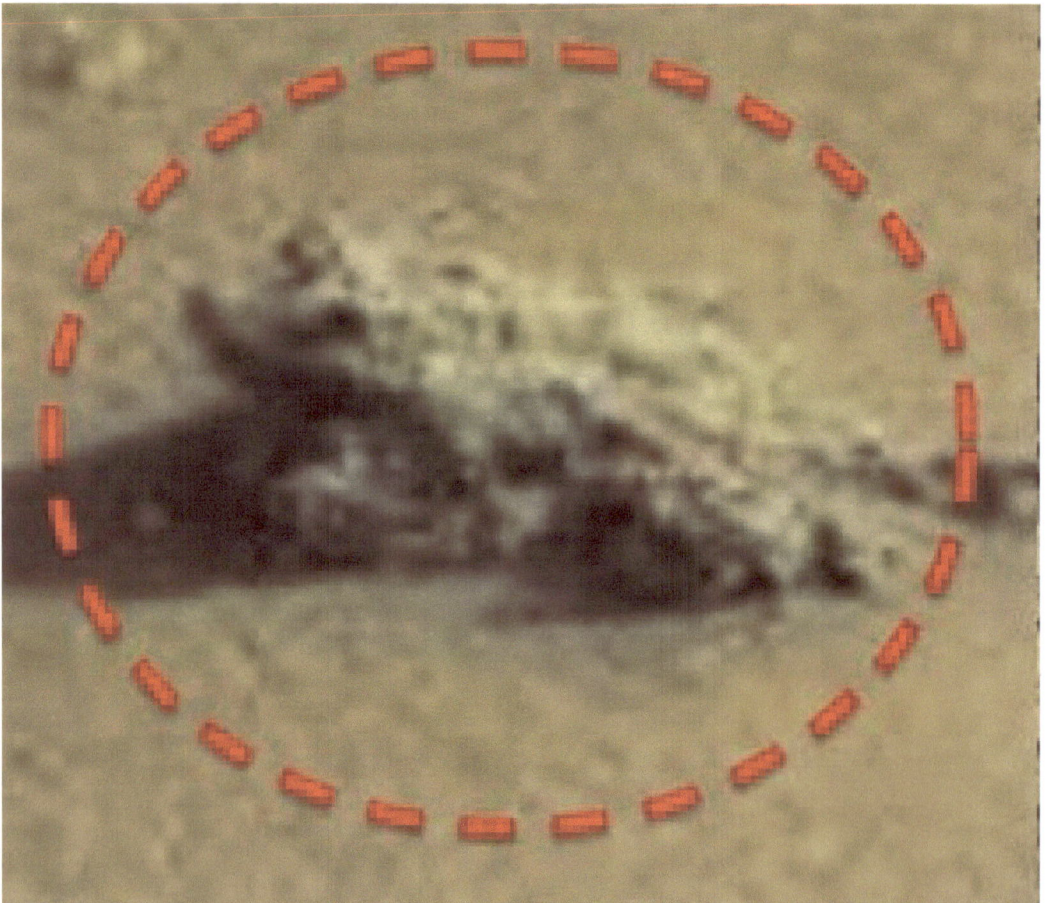

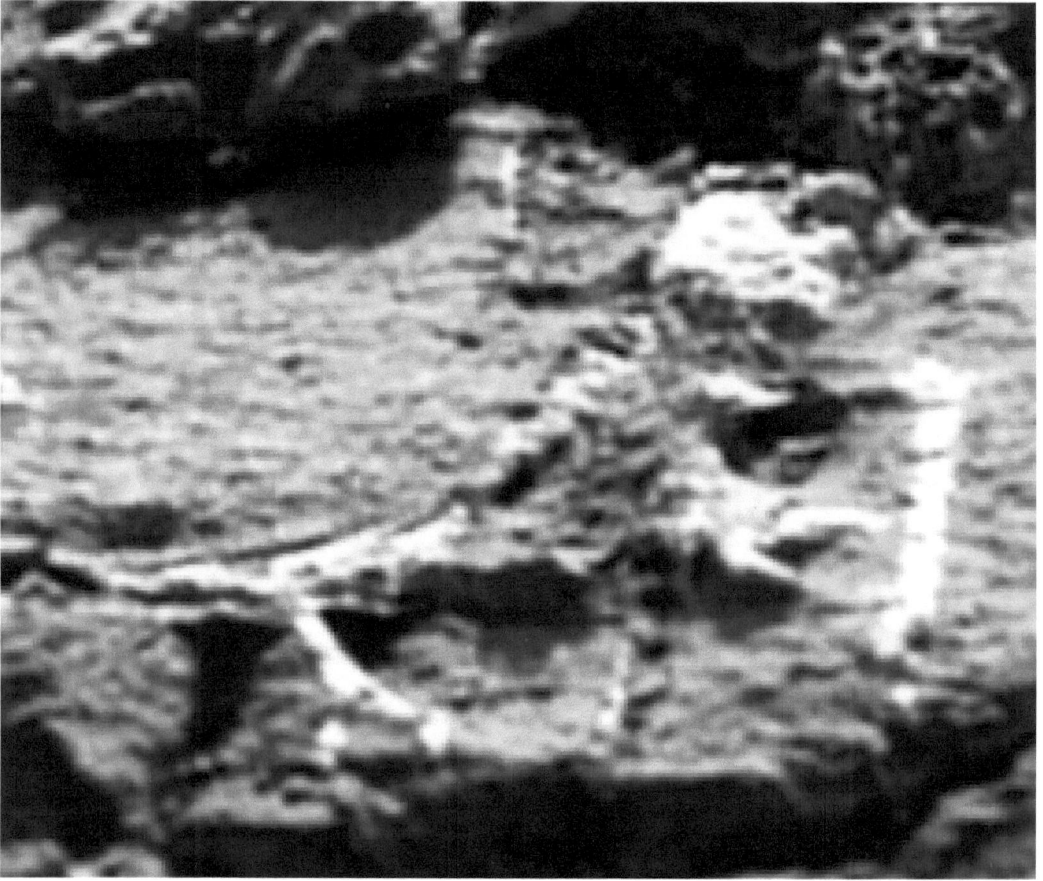

Life on Venus and Mars

263

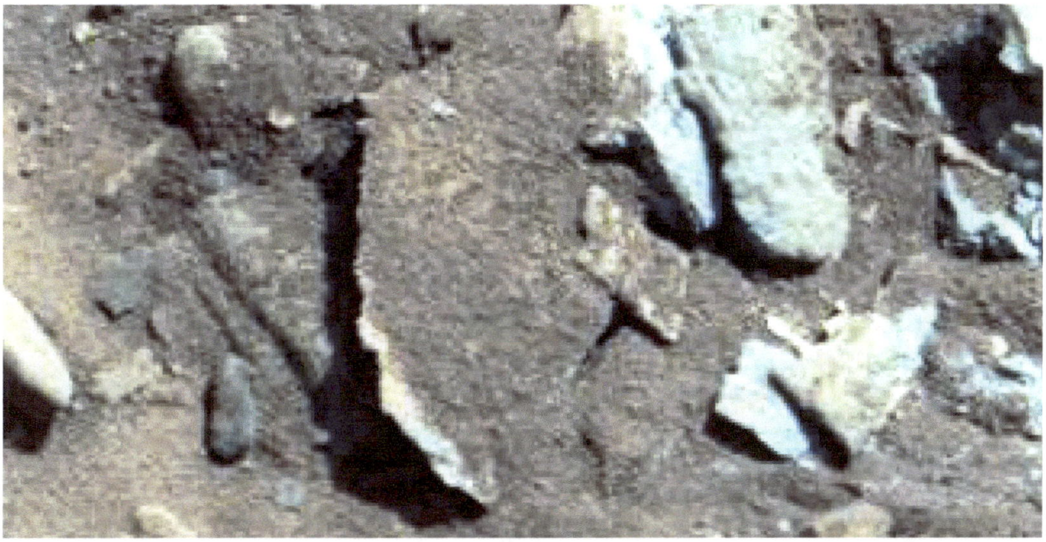

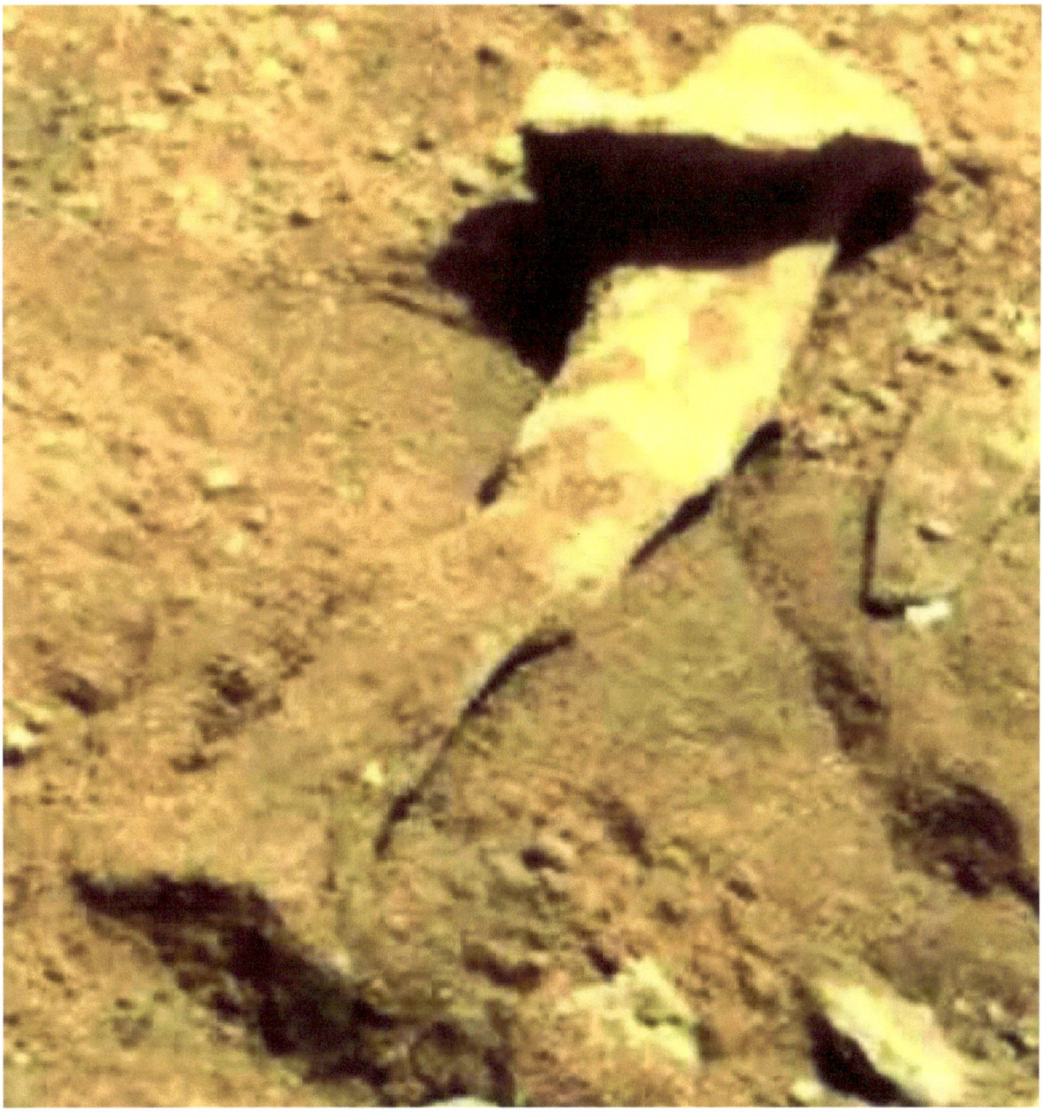

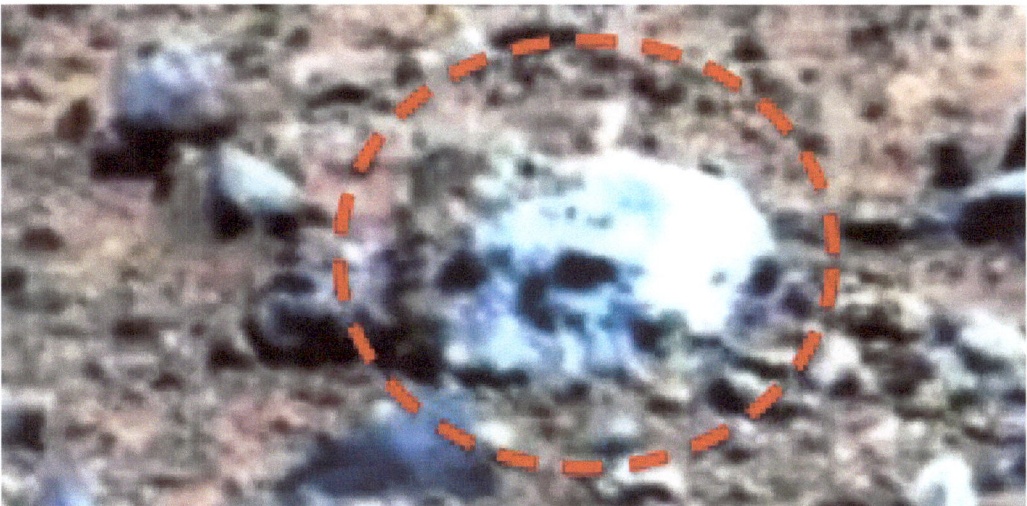

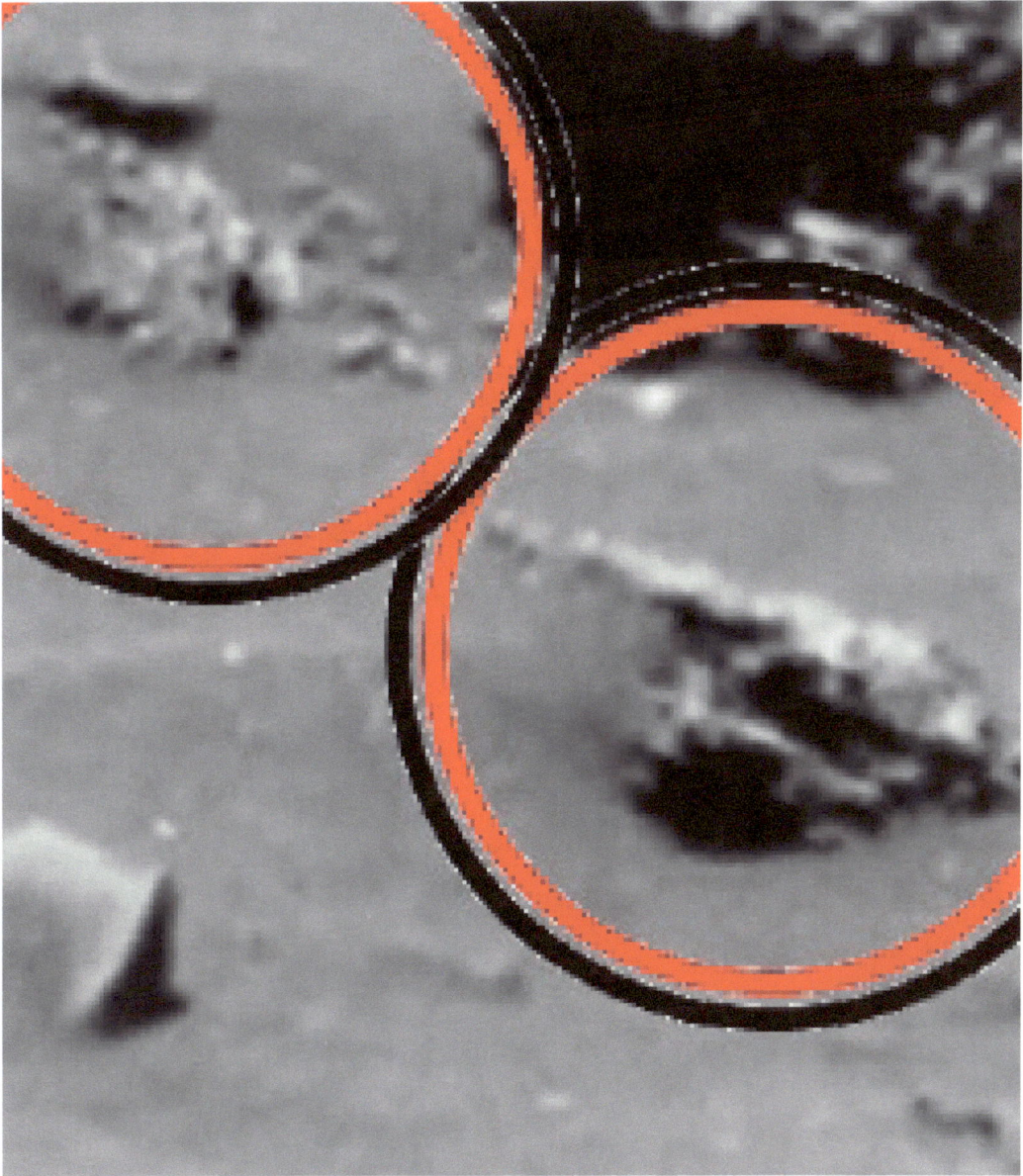

Figure Above: Skull set upon a grave?

Figure Above: A skull with nose, mouth, eyes, with the right hand and arm is laying across the chest, and two legs extended; laying upon a cushion (ejection seat?), near a circle of stones?

16. References

Abyzov, S. et al., (1998). Microbiologiya, 67, 547.

Ackermann HW, et al., (1987). Viruses of prokaryotes: General properties of bacteriophages. Boca Raton, Florida: CRC Press, Inc.

Ackermann HW (2007). 5500 Phages examined in the electron microscope. Arch Virol 2007, 152:227-243.

Acuña, M. H., et al. (1999), Global distribution of crustal magnetization discovered by the Mars Global Surveyor MAG/ER experiment, Science, 284, 790–793.

Acuña, M. H., et al. (2001), Magnetic field of Mars: Summary of results from the aerobraking and mapping orbits, J. Geophys. Res., 106, 23403–23417.

Agrawal A, Eastman QM, Schatz DG (1998) Transposition mediated by RAG1 and RAG2 and its implications for the evolution of the immune system. Nature 394: 744Y751.

Akao, M., et. al., (2001). Mitochondrial ATP-Sensitive Potassium Channels Inhibit Apoptosis Induced by Oxidative Stress in Cardiac Cells Circulation Research. 88, 1267-1275.

Alvarez-Buylla E.R., et al., (2000). An ancestral MADS-box gene duplication occurred before the divergence of plants and animals. Proc Natl Acad Sci U S A. 9;97(10):5328-3

Amaral et al. (2008). The Eukaryotic Genome as an RNA Machine, Science, 319. 1787 - 1789.

Andrews-Hanna, J. C., Zuber, M. T., Banerdt, W. B. 2008. The Borealis basin and the origin of the martian crustal dichotomy Nature 453, 1212-1215.

Ash R. D., Knott S. F., and Turner G. (1996) A 4-Gyr shock age for a martian meteorite and implications for the cratering history of Mars. Nature, 380, 57-59.

Butterfield, NJ (2005a). Probable Proterozoic fungi. Paleobiology. 31, 165–182.

Butterfield, N.J. (2005b). Reconstructing a complex early Neoproterozoic eukaryote, Wynniatt formation, arctic Canada. Lethaia. 38, 155–169.

Butterfield, N.J, Knoll, A.H, & Swett, N. (1994). Paleobiology of the Neoproterozoic Svanbergfjellet formation, Spitsbergen. Fossils Strata. 34, 1–84.

Butterfield N.J, Rainbird R.H (1998). Diverse organic-walled fossils, including 'possible dinoflagellates', from the early Neoproterozoic of arctic Canada. Geology. 26, 963–966.

Claus, G., Nagy, B. (1961) A Microbiological Examination of Some Carbonaceous Chondrites. Nature 192, 594 - 596.

Connerney, J. E. P., M. H. Acuña, P. J. Wasilewski, G. Kletetschka, N. F. Ness, H. Rème, R. P. Lin, and D. L. Mitchell (2001), The global magnetic field of Mars and implications for crustal evolution, Geophys. Res. Lett., 28(21), 4015–4018, doi:10.1029/2001GL013619.

Conway Morris, S (2008). A Redescription of a Rare Chordate, Metaspriggina Walcotti Simonetta and Insom, from the Burgess Shale (Middle Cambrian), British Columbia, Canada. Journal of Paleontology 82: 424.

Life on Venus and Mars

Cook P.M, Shergold J.H (1986). Proterozoic and Cambrian phosphorites. Cambridge, UK:Cambridge University Press.

Hartmann, W. K., and G. Neukum (2001), Cratering chronology and the evolution of Mars, Space Sci. Rev., 96, 165–194.

Hood, L. L., K. P. Harrison, B. Langlais, R. J. Lillis, F. Poulet, and D. A. Williams (2010), Magnetic anomalies near Apollinaris Patera and the Medusae Fossae Formation in Lucus Planum, Mars, Icarus, 208(1), 118–131, doi:10.1016/j.icarus.2010.01.009.

Hoover, R. B., 2006. Comets, carbonaceous meteorites, and the origin of the biosphere. Biogeosciences Discussions, 3, 23-70.

Hoover, R. B., Rozanov, A., 2003. Microfossils, biominerals and chemical biomarkers in Meteorites, in: Instruments Methods and Missions for Astrobiology VI, edited by: Hoover, R. B., Rozanov, A. Yu., and Lipps, J. H., Proc. SPIE 4939, 10-27.

Joseph, R. (2000). Astrobiology, the Origins of Life, and the Death of Darwinism. University Press, Californian

Joseph, R. (2009). Life on Earth Came From Other Planets. Journal of Cosmology, 1, 1-40.

Joseph, R. (2014) Life on Mars: Lichens, Fungi, Algae, Cosmology, 22, 40-62.

Joseph, R. (2016) A High Probability of Life on Mars, Cosmology, 25, 1-25.

Joseph, R. (2017) Martian Organisms Attack, Damage Curiosity's Rover Wheels After Only 10 Miles. Cosmology, Vol 27, 3/25/17.

Joseph, R., & Rabb, H. (2016) Martian Fungi & Bacteria Damaged the Mars Rovers. Cosmology, Vol 26, December 2016.

Joseph R. Schild, R. 2010a. Biological Cosmology and the Origins of Life in the Universe. Journal of Cosmology, 5, 1040-1090.

Joseph, R (2019). Life on Venus and the Interplanetary Transfer of Biota from Earth. Astrophysics and Space Science, 364(11), DOI: 10.1007/s10509-019-3678-x

Joseph, R. G, Dass, R. S., Rizzo, V., Cantasano, N., Bianciardi, G. (2019), Evidence of Life on Mars? Journal of Astrobiology and Space Science Reviews, 1, 40–81.

Joseph, R. Graham, L., Budel, B., Jung, P., Kidron, G. J., Latif, K., Armstrong, R. A., Mansour, H. A., Ray, J. G., Ramos, G.J.P., Consorti, L., et al. (2020a). Mars: Algae (Cyanobacteria), Lichens, Fungi, Microbial Mats, Stromatolites, and Trace Fossils in Gale Crater. Journal of Astrobiology, 3, 40-111.

Joseph, R. Gibson, C., Schild, R. (2020b). Water, Ice and Mud in the Gale Crater: Implications for Life on Mars. Submitted and Under Review.

Knoll, A. H. (1996). Archean and Proterozoic paleontology. In Palynology: Principles and applications, vol. 1 (ed. J. Jansonius and D. C. McGregor), pp. 51-80. American Association of Palynologists Foundation.

Knoll, A. H. and Carroll, S. B. (1999). Early animal evolution: Emerging views from comparative biology and geology. Science 184, 2129-2137.

Knoll, A. H., et al., (2004). A New Period for the Geologic Time Scale Science, 305. 621 - 622.

Knoll AH et al., (2006). Eukaryotic organisms in Proterozoic oceans. Philos Trans R Soc Lond B Biol Sci. 36, 1023-1038.

Kohler, M., et al., (1996). Small-Conductance, Calcium-Activated Potassium Channels from Mammalian Brain Science, 273, 1709-1714.

Konstantinidis KT, Tiedje JM. (2004). Trends between gene content and genome size in prokaryotic species with larger genomes. Proc. Natl Acad. Sci. USA, 101:3160–3165.

Koonin, EV. (2003) Comparative genomics, minimal gene-sets and the last universal common ancestor. Nature Rev. Microbiol. 1:127–136.

Koonin, E.V., et al. (2004). A comprehensive evolutionary classification of proteins encoded in complete eukaryotic genomes. Genome Biol. 5, R7.

Koonin EV. (2006). The origin of introns and their role in eukaryogenesis: a compromise solution to the introns-early versus introns-late debate? Biol Direct. Aug 14;1:22.

Koonin, E.V., & Wolf, Y.I. (2008). Genomics of bacteria and archaea: the emerging generalizations after 13 years. Nucleic Acids Res. 36, 6688–6719.

Lillis, R. J., H. V. Frey, M. Manga, D. L. Mitchell, R. P. Lin, M. H. Acuña, and S. W. Bougher (2008a), An improved crustal magnetic field map of Mars from electron reflectometry: Highland volcano magmatic history and the end of the Martian dynamo, Icarus, 194, 575–196.

Lillis, R. J., H. V. Frey, and M. Manga (2008b), Rapid decrease in Martian crustal magnetization in the Noachian era: Implications for the dynamo and climate of early Mars, Geophys. Res. Lett., 35, L14203, doi:10.1029/2008GL034338.

McKay, D.S., Gibson, E.K., Thomas-Keprta, K.L., Hojatollah, V., Romanek, C.S., Clemmett, S.J., Chillier, X.D.F., Maechling, C.R., and Zare, R.N. (1996) Search for past life on Mars: possible relic biogenic activity in Martian meteorite ALH84001. Science 273: 924-930.

McKay, G., Mikouchi, T., Schwandt, C. & Lofgren, G. (1998) Fracture fillings in ALH84001. Feldspathic glass : carbonatic and silica. 29 th Annual Lunar and Planetary Science Conference held March 16-20, 1998 in Houston, Texas. LPI Contribution No. 1998, Abstract no. 1944.

McKay, D.S., Thomas-Keprta, K.L., Clemett, S.J., Gibson Jr, E.K., Spencer, L. and Wentworth, S.J. (2009) Life on Mars: new evidence from martian meteorites. In, Instruments and Methods for Astrobiology and Planetary Missions, 7441, 744102.

Mitchell, D. L., R. J. Lillis, R. P. Lin, J. E. P. Connerney, and M. H. Acuña (2007), A global map of Mars' crustal magnetic field based on electron reflectometry, J. Geophys. Res., 112, E01002, doi:10.1029/2005JE002564.

Mojzsis, S.J., Arrhenius, G., McKeegan, K.D., Harrison, T.M., Nutman, A.P., Friend, C.R.L., 1996. Evidence for life on Earth before 3,800 million years ago. Nature 384, 55-59.

Moser, D. P. et al., 2005. Desulfotomaculum and Methanobacterium spp. Dominate a 4- to 5-Kilometer-Deep Fault. Applied and Environmental Microbiology, 71, 8773-8783.

Nagy, B., Meinschein, W. G. Hennessy, D, J. 1961, Mass-spectroscopic analysis of the Orgueil meteorite: evidence for biogenic hydrocarbons. Annals of the New York Academy of Sciences 93, 25-35.

Nagy, B., Claus, G., Hennessy, D, J., 1962, Organic Particles embedded in Minerals in the Orgueil and Ivuna Carbonaceous Chondrites. Nature 193, 1129 - 1133.

Nemchin, A. A., Whitehouse, M.J., Menneken, M., Geisler, T., Pidgeon, R.T., Wilde, S. A. 2008. A light carbon reservoir recorded in zircon-hosted diamond from the Jack Hills. Nature 454, 92-95.

Nutman, Allen P.; Bennett, Vickie C.; Friend, Clark R. L.; Kranendonk, Martin J. Van; Chivas, Allan R. (2016). "Rapid emergence of life shown by discovery of 3,700-million-year-old microbial structures". Nature. 537: 535–538.

O'Neil, J., Carlson, R. W., Francis, E., Stevenson, R. K. 2008. Neodymium-142 Evidence for Hadean Mafic Crust Science 321, 1828 - 1831.

Perron, J. Taylor; Jerry X. Mitrovica; Michael Manga; Isamu Matsuyama & Mark A. Richards (2007). Evidence for an ancient martian ocean in the topography of deformed shorelines Nature. 447: 840-843.

Pflug, H. D, 1978 Yeast-like microfossils detected in oldest sediments of the earth Journal Naturwissenschaften 65, 121-134.

Pflug, H.D. 1984. Microvesicles in meteorites, a model of pre-biotic evolution. Journal Naturwissenschaften, 71, 531-533.

Renno, N. O., and 22 colleagues (2009) Physical and Thermodynamical Evidence for Liquid Water on Mars, Lunar and Planetary Science Conference, Houston, March 23-27.

Rizzo, V., & Cantasano, N. (2009) Possible organosedimentary structures on Mars. International Journal of Astrobiology 8 (4): 267-280.

Rizzo, V., & Cantasano, N. (2011), Cyanobacteria on Terrestrial Meteorites and Stromatolites on Mars, Journal of Cosmology, 13, 15.

Rosing, M. T., 1999. C-13-depleted carbon microparticles in > 3700-Ma sea-floor sedimentary rocks from west Greenland. Science 283, 674-676.

Rosing, M. T., Frei, R., 2004. U-rich Archaean sea-floor sediments from Greenland - indications of > 3700 Ma oxygenic photosynthesis. Earth and Planetary Science Letters 217, 237-244.

Squyres, S. W. and 18 colleagues (2004) In Situ Evidence for an Ancient Aqueous Environment at Meridiani Planum, Mars, Science 306 (5702), 1709-1714. 2004 Dec 03.

Thomas-Keprta K.L., Clemett S.J., Bazylinski D.A., Kirschvink J.L., McKay D.S., Wentworth S.J., Vali H., and Gibson E.K., Romanek C.S. (2002) "Magnetofossils from Ancient Mars: A Robust Biosignature in the Martian Meteorite ALH84001." Applied and Environmental Microbiology 68, 3663-3672.

Thomas-Keprta, K. L., et al., (2009). Origins of magnetite nanocrystals in Martian meteorite ALH84001. Geochimica et Cosmochimica Acta, 73, 6631-6677.

Zhmur, S. I., Gerasimenko, L. M. 1999. Biomorphic forms in carbonaceous meteorite Alliende and possible ecological system - producer of organic matter hondrites" in Instruments, Methods and Missions for Astrobiology II, RB. Hoover, Editor, Proceedings of SPIE Vol. 3755 p. 48-58.   Zhmur, S. I., Rozanov, A. Yu., Gorlenko, V. M. 1997. Lithified remnants of microorganisms in carbonaceous chondrites, Geochemistry International, 35, 58-60

www.ingramcontent.com/pod-product-compliance
Lightning Source LLC
Chambersburg PA
CBHW061134030426
42334CB00003B/36